PRAISE FOR *MY FRIEND DAHMER*

"A well-told, powerful story." –R. Crumb

"Stunning. Horrifying. Beautifully done." –Alison Bechdel

"A solid job. Putrid serial killer Jeffrey Dahmer's origins are explored in this
fine book. Dig it—it'll hang you out to dry."
–James Ellroy

"This one's still haunting me." –Brad Meltzer

"A brilliant graphic novel and surely ranks among the very best of the form."
–Dan Chaon

"It wasn't easy reading this book, but I'm glad I did."
–David Small

"*My Friend Dahmer* will certainly quench your dark little desires."
–Chuck Klosterman

"Wow. Reading this is unlike any other reading experience I've ever had. Do it."
–Rainn Wilson

★ "A small, dark classic." –*Publishers Weekly* (starred review)

★ "An exemplary demonstration of the transformative possibilities of
graphic narrative."
–*Kirkus* (starred review)

"One of the most thought-provoking comics released in a long time."
–Slate.com

"Astounding." –Lev Grossman, *TIME*

"One of the best graphic novels I've read this year." –*USA Today*

"Masterful . . . a rich tale full of complexity and sensitivity."
–*The Cleveland Plain Dealer*

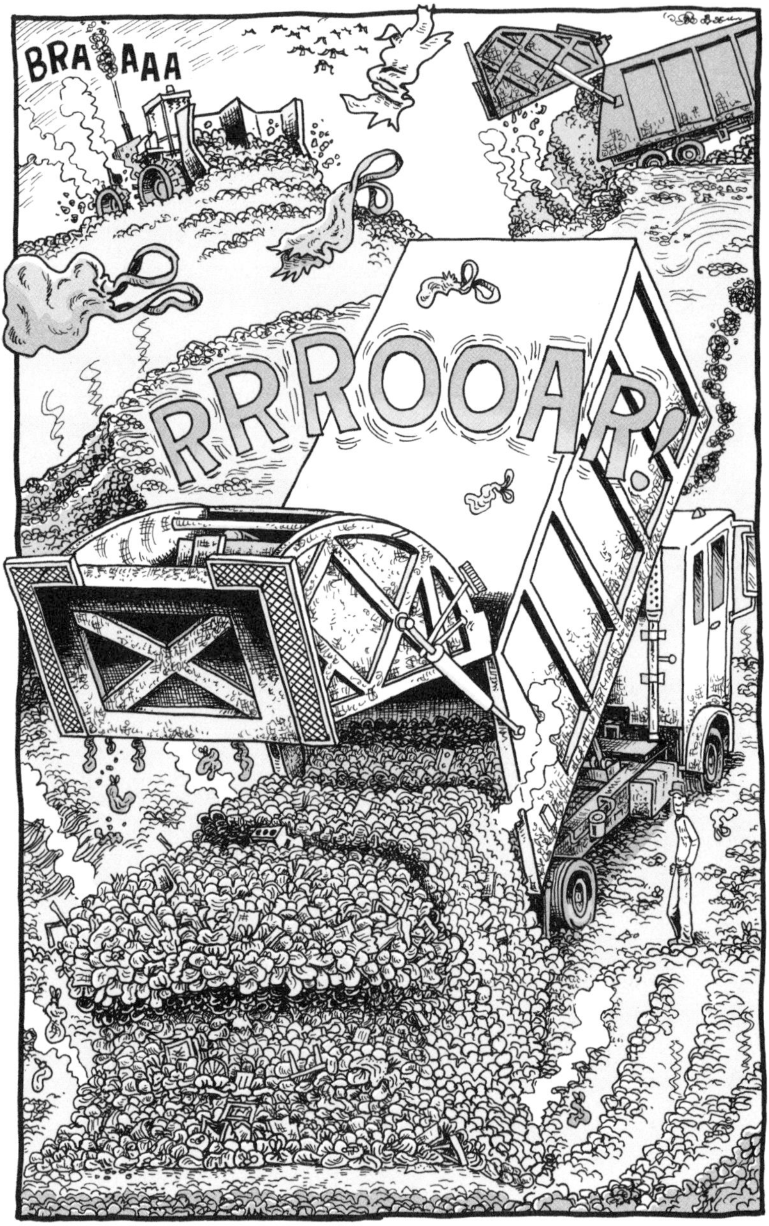
BRAAAA
RRROOAR!

ALSO BY DERF BACKDERF

My Friend Dahmer

TRASHED

a graphic novel by
DERF BACKDERF

Abrams ComicArts, New York

EDITOR: Charles Kochman
DESIGNER: Pamela Notarantonio
MANAGING EDITOR: Jen Graham
PRODUCTION MANAGER: Kathy Lovisolo

Library of Congress Cataloging-in-Publication Data
Backderf, Derf.
  Trashed : a graphic novel / by Derf Backderf.
    pages cm
  Summary: Graphic novel based on the author's time spent working on a garbage truck.
  ISBN 978-1-4197-1453-5 (hardcover) — ISBN 978-1-4197-1454-2 (pbk.) —
  ISBN 978-1-61312-865-7 (ebook)
  1. Sanitation workers—Fiction. 2. Refuse and refuse disposal—Fiction.
  3. Graphic novels. I. Title.
  PN6727.D466T73 2015
  741.5'973—dc23
                        2015011115

Printed and bound in the United States
10 9 8 7 6 5 4 3 2 1

Abrams ComicArts books are available at special discounts when purchased in quan-
tity for premiums and promotions as well as fundraising or educational use. Special
editions can also be created to specification. For details, contact specialsales@abrams-
books.com or the address below.

ABRAMS
THE ART OF BOOKS SINCE 1949
115 West 18th Street
New York, NY 10011
www.abramsbooks.com

Contents

PREFACE

My first *Trashed* stories were memoir, the tale of my time as a garbageman in 1979 and 1980. These were published as a fifty-page floppy comic book in 2002. It was my first attempt at long-form comics storytelling. I also received my first Eisner nomination for the *Trashed* floppy, so it's always been a special project for me.

I returned to *Trashed* in 2010. I decided to experiment with a webcomic, and thought a new *Trashed* would be a fun venture in this format. I did one forty-page *Trashed* story in 2010, then another installment in 2011. But as I worked on the narrative it became clear this was no longer a memoir. I brought the story up to the present day and decided to make it fiction. Portions of this book are taken from those two webcomics. This book is inspired by my experiences, but none of the characters or places are real.

Thanks to my beloved editor, Charlie Kochman, for suggesting this book, and to my awesome agent, Matthew Carnicelli. You guys are the best.

Thanks also to Pam Notarantonio, who designed and colored the cover and babysat this book all the way to print.

And to Jen Graham, who copyedited my shoddy prose and saved me from misspelling every word longer than two syllables.

Thanks to my old pal Mike, who made that year on the truck bearable.

Thanks to all the fans and readers around the globe who have made my late-career reboot as a graphic novelist so fabulous.

This book is dedicated to my wife, Sheryl, who suffered nobly as a "comix widow" for the fourteen months it took me to make it. She is, as she so often reminds me, my muse.

derf

Derf Backderf
Shaker Heights, Ohio
June 2015

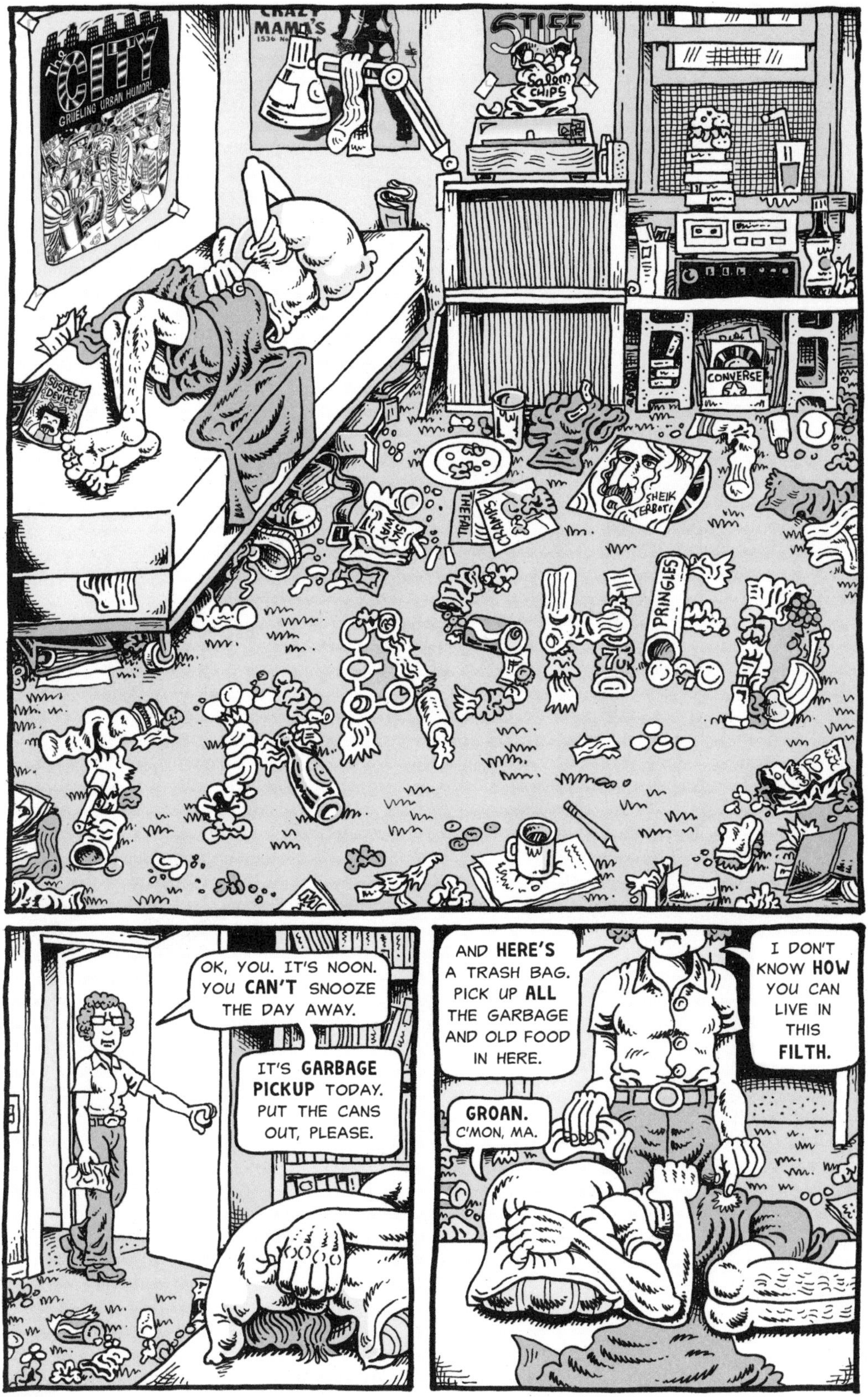
The CITY
GRUELING URBAN HUMOR!
STIFF
Salem CHIPS
SUSPECT DEVICE
CONVERSE
THE FALL
CRAMPS
SHEIK YERBOTI
PRINGLES
TRASHED
OK, YOU. IT'S NOON. YOU CAN'T SNOOZE THE DAY AWAY.
IT'S GARBAGE PICKUP TODAY. PUT THE CANS OUT, PLEASE.
AND HERE'S A TRASH BAG. PICK UP ALL THE GARBAGE AND OLD FOOD IN HERE.
I DON'T KNOW HOW YOU CAN LIVE IN THIS FILTH.
GROAN. C'MON, MA.

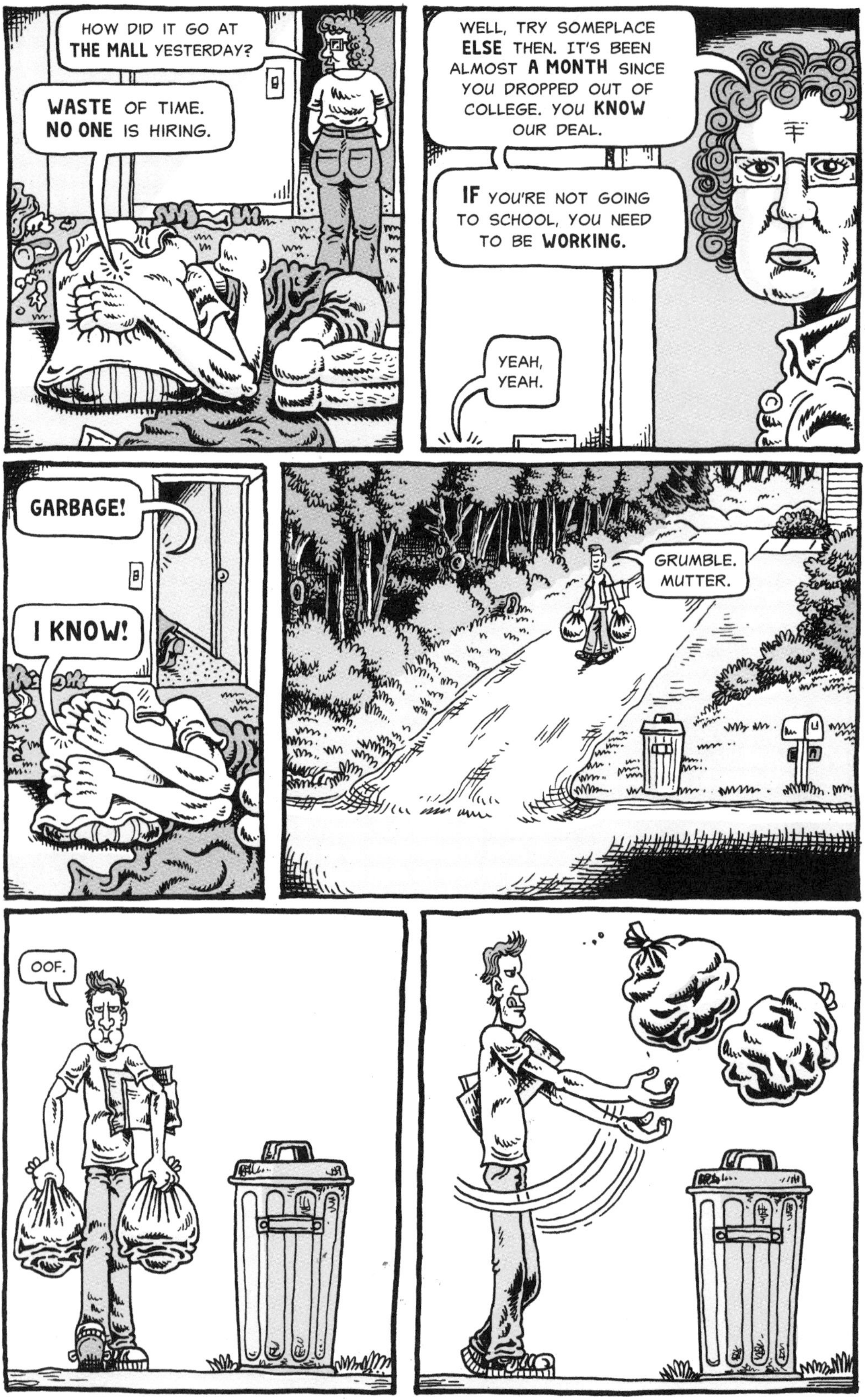
HOW DID IT GO AT THE MALL YESTERDAY?
WASTE OF TIME. NO ONE IS HIRING.
WELL, TRY SOMEPLACE ELSE THEN. IT'S BEEN ALMOST A MONTH SINCE YOU DROPPED OUT OF COLLEGE. YOU KNOW OUR DEAL.
IF YOU'RE NOT GOING TO SCHOOL, YOU NEED TO BE WORKING.
YEAH, YEAH.
GARBAGE!
I KNOW!
GRUMBLE. MUTTER.
OOF.

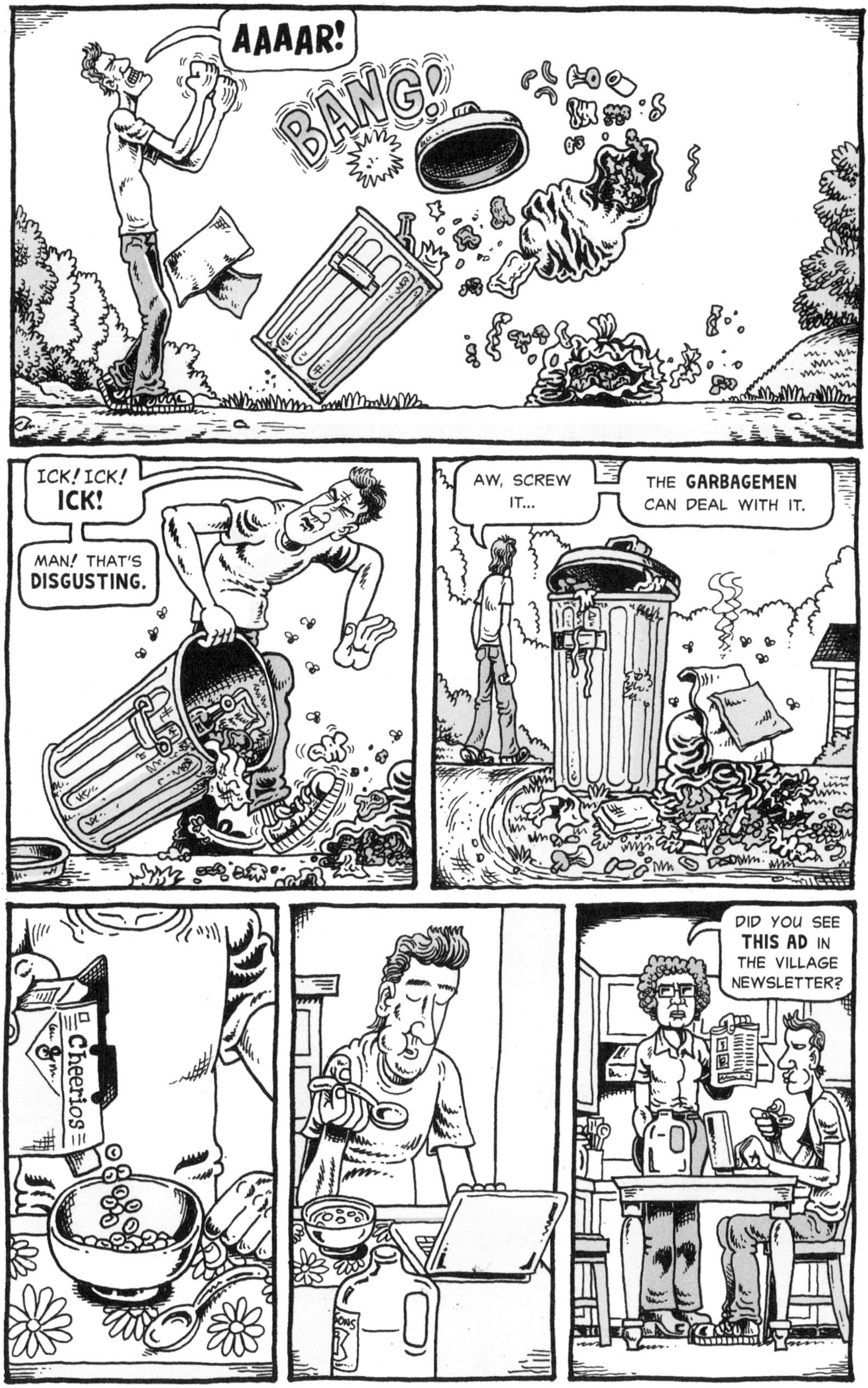
AAAAR!
BANG!
ICK! ICK! ICK!
MAN! THAT'S DISGUSTING.
AW, SCREW IT...
THE GARBAGEMEN CAN DEAL WITH IT.
Cheerios
DID YOU SEE THIS AD IN THE VILLAGE NEWSLETTER?

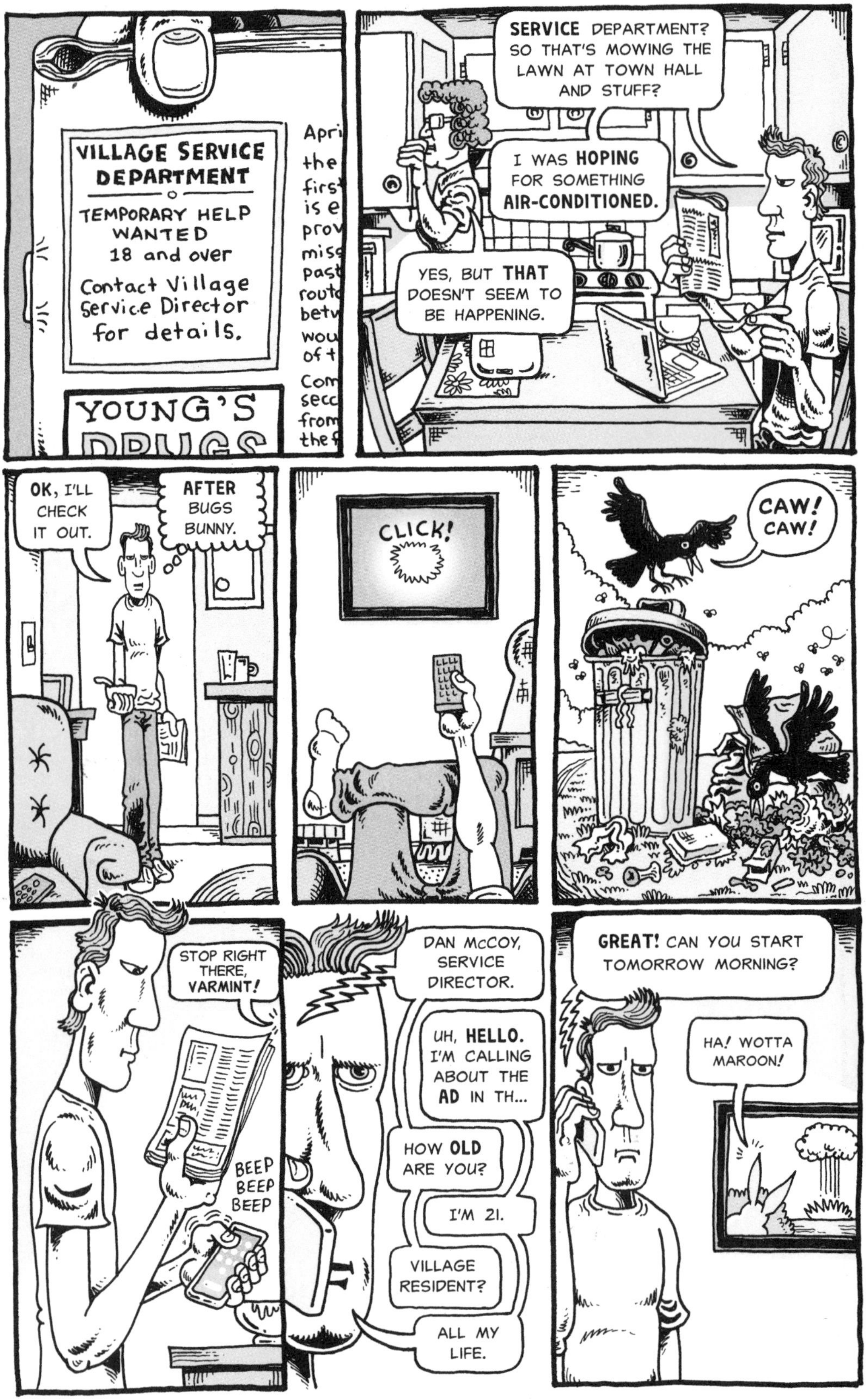
VILLAGE SERVICE DEPARTMENT
TEMPORARY HELP WANTED
18 and over
Contact Village Service Director for details.
YOUNG'S
SERVICE DEPARTMENT? SO THAT'S MOWING THE LAWN AT TOWN HALL AND STUFF?
I WAS HOPING FOR SOMETHING AIR-CONDITIONED.
YES, BUT THAT DOESN'T SEEM TO BE HAPPENING.
OK, I'LL CHECK IT OUT.
AFTER BUGS BUNNY.
CLICK!
CAW! CAW!
STOP RIGHT THERE, VARMINT!
BEEP BEEP BEEP
DAN McCOY, SERVICE DIRECTOR.
UH, HELLO. I'M CALLING ABOUT THE AD IN TH...
HOW OLD ARE YOU?
I'M 21.
VILLAGE RESIDENT?
ALL MY LIFE.
GREAT! CAN YOU START TOMORROW MORNING?
HA! WOTTA MAROON!

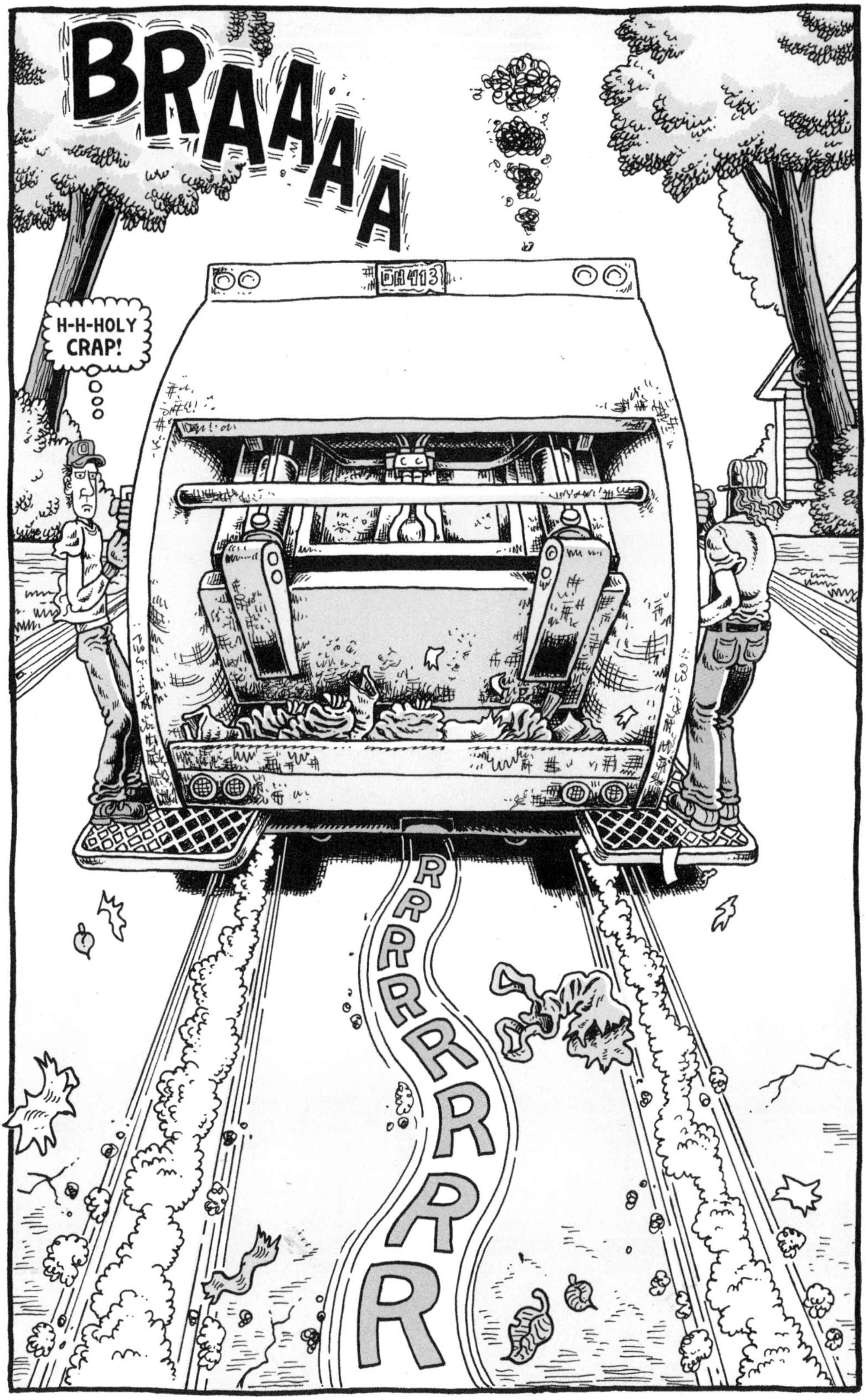

BRAAAAA
DH 413
H-H-HOLY CRAP!
RRRRRRRRR

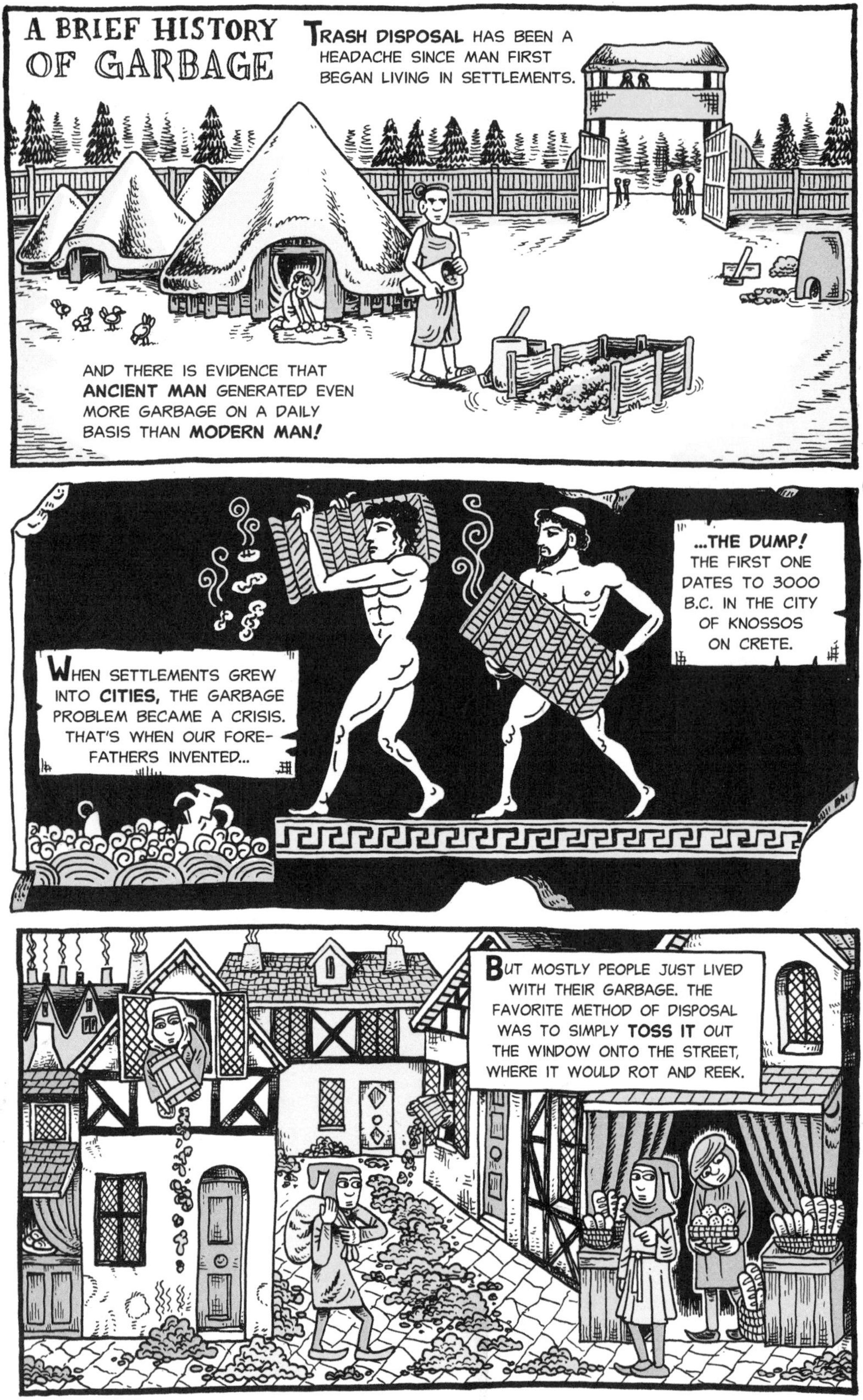
A BRIEF HISTORY OF GARBAGE
TRASH DISPOSAL HAS BEEN A HEADACHE SINCE MAN FIRST BEGAN LIVING IN SETTLEMENTS.
AND THERE IS EVIDENCE THAT ANCIENT MAN GENERATED EVEN MORE GARBAGE ON A DAILY BASIS THAN MODERN MAN!
WHEN SETTLEMENTS GREW INTO CITIES, THE GARBAGE PROBLEM BECAME A CRISIS. THAT'S WHEN OUR FOREFATHERS INVENTED...
...THE DUMP! THE FIRST ONE DATES TO 3000 B.C. IN THE CITY OF KNOSSOS ON CRETE.
BUT MOSTLY PEOPLE JUST LIVED WITH THEIR GARBAGE. THE FAVORITE METHOD OF DISPOSAL WAS TO SIMPLY TOSS IT OUT THE WINDOW ONTO THE STREET, WHERE IT WOULD ROT AND REEK.

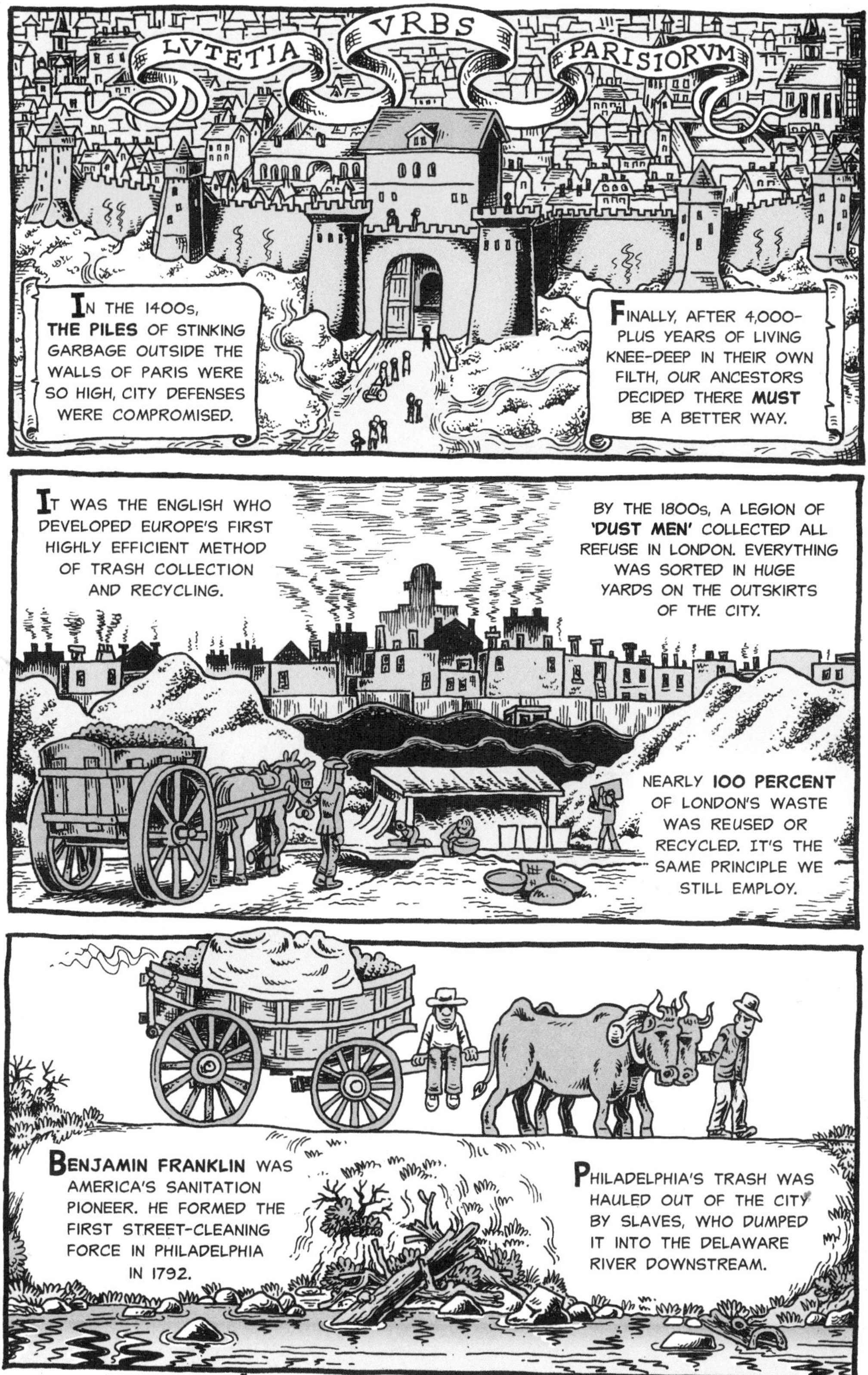
LVTETIA VRBS PARISIORVM
IN THE 1400s, THE PILES OF STINKING GARBAGE OUTSIDE THE WALLS OF PARIS WERE SO HIGH, CITY DEFENSES WERE COMPROMISED.
FINALLY, AFTER 4,000-PLUS YEARS OF LIVING KNEE-DEEP IN THEIR OWN FILTH, OUR ANCESTORS DECIDED THERE MUST BE A BETTER WAY.
IT WAS THE ENGLISH WHO DEVELOPED EUROPE'S FIRST HIGHLY EFFICIENT METHOD OF TRASH COLLECTION AND RECYCLING.
BY THE 1800s, A LEGION OF 'DUST MEN' COLLECTED ALL REFUSE IN LONDON. EVERYTHING WAS SORTED IN HUGE YARDS ON THE OUTSKIRTS OF THE CITY.
NEARLY 100 PERCENT OF LONDON'S WASTE WAS REUSED OR RECYCLED. IT'S THE SAME PRINCIPLE WE STILL EMPLOY.
BENJAMIN FRANKLIN WAS AMERICA'S SANITATION PIONEER. HE FORMED THE FIRST STREET-CLEANING FORCE IN PHILADELPHIA IN 1792.
PHILADELPHIA'S TRASH WAS HAULED OUT OF THE CITY BY SLAVES, WHO DUMPED IT INTO THE DELAWARE RIVER DOWNSTREAM.

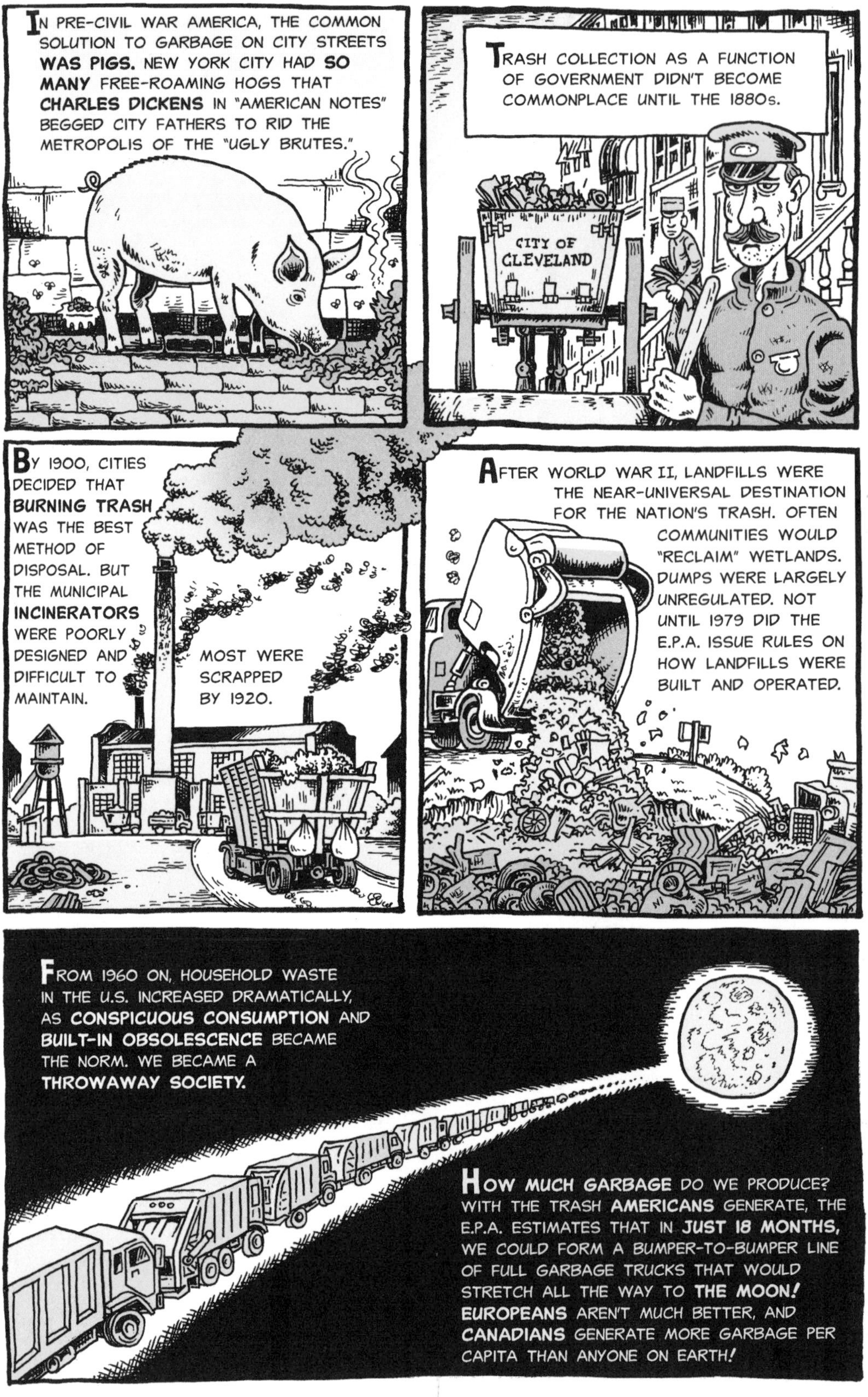
IN PRE-CIVIL WAR AMERICA, THE COMMON SOLUTION TO GARBAGE ON CITY STREETS WAS PIGS. NEW YORK CITY HAD SO MANY FREE-ROAMING HOGS THAT CHARLES DICKENS IN "AMERICAN NOTES" BEGGED CITY FATHERS TO RID THE METROPOLIS OF THE "UGLY BRUTES."
TRASH COLLECTION AS A FUNCTION OF GOVERNMENT DIDN'T BECOME COMMONPLACE UNTIL THE 1880s.
CITY OF CLEVELAND
BY 1900, CITIES DECIDED THAT BURNING TRASH WAS THE BEST METHOD OF DISPOSAL. BUT THE MUNICIPAL INCINERATORS WERE POORLY DESIGNED AND DIFFICULT TO MAINTAIN.
MOST WERE SCRAPPED BY 1920.
AFTER WORLD WAR II, LANDFILLS WERE THE NEAR-UNIVERSAL DESTINATION FOR THE NATION'S TRASH. OFTEN COMMUNITIES WOULD "RECLAIM" WETLANDS. DUMPS WERE LARGELY UNREGULATED. NOT UNTIL 1979 DID THE E.P.A. ISSUE RULES ON HOW LANDFILLS WERE BUILT AND OPERATED.
FROM 1960 ON, HOUSEHOLD WASTE IN THE U.S. INCREASED DRAMATICALLY, AS CONSPICUOUS CONSUMPTION AND BUILT-IN OBSOLESCENCE BECAME THE NORM. WE BECAME A THROWAWAY SOCIETY.
HOW MUCH GARBAGE DO WE PRODUCE? WITH THE TRASH AMERICANS GENERATE, THE E.P.A. ESTIMATES THAT IN JUST 18 MONTHS, WE COULD FORM A BUMPER-TO-BUMPER LINE OF FULL GARBAGE TRUCKS THAT WOULD STRETCH ALL THE WAY TO THE MOON! EUROPEANS AREN'T MUCH BETTER, AND CANADIANS GENERATE MORE GARBAGE PER CAPITA THAN ANYONE ON EARTH!

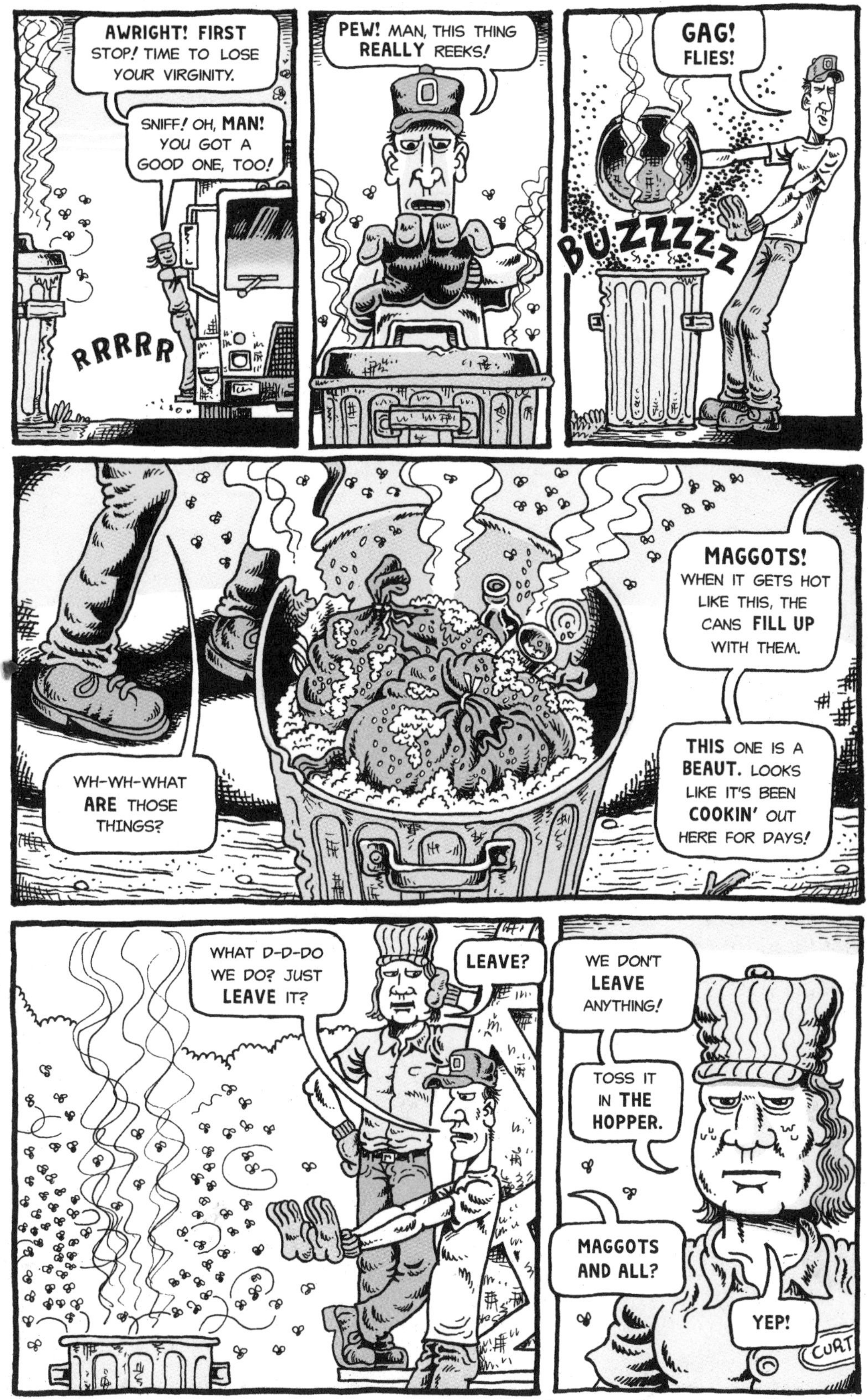
AWRIGHT! FIRST STOP! TIME TO LOSE YOUR VIRGINITY.
SNIFF! OH, MAN! YOU GOT A GOOD ONE, TOO!
RRRRR
PEW! MAN, THIS THING REALLY REEKS!
GAG! FLIES!
BUZZZZZ
MAGGOTS! WHEN IT GETS HOT LIKE THIS, THE CANS FILL UP WITH THEM.
WH-WH-WHAT ARE THOSE THINGS?
THIS ONE IS A BEAUT. LOOKS LIKE IT'S BEEN COOKIN' OUT HERE FOR DAYS!
WHAT D-D-DO WE DO? JUST LEAVE IT?
LEAVE?
WE DON'T LEAVE ANYTHING!
TOSS IT IN THE HOPPER.
MAGGOTS AND ALL?
YEP!
CURT

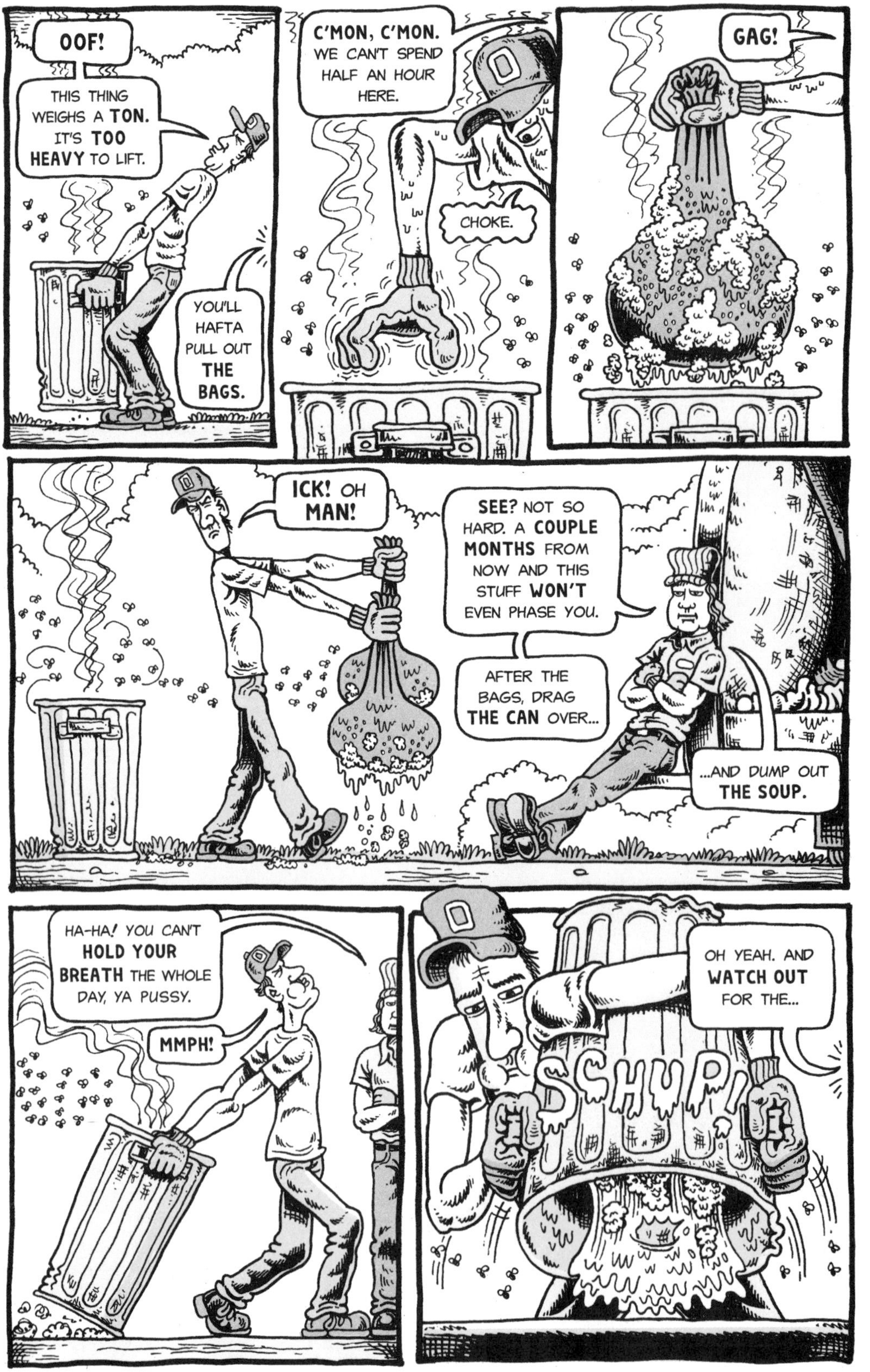
OOF!
THIS THING WEIGHS A TON. IT'S TOO HEAVY TO LIFT.
YOU'LL HAFTA PULL OUT THE BAGS.
C'MON, C'MON. WE CAN'T SPEND HALF AN HOUR HERE.
CHOKE.
GAG!
ICK! OH MAN!
SEE? NOT SO HARD. A COUPLE MONTHS FROM NOW AND THIS STUFF WON'T EVEN PHASE YOU.
AFTER THE BAGS, DRAG THE CAN OVER...
...AND DUMP OUT THE SOUP.
HA-HA! YOU CAN'T HOLD YOUR BREATH THE WHOLE DAY, YA PUSSY.
MMPH!
OH YEAH. AND WATCH OUT FOR THE...
SCHUP!

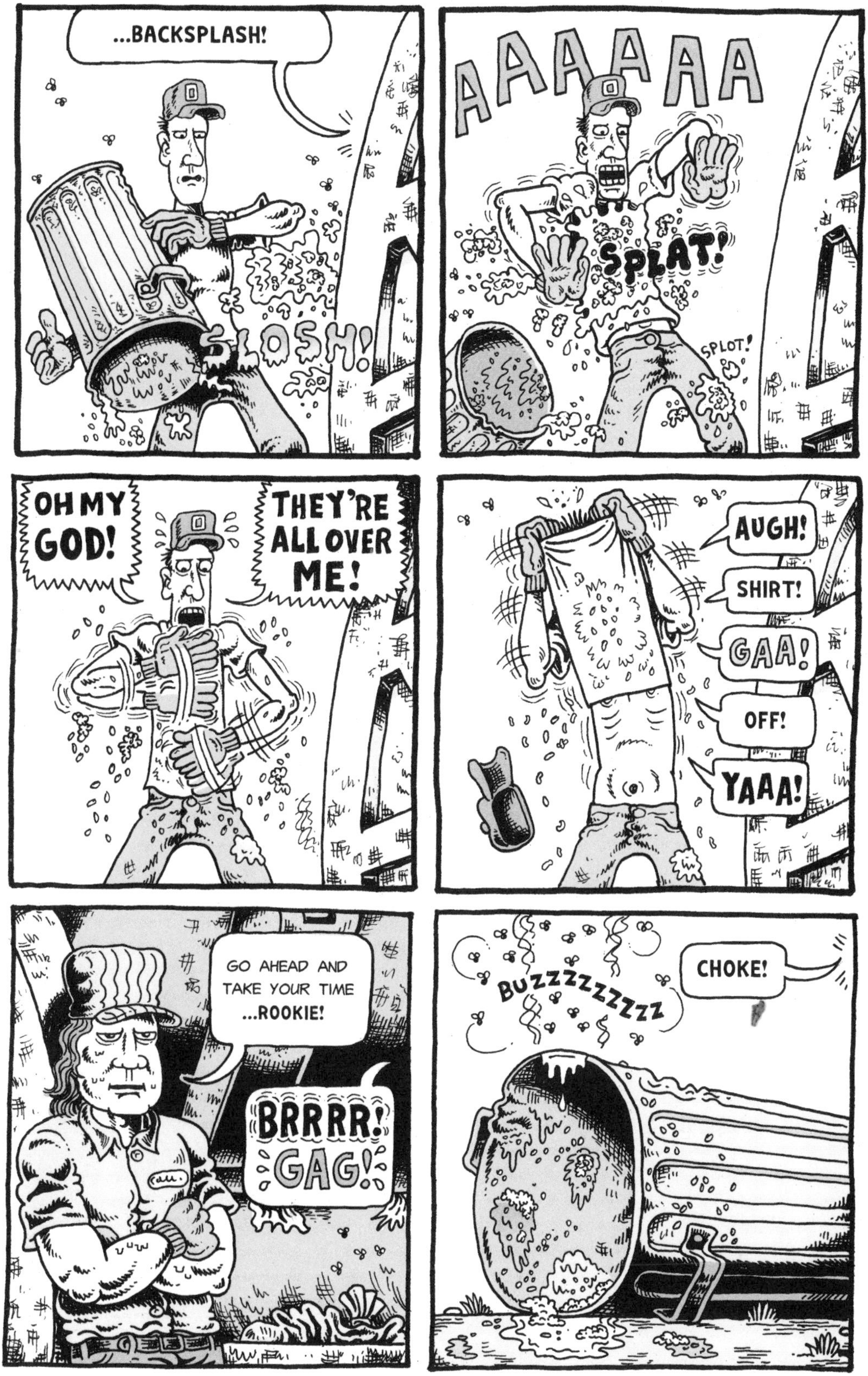
...BACKSPLASH!
SLOSH!
AAAAAA
SPLAT!
SPLOT!
OH MY GOD!
THEY'RE ALL OVER ME!
AUGH!
SHIRT!
GAA!
OFF!
YAAA!
GO AHEAD AND TAKE YOUR TIME ...ROOKIE!
BRRRR!
GAG!
BUZZZZZZZZZ
CHOKE!

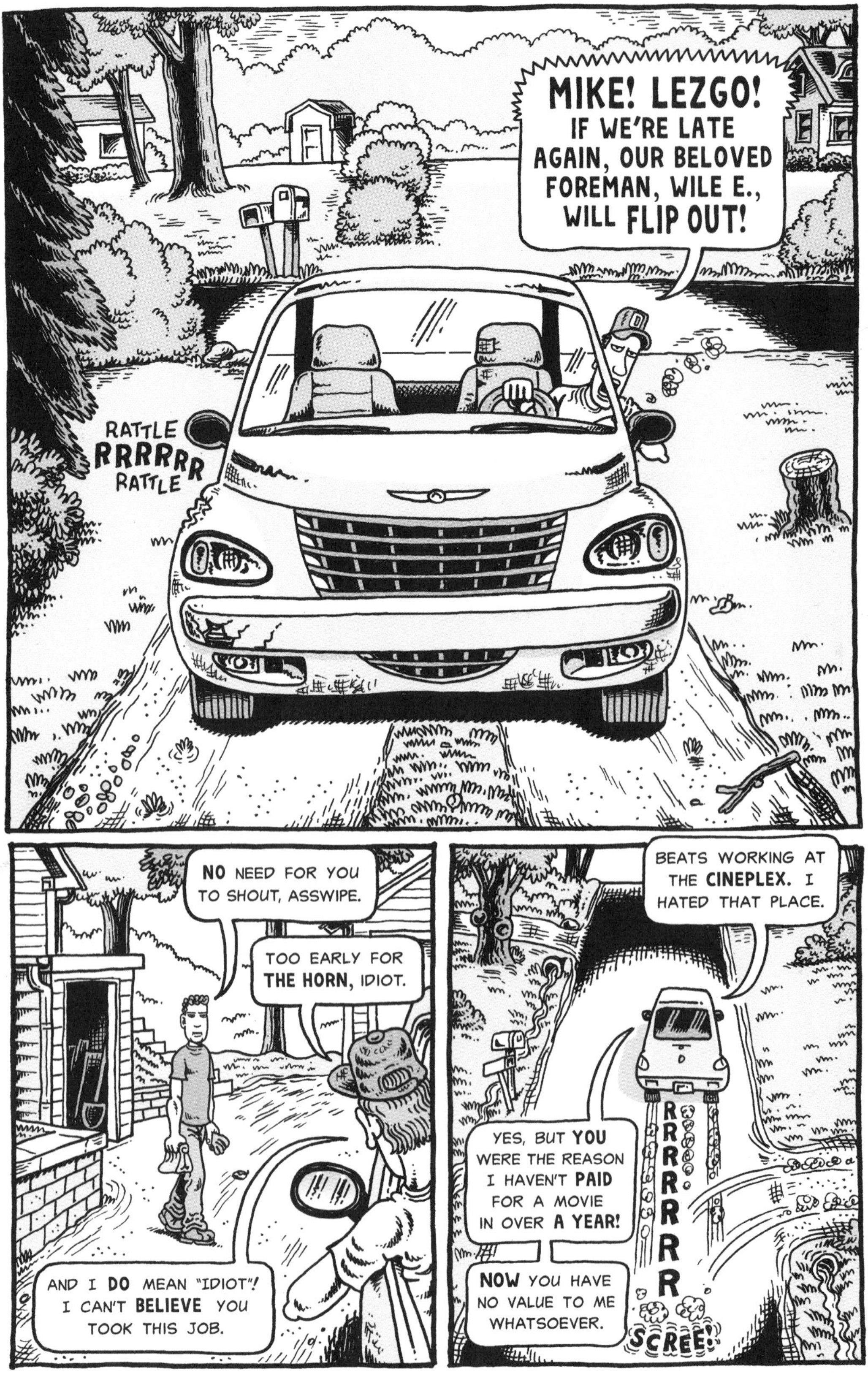
MIKE! LEZGO! IF WE'RE LATE AGAIN, OUR BELOVED FOREMAN, WILE E., WILL FLIP OUT!
RATTLE RRRRRR RATTLE
NO NEED FOR YOU TO SHOUT, ASSWIPE.
TOO EARLY FOR THE HORN, IDIOT.
AND I DO MEAN "IDIOT"! I CAN'T BELIEVE YOU TOOK THIS JOB.
BEATS WORKING AT THE CINEPLEX. I HATED THAT PLACE.
YES, BUT YOU WERE THE REASON I HAVEN'T PAID FOR A MOVIE IN OVER A YEAR!
NOW YOU HAVE NO VALUE TO ME WHATSOEVER.
RRRRRRR
SCREE!

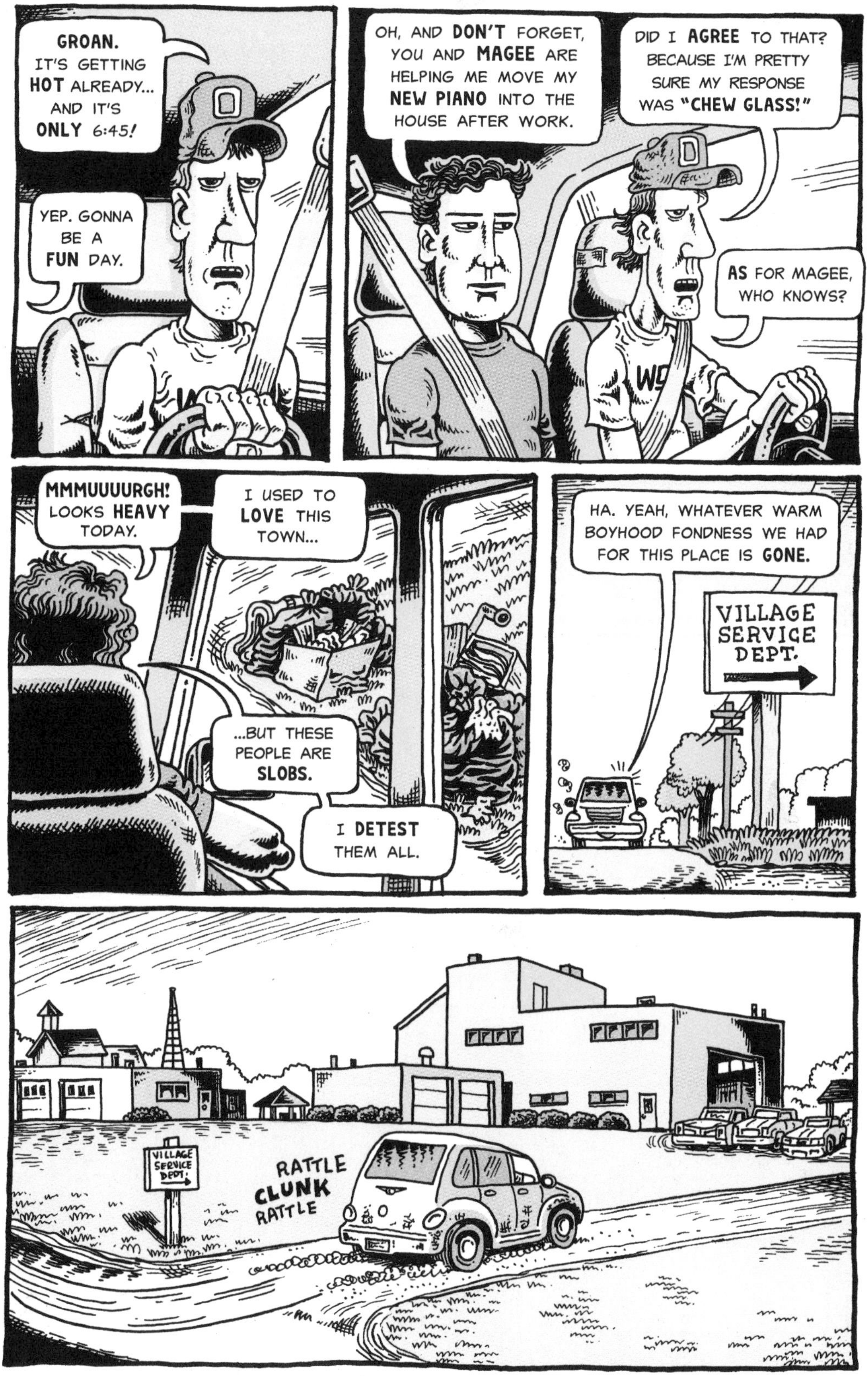
GROAN. IT'S GETTING HOT ALREADY... AND IT'S ONLY 6:45!
YEP. GONNA BE A FUN DAY.
OH, AND DON'T FORGET, YOU AND MAGEE ARE HELPING ME MOVE MY NEW PIANO INTO THE HOUSE AFTER WORK.
DID I AGREE TO THAT? BECAUSE I'M PRETTY SURE MY RESPONSE WAS "CHEW GLASS!"
AS FOR MAGEE, WHO KNOWS?
MMMUUUURGH! LOOKS HEAVY TODAY.
I USED TO LOVE THIS TOWN...
...BUT THESE PEOPLE ARE SLOBS.
I DETEST THEM ALL.
HA. YEAH, WHATEVER WARM BOYHOOD FONDNESS WE HAD FOR THIS PLACE IS GONE.
VILLAGE SERVICE DEPT.
VILLAGE SERVICE DEPT.
RATTLE
CLUNK
RATTLE

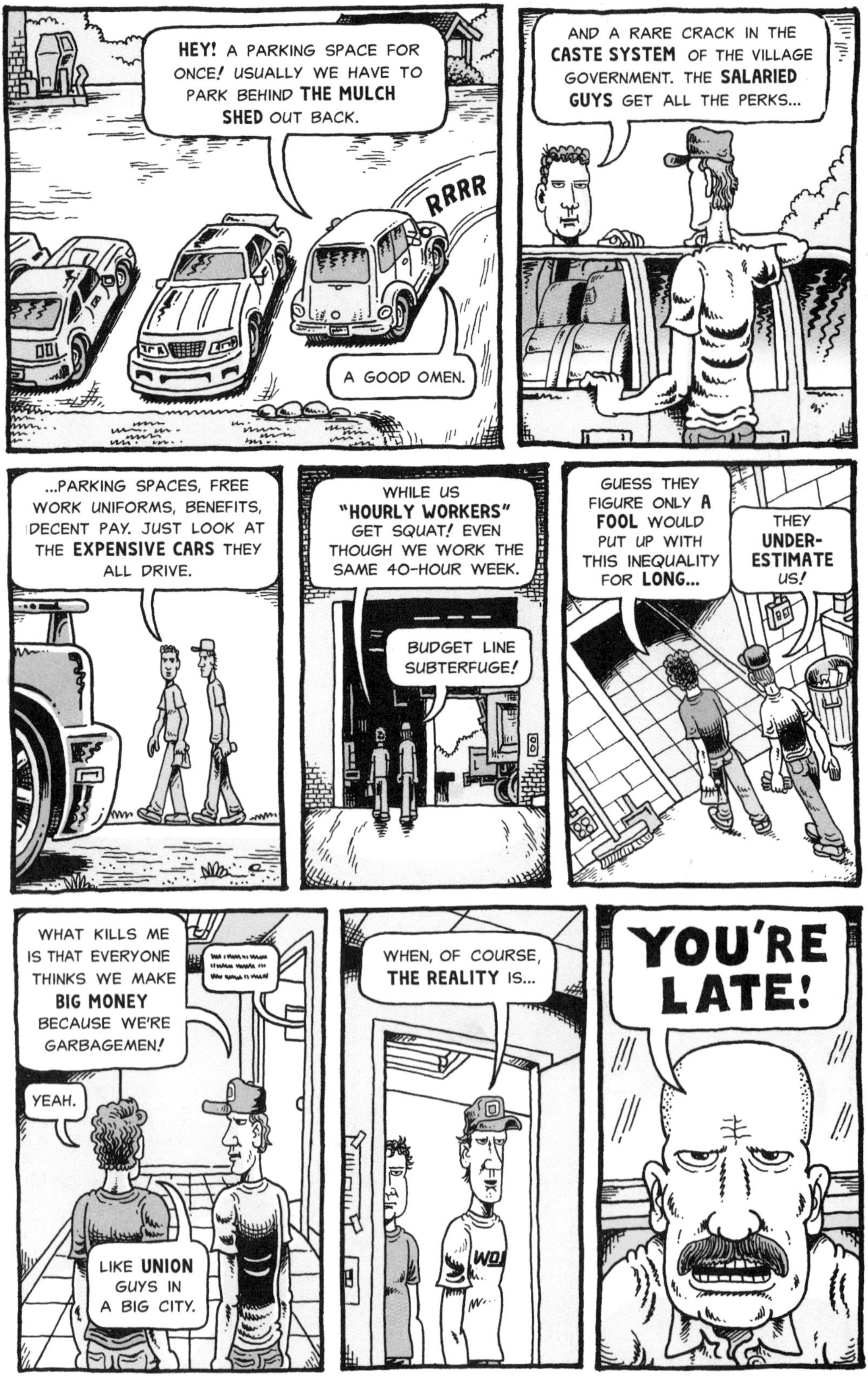
HEY! A PARKING SPACE FOR ONCE! USUALLY WE HAVE TO PARK BEHIND THE MULCH SHED OUT BACK.
RRRR
A GOOD OMEN.
AND A RARE CRACK IN THE CASTE SYSTEM OF THE VILLAGE GOVERNMENT. THE SALARIED GUYS GET ALL THE PERKS...
...PARKING SPACES, FREE WORK UNIFORMS, BENEFITS, DECENT PAY. JUST LOOK AT THE EXPENSIVE CARS THEY ALL DRIVE.
WHILE US "HOURLY WORKERS" GET SQUAT! EVEN THOUGH WE WORK THE SAME 40-HOUR WEEK.
BUDGET LINE SUBTERFUGE!
GUESS THEY FIGURE ONLY A FOOL WOULD PUT UP WITH THIS INEQUALITY FOR LONG...
THEY UNDER-ESTIMATE US!
WHAT KILLS ME IS THAT EVERYONE THINKS WE MAKE BIG MONEY BECAUSE WE'RE GARBAGEMEN!
YEAH.
LIKE UNION GUYS IN A BIG CITY.
WHEN, OF COURSE, THE REALITY IS...
YOU'RE LATE!

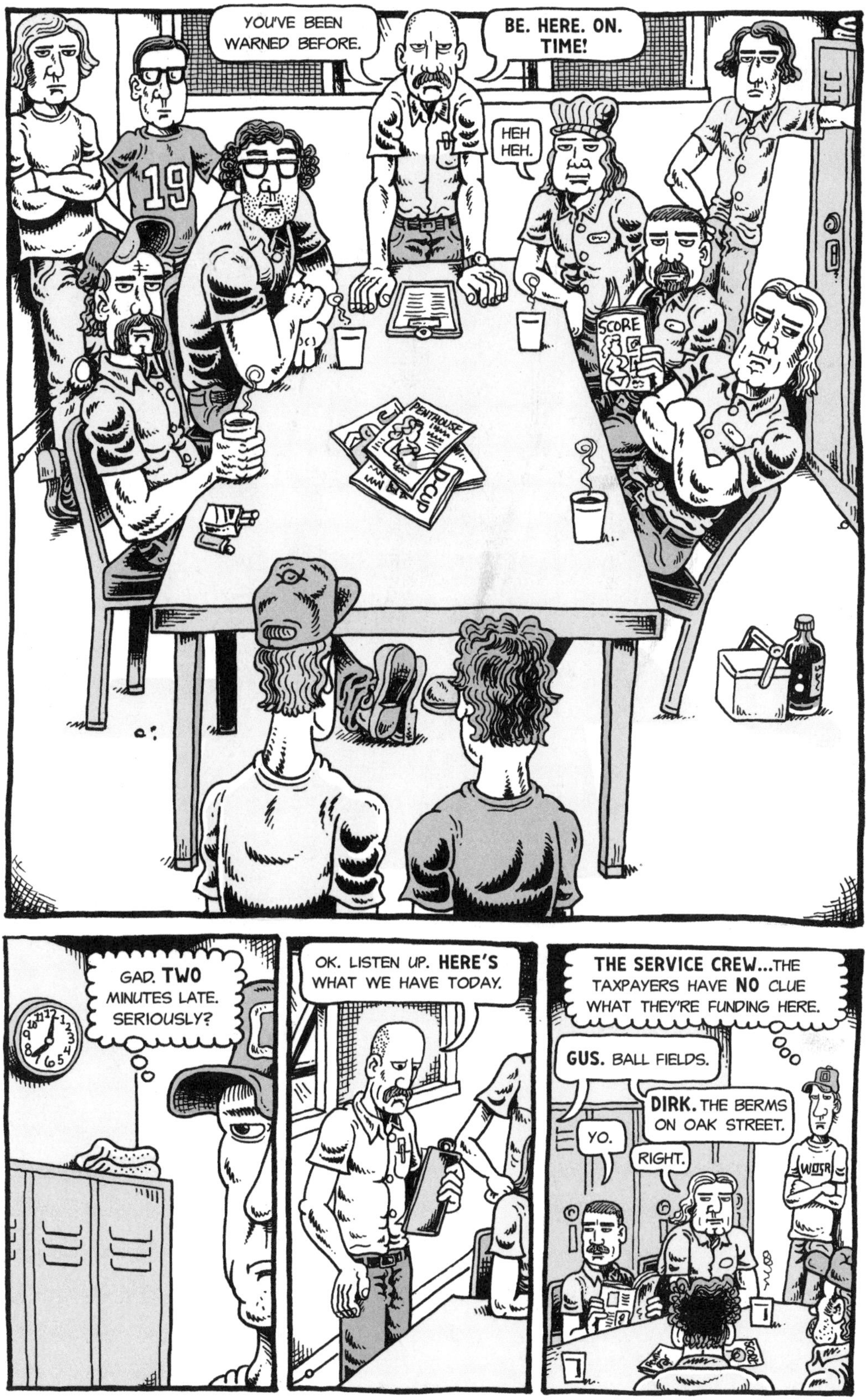
YOU'VE BEEN WARNED BEFORE.
BE. HERE. ON. TIME!
HEH HEH.
SCORE
PENTHOUSE
D-CUP
GAD. TWO MINUTES LATE. SERIOUSLY?
OK. LISTEN UP. HERE'S WHAT WE HAVE TODAY.
THE SERVICE CREW...THE TAXPAYERS HAVE NO CLUE WHAT THEY'RE FUNDING HERE.
GUS. BALL FIELDS.
YO.
DIRK. THE BERMS ON OAK STREET.
RIGHT.

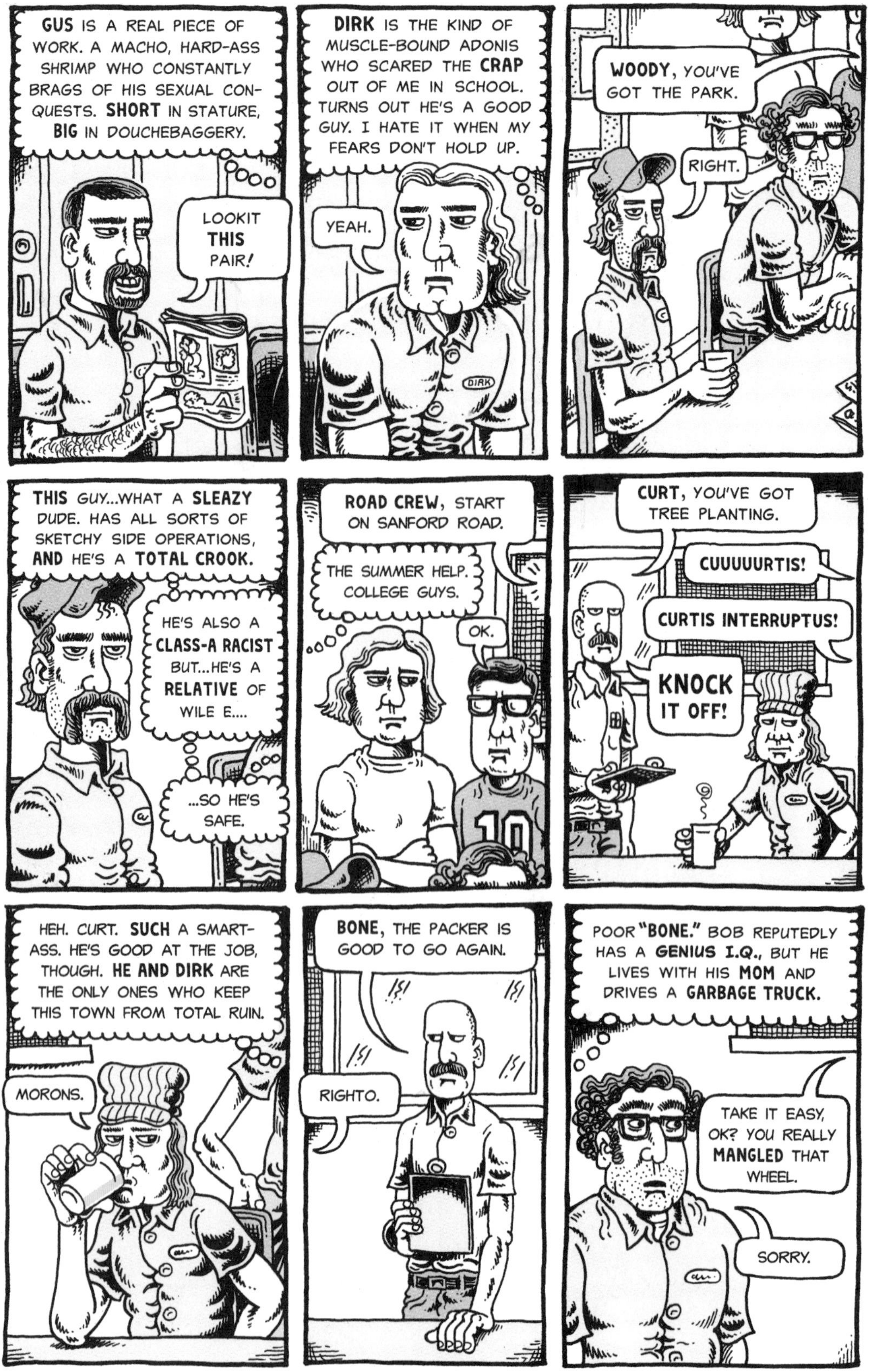
GUS IS A REAL PIECE OF WORK. A MACHO, HARD-ASS SHRIMP WHO CONSTANTLY BRAGS OF HIS SEXUAL CONQUESTS. SHORT IN STATURE, BIG IN DOUCHEBAGGERY.
LOOKIT THIS PAIR!
DIRK IS THE KIND OF MUSCLE-BOUND ADONIS WHO SCARED THE CRAP OUT OF ME IN SCHOOL. TURNS OUT HE'S A GOOD GUY. I HATE IT WHEN MY FEARS DON'T HOLD UP.
YEAH.
DIRK
WOODY, YOU'VE GOT THE PARK.
RIGHT.
THIS GUY...WHAT A SLEAZY DUDE. HAS ALL SORTS OF SKETCHY SIDE OPERATIONS, AND HE'S A TOTAL CROOK.
HE'S ALSO A CLASS-A RACIST BUT...HE'S A RELATIVE OF WILE E....
...SO HE'S SAFE.
ROAD CREW, START ON SANFORD ROAD.
THE SUMMER HELP. COLLEGE GUYS.
OK.
CURT, YOU'VE GOT TREE PLANTING.
CUUUUURTIS!
CURTIS INTERRUPTUS!
KNOCK IT OFF!
HEH. CURT. SUCH A SMART-ASS. HE'S GOOD AT THE JOB, THOUGH. HE AND DIRK ARE THE ONLY ONES WHO KEEP THIS TOWN FROM TOTAL RUIN.
MORONS.
BONE, THE PACKER IS GOOD TO GO AGAIN.
RIGHTO.
POOR "BONE." BOB REPUTEDLY HAS A GENIUS I.Q., BUT HE LIVES WITH HIS MOM AND DRIVES A GARBAGE TRUCK.
TAKE IT EASY, OK? YOU REALLY MANGLED THAT WHEEL.
SORRY.

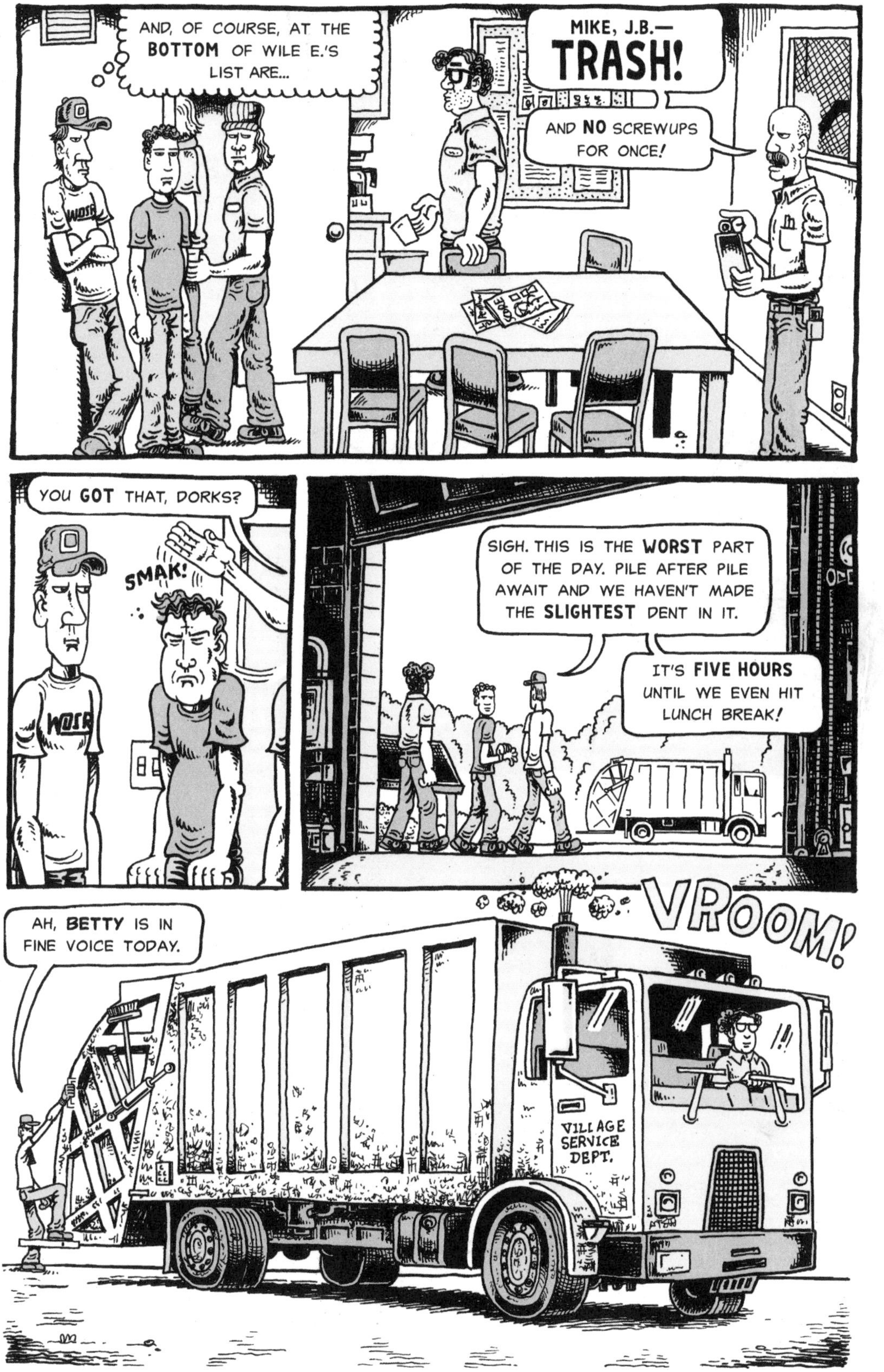

AND, OF COURSE, AT THE BOTTOM OF WILE E.'S LIST ARE...
MIKE, J.B.— TRASH!
AND NO SCREWUPS FOR ONCE!
YOU GOT THAT, DORKS?
SMAK!
SIGH. THIS IS THE WORST PART OF THE DAY. PILE AFTER PILE AWAIT AND WE HAVEN'T MADE THE SLIGHTEST DENT IN IT.
IT'S FIVE HOURS UNTIL WE EVEN HIT LUNCH BREAK!
AH, BETTY IS IN FINE VOICE TODAY.
VROOM!
VILLAGE SERVICE DEPT.

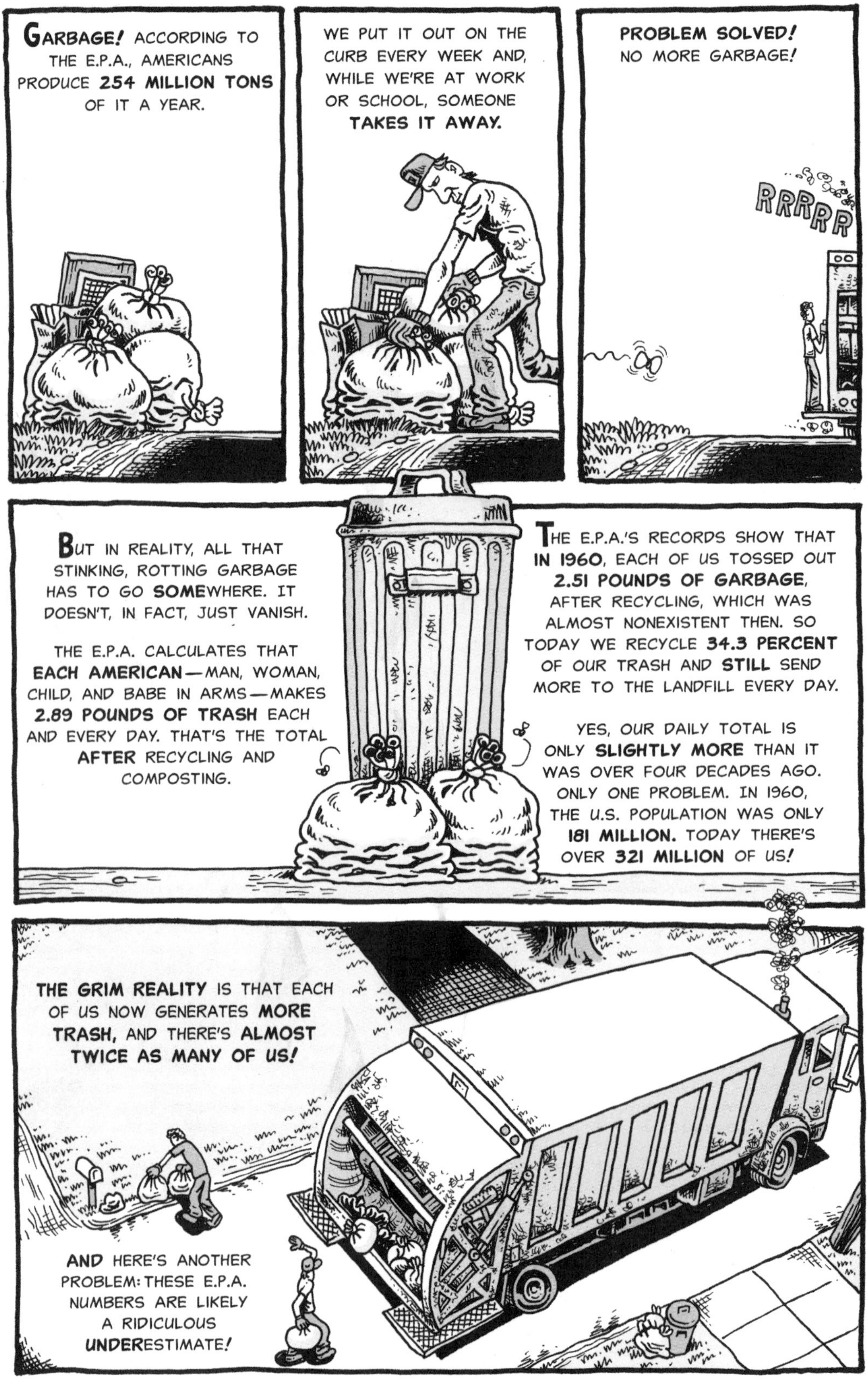
GARBAGE! ACCORDING TO THE E.P.A., AMERICANS PRODUCE 254 MILLION TONS OF IT A YEAR.
WE PUT IT OUT ON THE CURB EVERY WEEK AND, WHILE WE'RE AT WORK OR SCHOOL, SOMEONE TAKES IT AWAY.
PROBLEM SOLVED! NO MORE GARBAGE!
RRRRR
BUT IN REALITY, ALL THAT STINKING, ROTTING GARBAGE HAS TO GO SOMEWHERE. IT DOESN'T, IN FACT, JUST VANISH.
THE E.P.A. CALCULATES THAT EACH AMERICAN—MAN, WOMAN, CHILD, AND BABE IN ARMS—MAKES 2.89 POUNDS OF TRASH EACH AND EVERY DAY. THAT'S THE TOTAL AFTER RECYCLING AND COMPOSTING.
THE E.P.A.'S RECORDS SHOW THAT IN 1960, EACH OF US TOSSED OUT 2.51 POUNDS OF GARBAGE, AFTER RECYCLING, WHICH WAS ALMOST NONEXISTENT THEN. SO TODAY WE RECYCLE 34.3 PERCENT OF OUR TRASH AND STILL SEND MORE TO THE LANDFILL EVERY DAY.
YES, OUR DAILY TOTAL IS ONLY SLIGHTLY MORE THAN IT WAS OVER FOUR DECADES AGO. ONLY ONE PROBLEM. IN 1960, THE U.S. POPULATION WAS ONLY 181 MILLION. TODAY THERE'S OVER 321 MILLION OF US!
THE GRIM REALITY IS THAT EACH OF US NOW GENERATES MORE TRASH, AND THERE'S ALMOST TWICE AS MANY OF US!
AND HERE'S ANOTHER PROBLEM: THESE E.P.A. NUMBERS ARE LIKELY A RIDICULOUS UNDERESTIMATE!

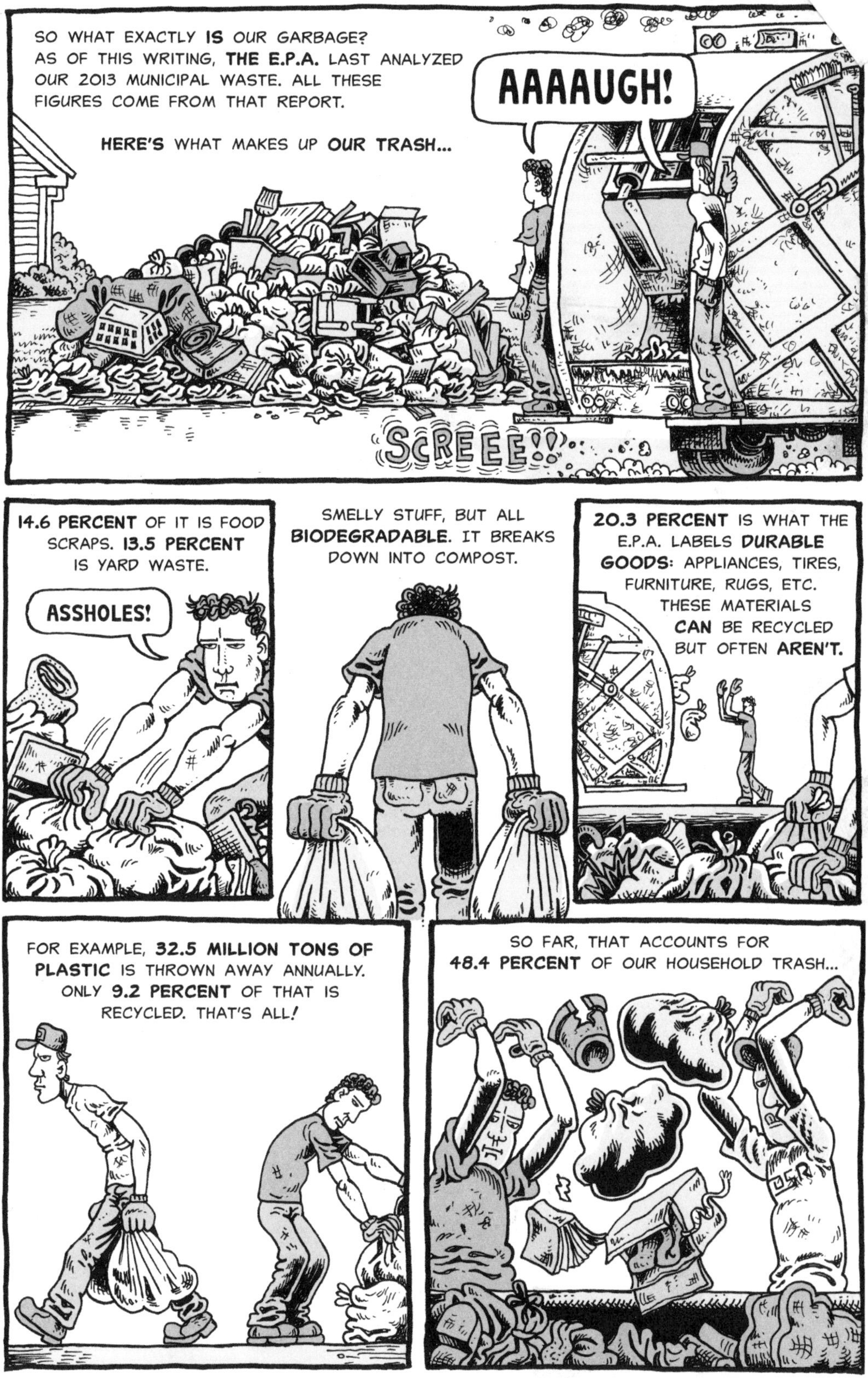
SO WHAT EXACTLY IS OUR GARBAGE? AS OF THIS WRITING, THE E.P.A. LAST ANALYZED OUR 2013 MUNICIPAL WASTE. ALL THESE FIGURES COME FROM THAT REPORT.
HERE'S WHAT MAKES UP OUR TRASH...
AAAAUGH!
SCREEE!!
14.6 PERCENT OF IT IS FOOD SCRAPS. 13.5 PERCENT IS YARD WASTE.
ASSHOLES!
SMELLY STUFF, BUT ALL BIODEGRADABLE. IT BREAKS DOWN INTO COMPOST.
20.3 PERCENT IS WHAT THE E.P.A. LABELS DURABLE GOODS: APPLIANCES, TIRES, FURNITURE, RUGS, ETC. THESE MATERIALS CAN BE RECYCLED BUT OFTEN AREN'T.
FOR EXAMPLE, 32.5 MILLION TONS OF PLASTIC IS THROWN AWAY ANNUALLY. ONLY 9.2 PERCENT OF THAT IS RECYCLED. THAT'S ALL!
SO FAR, THAT ACCOUNTS FOR 48.4 PERCENT OF OUR HOUSEHOLD TRASH...

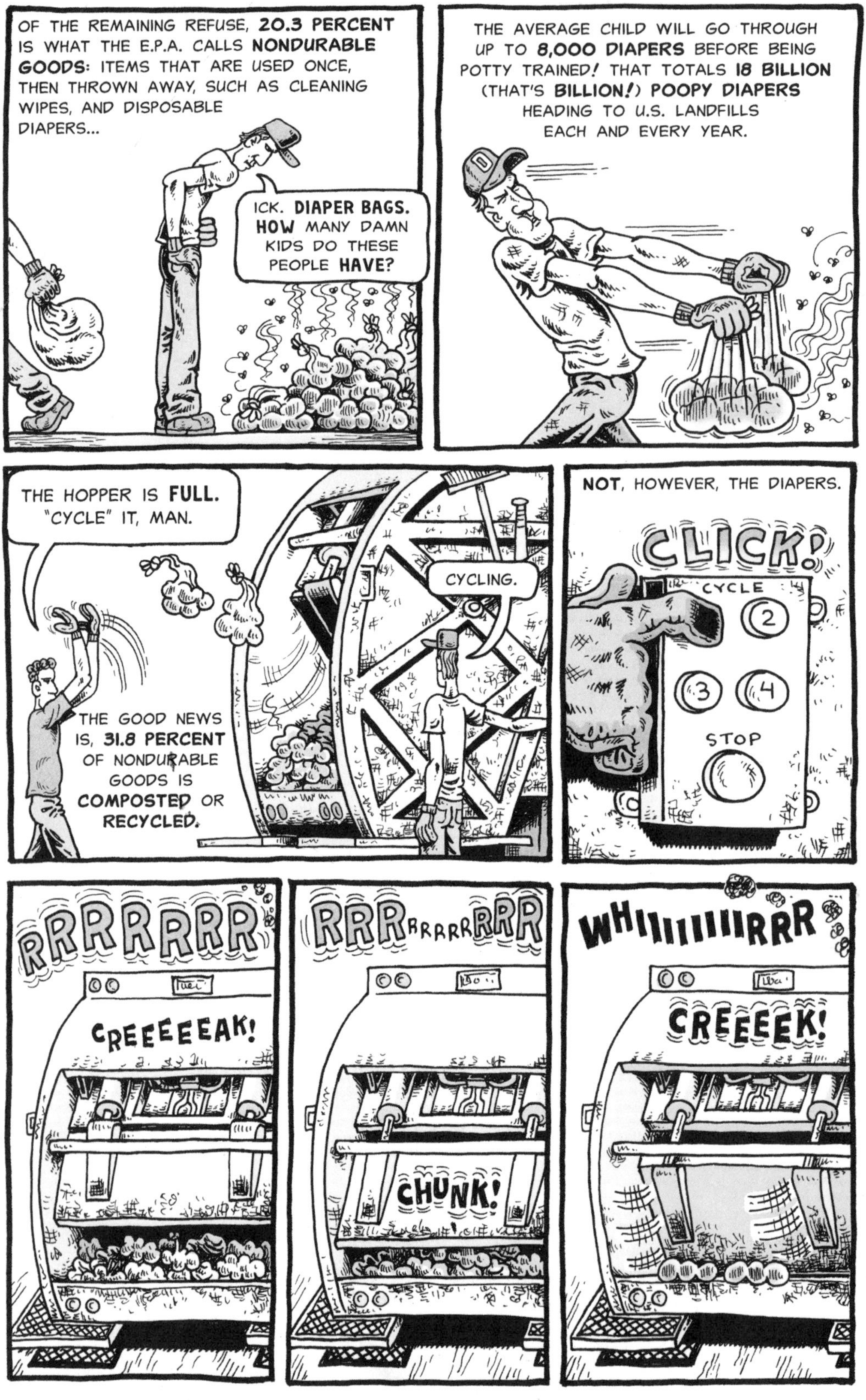
OF THE REMAINING REFUSE, 20.3 PERCENT IS WHAT THE E.P.A. CALLS NONDURABLE GOODS: ITEMS THAT ARE USED ONCE, THEN THROWN AWAY, SUCH AS CLEANING WIPES, AND DISPOSABLE DIAPERS...
ICK. DIAPER BAGS. HOW MANY DAMN KIDS DO THESE PEOPLE HAVE?
THE AVERAGE CHILD WILL GO THROUGH UP TO 8,000 DIAPERS BEFORE BEING POTTY TRAINED! THAT TOTALS 18 BILLION (THAT'S BILLION!) POOPY DIAPERS HEADING TO U.S. LANDFILLS EACH AND EVERY YEAR.
THE HOPPER IS FULL. "CYCLE" IT, MAN.
CYCLING.
THE GOOD NEWS IS, 31.8 PERCENT OF NONDURABLE GOODS IS COMPOSTED OR RECYCLED.
NOT, HOWEVER, THE DIAPERS.
CLICK!
CYCLE
2
3
4
STOP
RRRRRRR
CREEEEEAK!
RRRRRRRRRRR
CHUNK!
WHIIIIIIIIIRRR
CREEEEK!

AND THE REMAINING **29.8 PERCENT** OF OUR GARBAGE? IT'S **PACKAGING AND CONTAINERS.**

THAT'S RIGHT. THE **LARGEST PART** OF OUR CRAP IS THE CRAP OUR CRAP COMES IN!

RRRRRRRRRR

HEADS UP!

DIAPER BOMBS!

BLAM! BLAM!

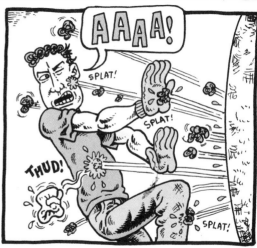

AAAA!

SPLAT!

SPLAT!

THUD!

SPLAT!

AWWWW, **MAN!**

HA! DIRECT HIT!

ICK.

WELL, **DON'T** STAND IN BETTY'S **LINE OF FIRE,** YOU FOOL.

THIS IS TH' **LAST** OF IT.

THAT'S OUR TRASH.

RRRRRRRR

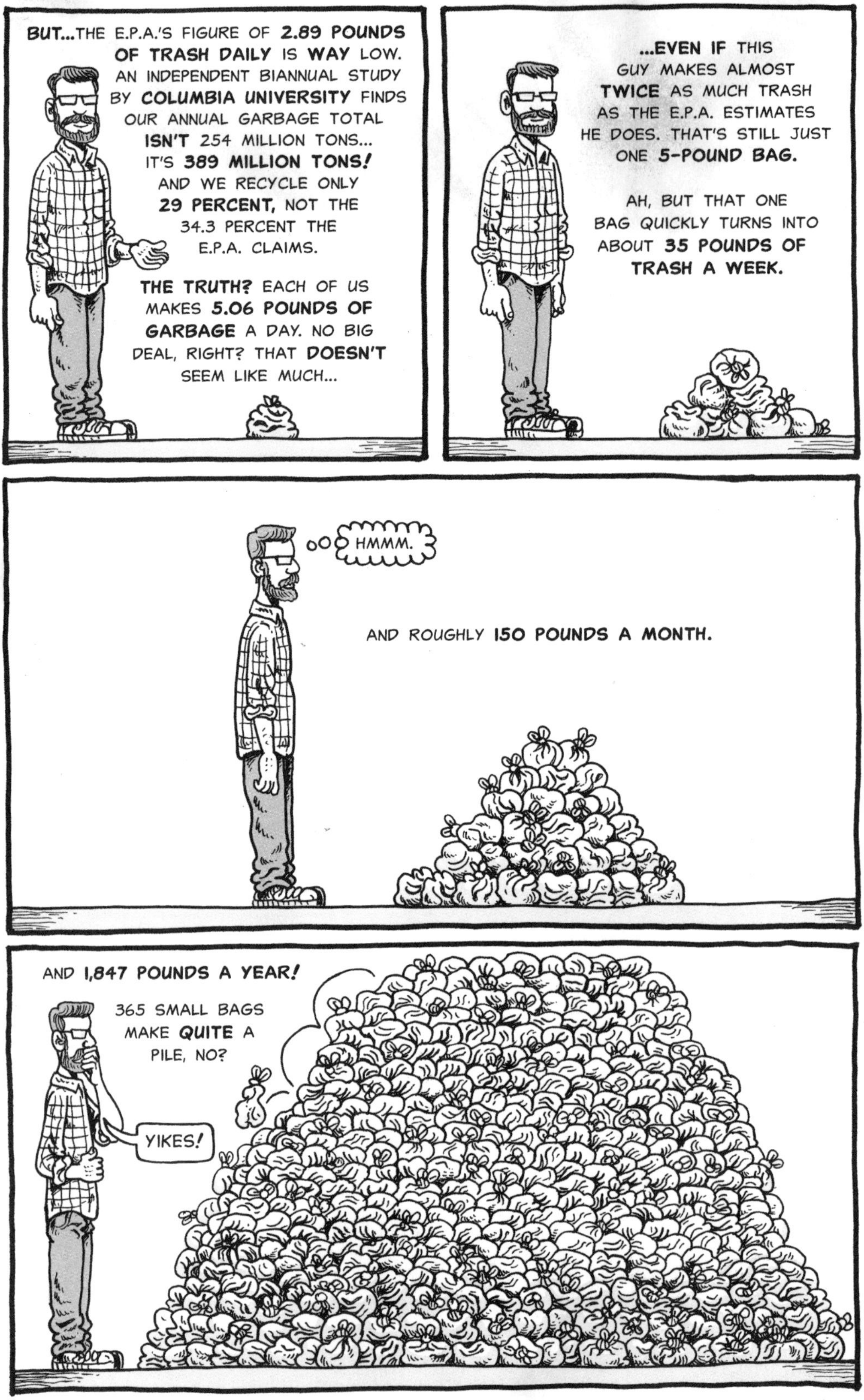
BUT...THE E.P.A.'S FIGURE OF 2.89 POUNDS OF TRASH DAILY IS WAY LOW. AN INDEPENDENT BIANNUAL STUDY BY COLUMBIA UNIVERSITY FINDS OUR ANNUAL GARBAGE TOTAL ISN'T 254 MILLION TONS... IT'S 389 MILLION TONS! AND WE RECYCLE ONLY 29 PERCENT, NOT THE 34.3 PERCENT THE E.P.A. CLAIMS.
THE TRUTH? EACH OF US MAKES 5.06 POUNDS OF GARBAGE A DAY. NO BIG DEAL, RIGHT? THAT DOESN'T SEEM LIKE MUCH...
...EVEN IF THIS GUY MAKES ALMOST TWICE AS MUCH TRASH AS THE E.P.A. ESTIMATES HE DOES. THAT'S STILL JUST ONE 5-POUND BAG.
AH, BUT THAT ONE BAG QUICKLY TURNS INTO ABOUT 35 POUNDS OF TRASH A WEEK.
HMMM.
AND ROUGHLY 150 POUNDS A MONTH.
AND 1,847 POUNDS A YEAR!
365 SMALL BAGS MAKE QUITE A PILE, NO?
YIKES!

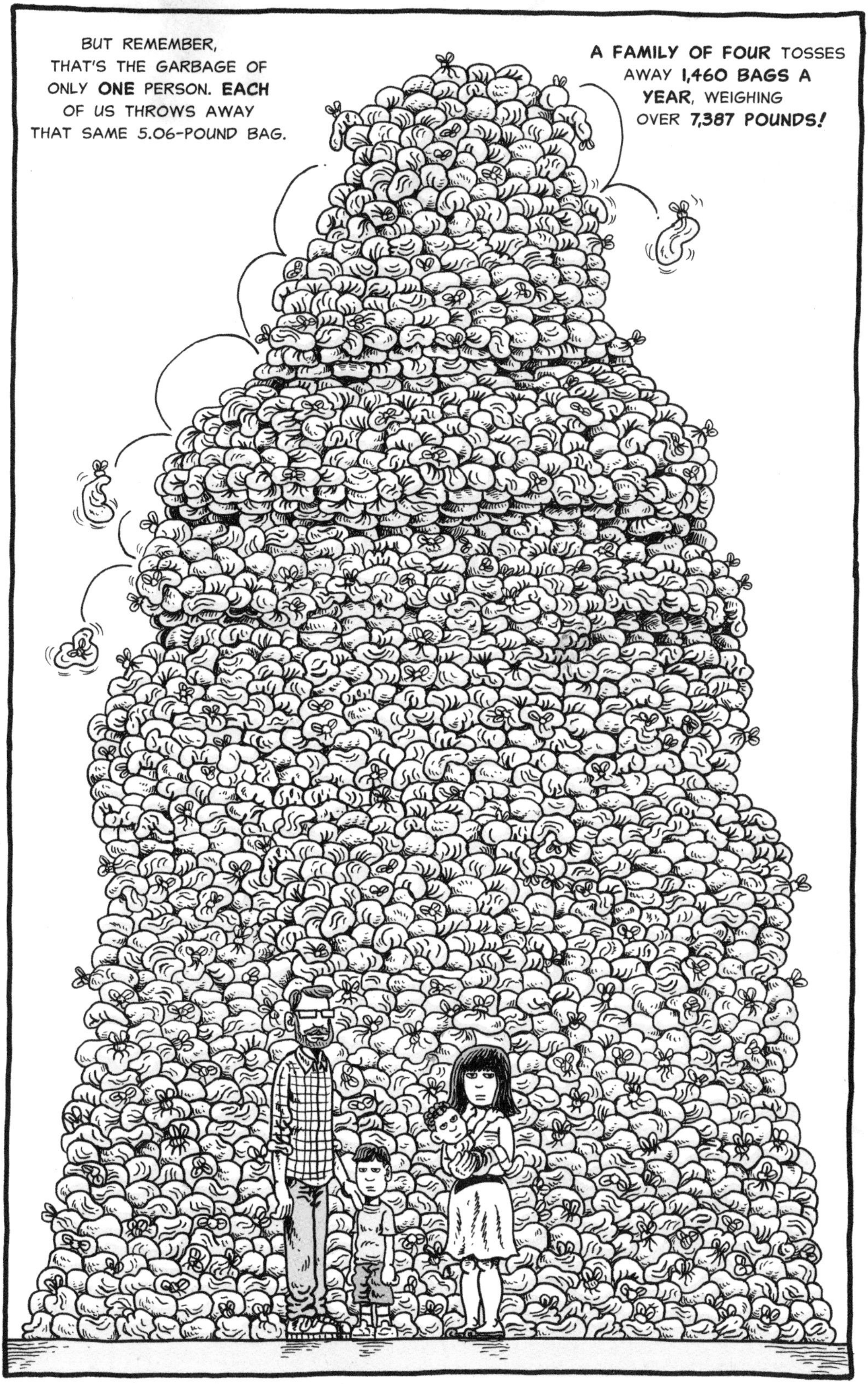
BUT REMEMBER, THAT'S THE GARBAGE OF ONLY ONE PERSON. EACH OF US THROWS AWAY THAT SAME 5.06-POUND BAG.
A FAMILY OF FOUR TOSSES AWAY 1,460 BAGS A YEAR, WEIGHING OVER 7,387 POUNDS!

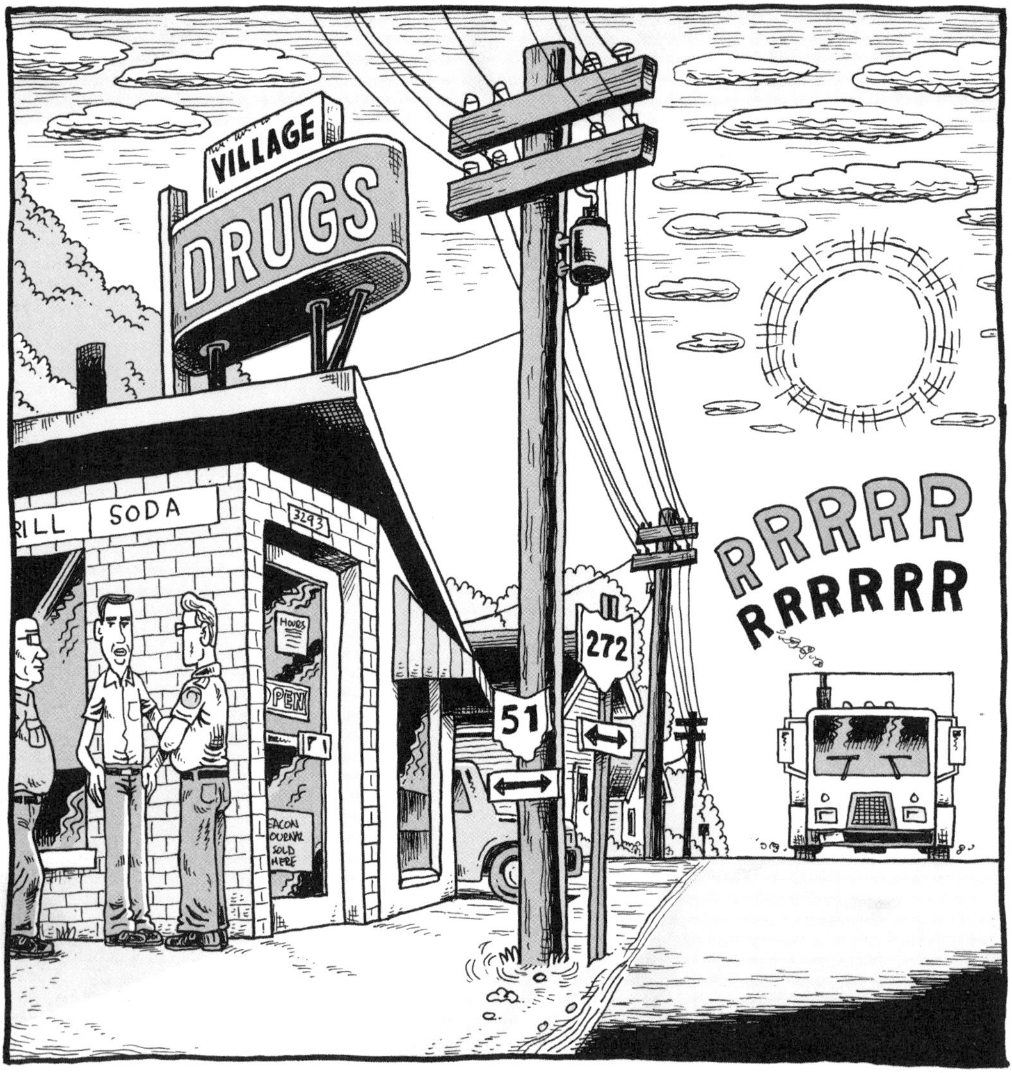
VILLAGE
DRUGS
ILL SODA
3293
HOURS
OPEN
BEACON JOURNAL SOLD HERE
272
51
RRRRR
RRRRRR

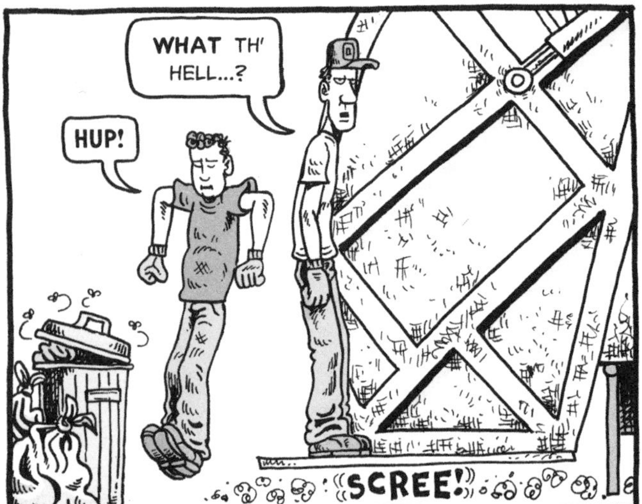
WHAT TH' HELL...?
HUP!
SCREE!

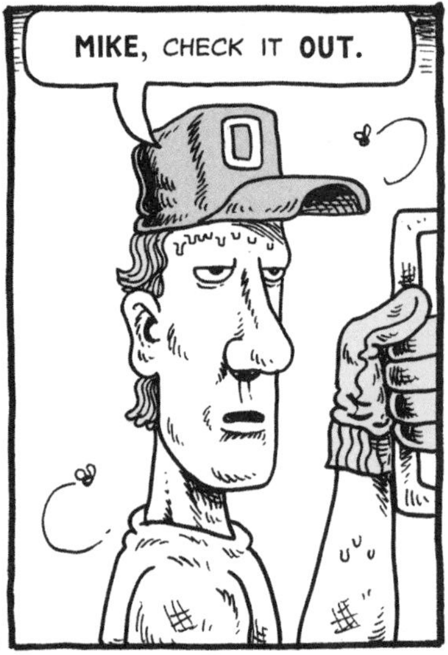
MIKE, CHECK IT OUT.

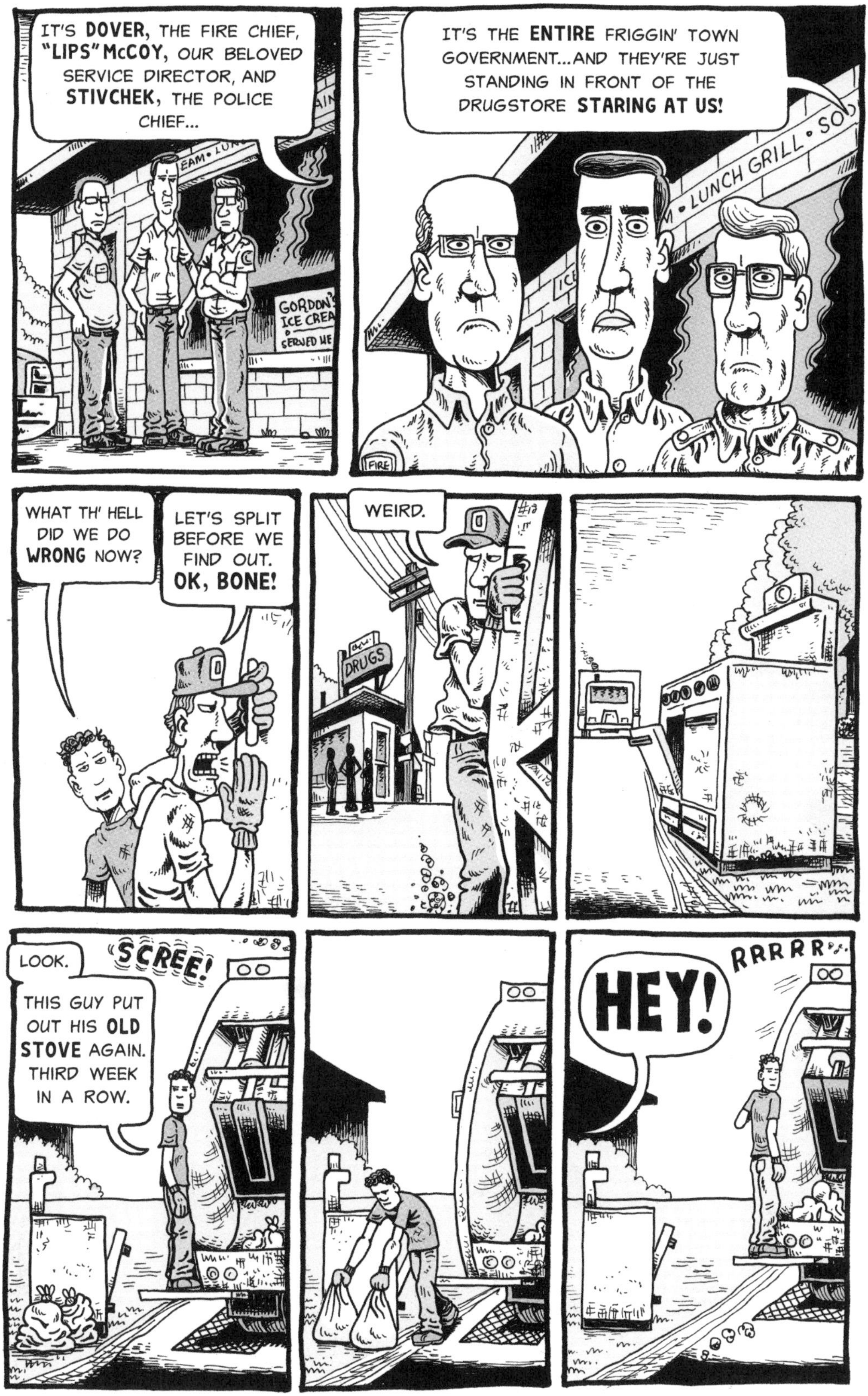
IT'S DOVER, THE FIRE CHIEF, "LIPS" McCOY, OUR BELOVED SERVICE DIRECTOR, AND STIVCHEK, THE POLICE CHIEF...
GORDON'S ICE CREA SERVED HE
IT'S THE ENTIRE FRIGGIN' TOWN GOVERNMENT...AND THEY'RE JUST STANDING IN FRONT OF THE DRUGSTORE STARING AT US!
LUNCH GRILL
FIRE
WHAT TH' HELL DID WE DO WRONG NOW?
LET'S SPLIT BEFORE WE FIND OUT. OK, BONE!
WEIRD.
DRUGS
LOOK.
SCREE!
THIS GUY PUT OUT HIS OLD STOVE AGAIN. THIRD WEEK IN A ROW.
HEY!
RRRRR

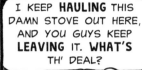

NOW, LOOK...

I KEEP **HAULING** THIS DAMN STOVE OUT HERE, AND YOU GUYS KEEP **LEAVING** IT. **WHAT'S** TH' DEAL?

WE ONLY TAKE APPLIANCES ON **HEAVY TRASH DAY.** VILLAGE RULES. SORRY.

WHEN'S THE **NEXT** HEAVY TRASH DAY?

MARCH.

WHAT?

AW, C,MON.

APPLIANCES FILL US UP **TOO FAST.**

MAN, IT **IS** HEAVY.

TELL YA **WHAT...**

CREEEAK!

THUMP!

IF YOU TAKE **THE DOORS** OFF, WE CAN PICK **THOSE** UP WHEN WE PASS BY AGAIN LATER TODAY.

CREAK!

THAT'S **ALL?** THE DOORS?

YEP.

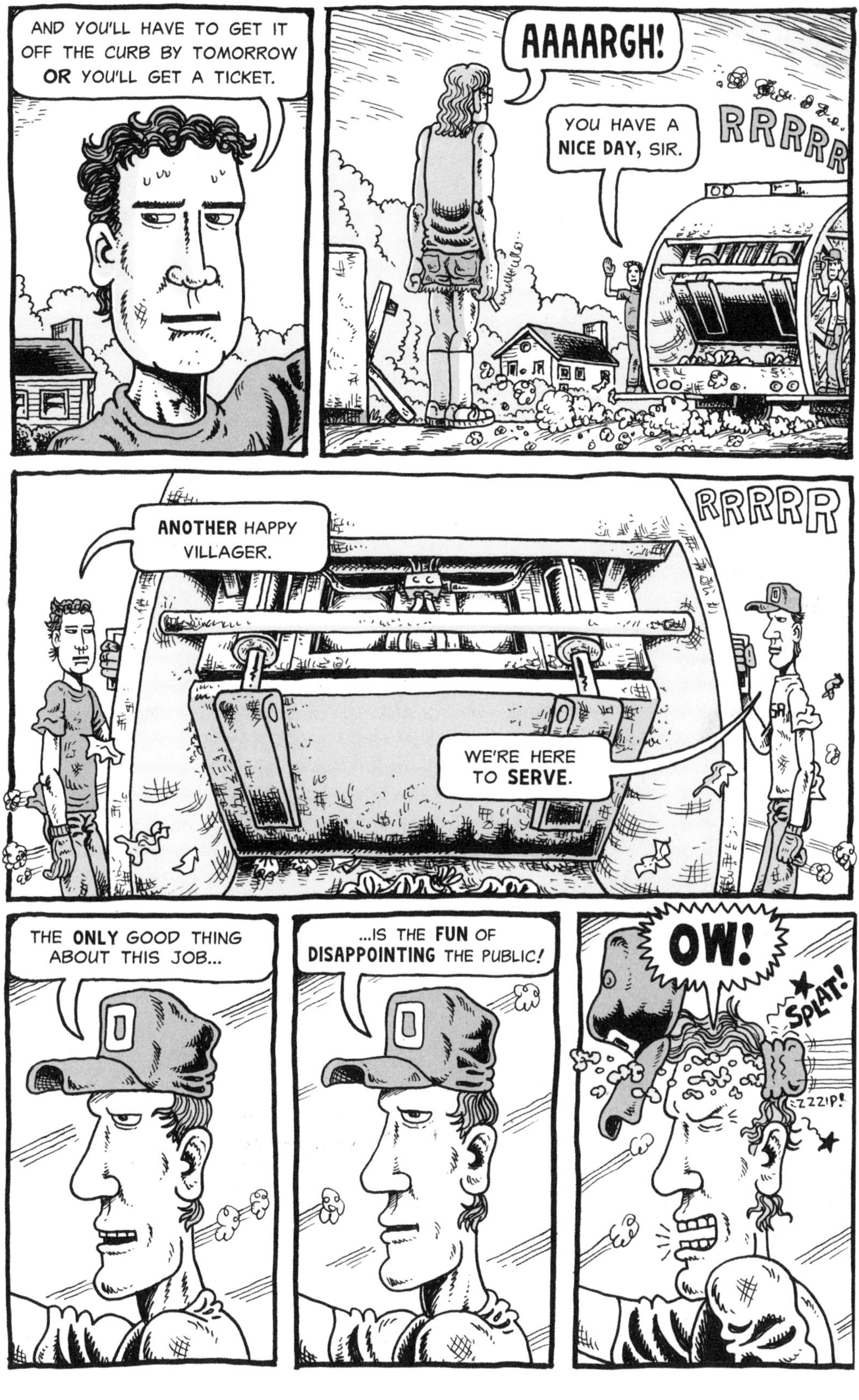
AND YOU'LL HAVE TO GET IT OFF THE CURB BY TOMORROW **OR** YOU'LL GET A TICKET.
AAAARGH!
YOU HAVE A **NICE DAY**, SIR.
RRRRRR
RRRRR
ANOTHER HAPPY VILLAGER.
WE'RE HERE TO **SERVE**.
THE **ONLY** GOOD THING ABOUT THIS JOB...
...IS THE **FUN** OF **DISAPPOINTING** THE PUBLIC!
OW!
SPLAT!
ZZZIP!

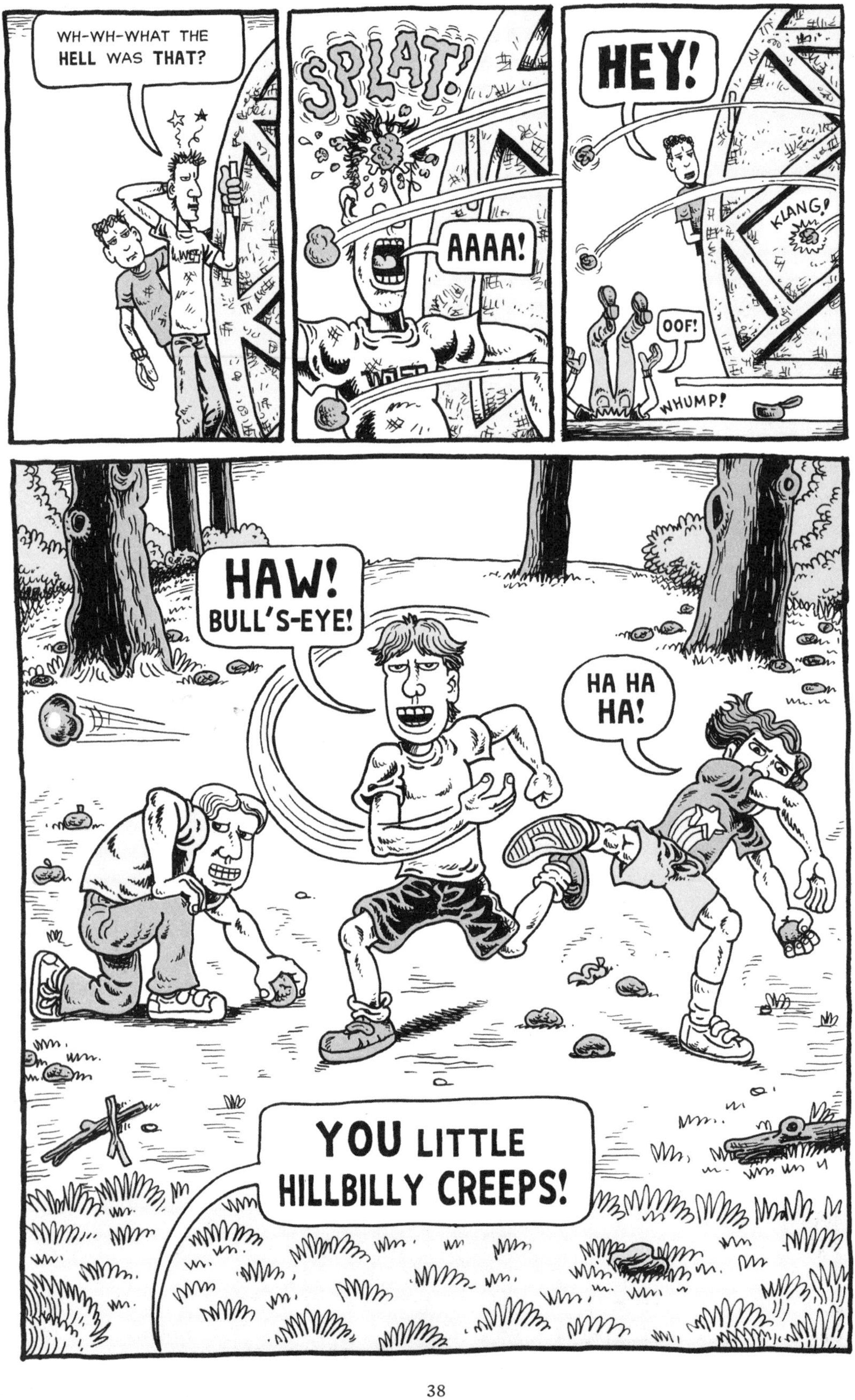
WH-WH-WHAT THE HELL WAS THAT?
SPLAT!
AAAA!
HEY!
KLANG!
OOF!
WHUMP!
HAW! BULL'S-EYE!
HA HA HA!
YOU LITTLE HILLBILLY CREEPS!

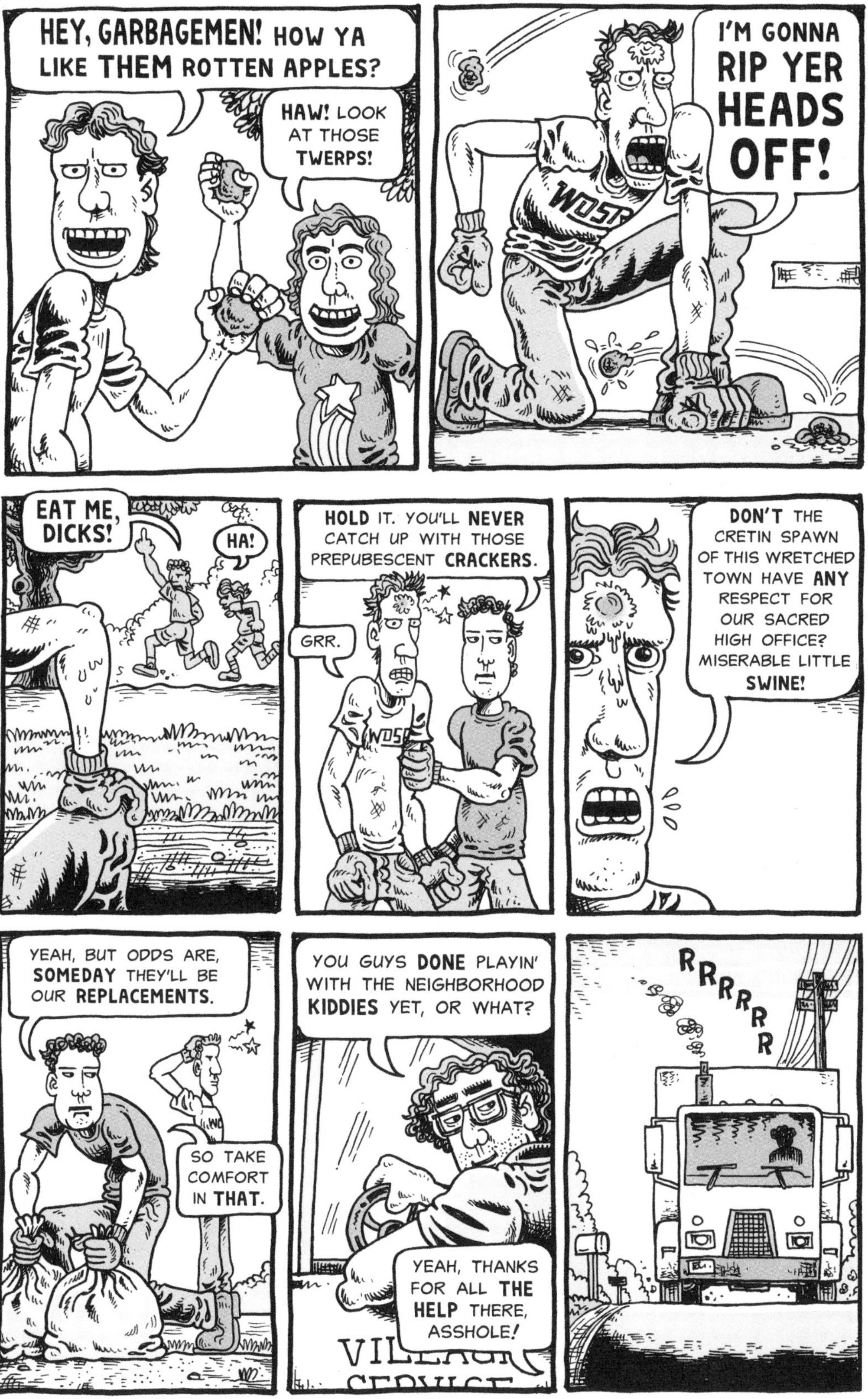
HEY, GARBAGEMEN! HOW YA LIKE THEM ROTTEN APPLES?
HAW! LOOK AT THOSE TWERPS!
I'M GONNA RIP YER HEADS OFF!
EAT ME, DICKS!
HA!
HOLD IT. YOU'LL NEVER CATCH UP WITH THOSE PREPUBESCENT CRACKERS.
GRR.
DON'T THE CRETIN SPAWN OF THIS WRETCHED TOWN HAVE ANY RESPECT FOR OUR SACRED HIGH OFFICE? MISERABLE LITTLE SWINE!
YEAH, BUT ODDS ARE, SOMEDAY THEY'LL BE OUR REPLACEMENTS.
SO TAKE COMFORT IN THAT.
YOU GUYS DONE PLAYIN' WITH THE NEIGHBORHOOD KIDDIES YET, OR WHAT?
YEAH, THANKS FOR ALL THE HELP THERE, ASSHOLE!
RRRRRR

SCRUNCH!

AAAAAR!

RIIIP!

DAMMIT!

THAT'S THE **THIRD** BAG THAT'S **SPLIT APART** TODAY!

TH' **GROCERY** MUSTA SWITCHED TO A **CHEAPER** GENERIC BRAND.

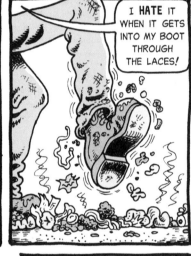

I **HATE** IT WHEN IT GETS INTO MY BOOT THROUGH THE LACES!

HANG ON. I'LL GET **THE SHOVEL.**

DON'T BOTHER.

OK, LET'S ROLL.

40

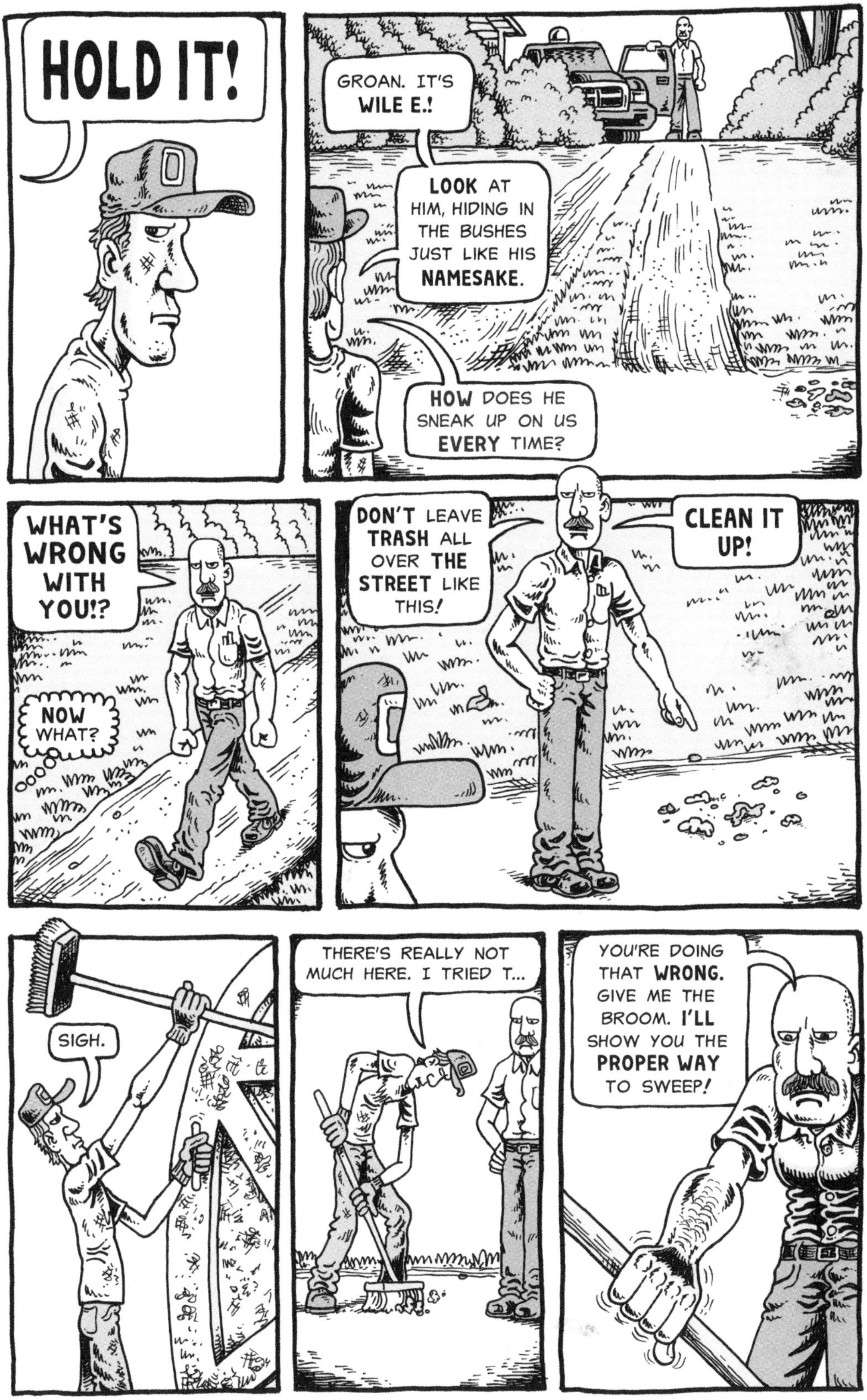
HOLD IT!
GROAN. IT'S WILE E.!
LOOK AT HIM, HIDING IN THE BUSHES JUST LIKE HIS NAMESAKE.
HOW DOES HE SNEAK UP ON US EVERY TIME?
WHAT'S WRONG WITH YOU!?
NOW WHAT?
DON'T LEAVE TRASH ALL OVER THE STREET LIKE THIS!
CLEAN IT UP!
SIGH.
THERE'S REALLY NOT MUCH HERE. I TRIED T...
YOU'RE DOING THAT WRONG. GIVE ME THE BROOM. I'LL SHOW YOU THE PROPER WAY TO SWEEP!

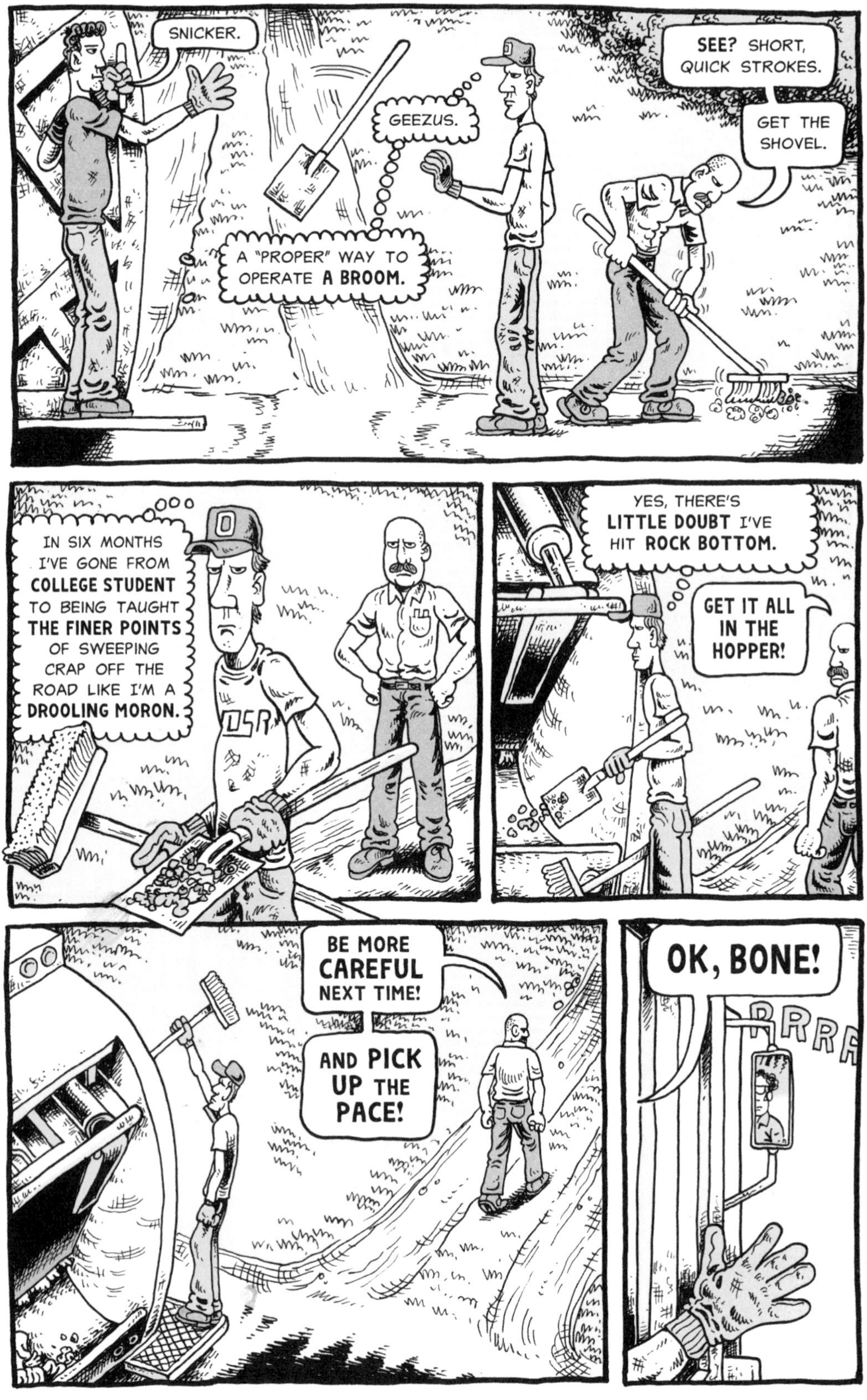
SNICKER.
GEEZUS.
A "PROPER" WAY TO OPERATE A BROOM.
SEE? SHORT, QUICK STROKES.
GET THE SHOVEL.
IN SIX MONTHS I'VE GONE FROM COLLEGE STUDENT TO BEING TAUGHT THE FINER POINTS OF SWEEPING CRAP OFF THE ROAD LIKE I'M A DROOLING MORON.
YES, THERE'S LITTLE DOUBT I'VE HIT ROCK BOTTOM.
GET IT ALL IN THE HOPPER!
BE MORE CAREFUL NEXT TIME!
AND PICK UP THE PACE!
OK, BONE!
RRRR

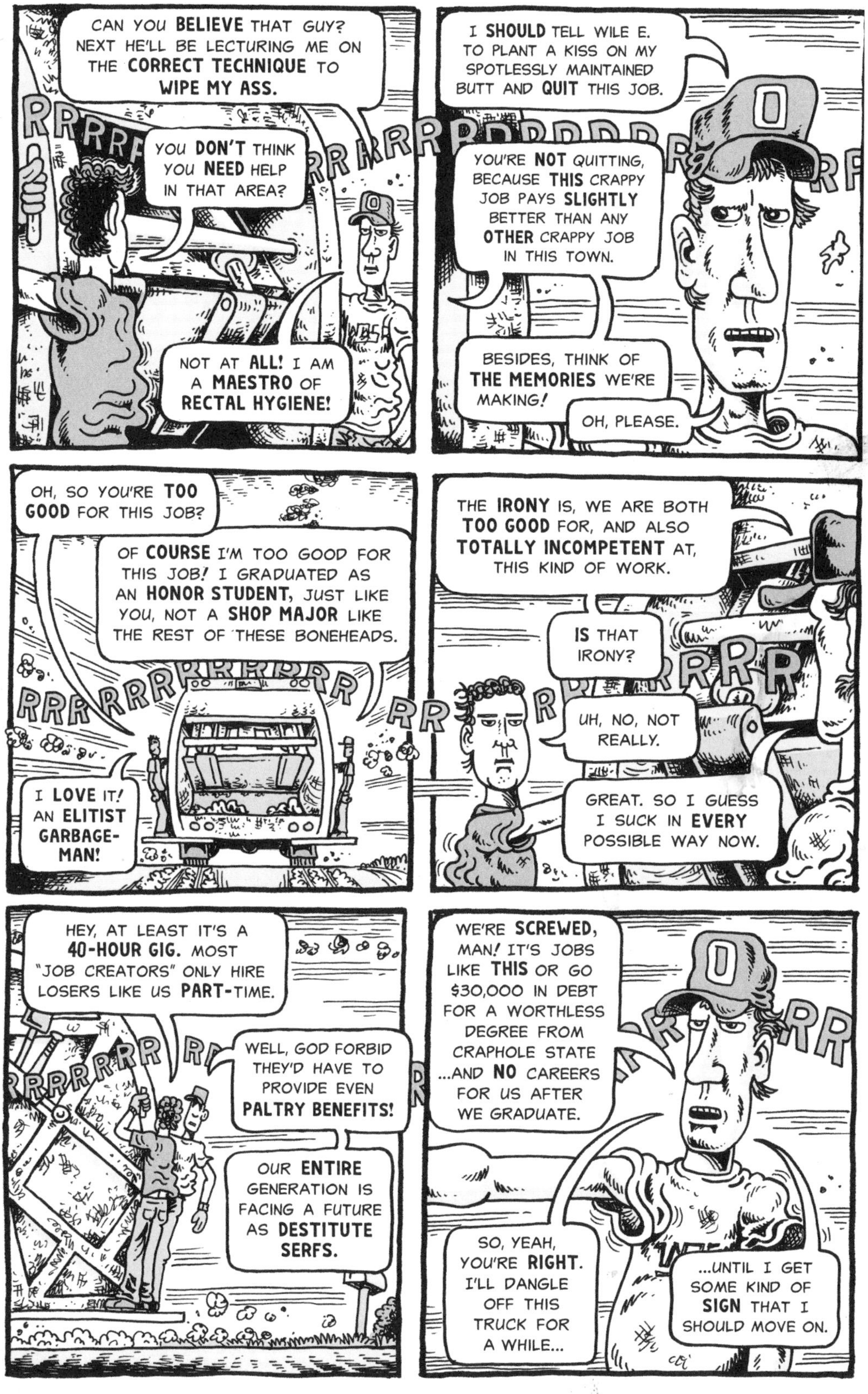
CAN YOU BELIEVE THAT GUY? NEXT HE'LL BE LECTURING ME ON THE CORRECT TECHNIQUE TO WIPE MY ASS.
YOU DON'T THINK YOU NEED HELP IN THAT AREA?
NOT AT ALL! I AM A MAESTRO OF RECTAL HYGIENE!
RRRRR
I SHOULD TELL WILE E. TO PLANT A KISS ON MY SPOTLESSLY MAINTAINED BUTT AND QUIT THIS JOB.
YOU'RE NOT QUITTING, BECAUSE THIS CRAPPY JOB PAYS SLIGHTLY BETTER THAN ANY OTHER CRAPPY JOB IN THIS TOWN.
BESIDES, THINK OF THE MEMORIES WE'RE MAKING!
OH, PLEASE.
OH, SO YOU'RE TOO GOOD FOR THIS JOB?
OF COURSE I'M TOO GOOD FOR THIS JOB! I GRADUATED AS AN HONOR STUDENT, JUST LIKE YOU, NOT A SHOP MAJOR LIKE THE REST OF THESE BONEHEADS.
I LOVE IT! AN ELITIST GARBAGE-MAN!
THE IRONY IS, WE ARE BOTH TOO GOOD FOR, AND ALSO TOTALLY INCOMPETENT AT, THIS KIND OF WORK.
IS THAT IRONY?
UH, NO, NOT REALLY.
GREAT. SO I GUESS I SUCK IN EVERY POSSIBLE WAY NOW.
HEY, AT LEAST IT'S A 40-HOUR GIG. MOST "JOB CREATORS" ONLY HIRE LOSERS LIKE US PART-TIME.
WELL, GOD FORBID THEY'D HAVE TO PROVIDE EVEN PALTRY BENEFITS!
OUR ENTIRE GENERATION IS FACING A FUTURE AS DESTITUTE SERFS.
WE'RE SCREWED, MAN! IT'S JOBS LIKE THIS OR GO $30,000 IN DEBT FOR A WORTHLESS DEGREE FROM CRAPHOLE STATE ...AND NO CAREERS FOR US AFTER WE GRADUATE.
SO, YEAH, YOU'RE RIGHT. I'LL DANGLE OFF THIS TRUCK FOR A WHILE...
...UNTIL I GET SOME KIND OF SIGN THAT I SHOULD MOVE ON.

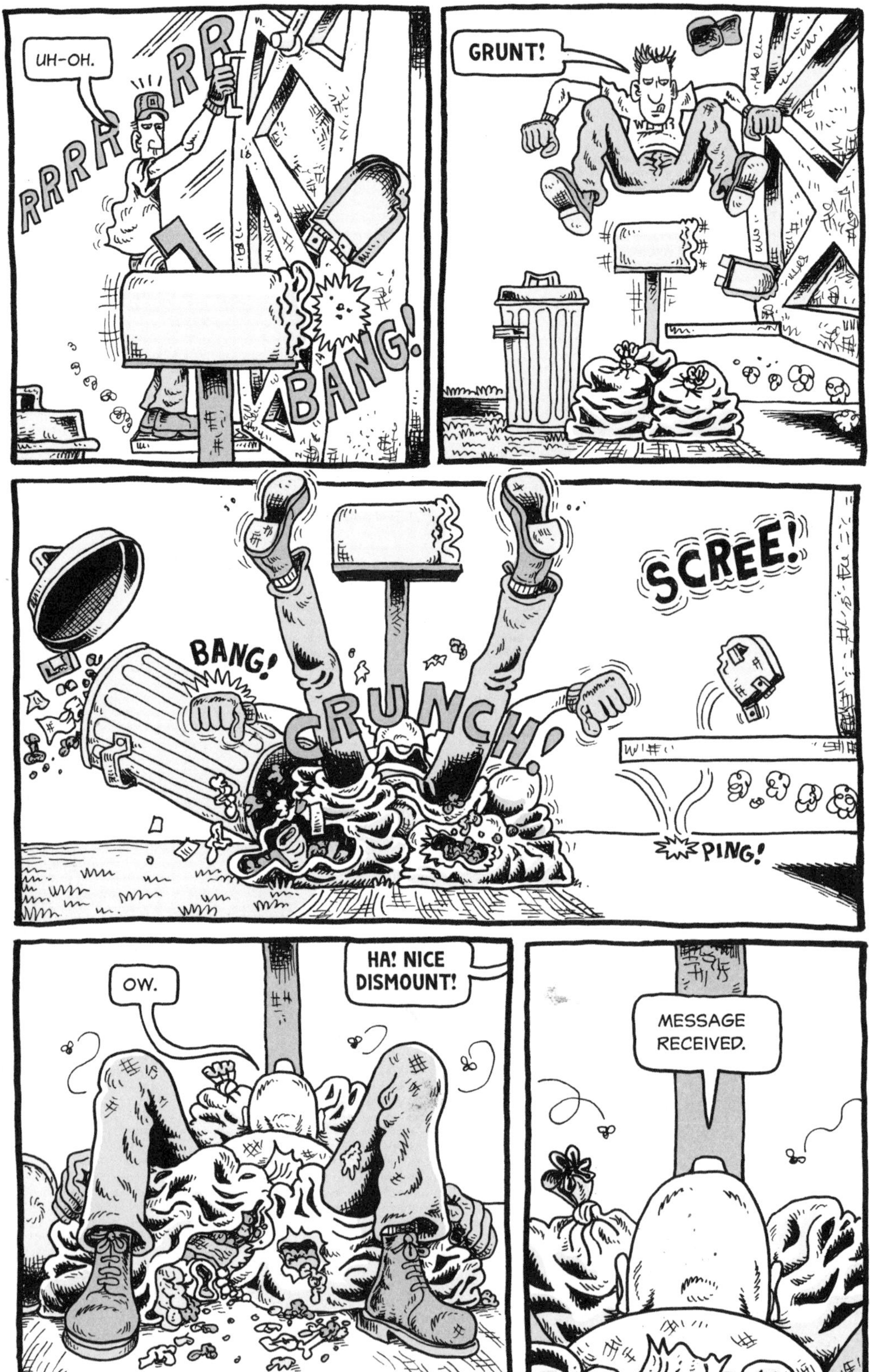
UH-OH.
RRRRRRR
BANG!
GRUNT!
BANG!
CRUNCH!
SCREE!
PING!
OW.
HA! NICE DISMOUNT!
MESSAGE RECEIVED.

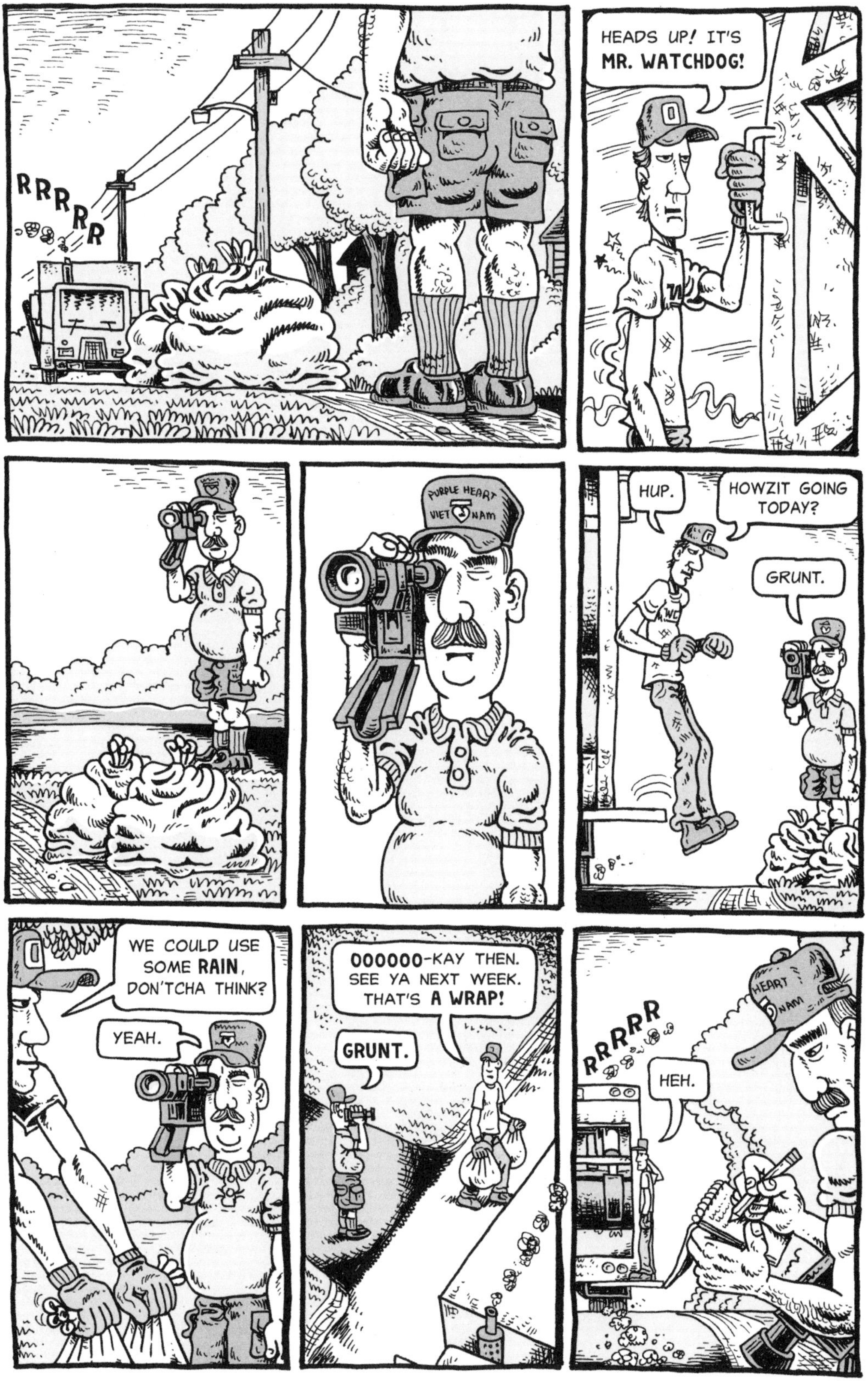
RRRRRR
HEADS UP! IT'S **MR. WATCHDOG!**
PURPLE HEART
VIET NAM
HUP.
HOWZIT GOING TODAY?
GRUNT.
WE COULD USE SOME **RAIN**, DON'TCHA THINK?
YEAH.
OOOOOO-KAY THEN. SEE YA NEXT WEEK. THAT'S **A WRAP!**
GRUNT.
RRRRR
HEH.
HEART
NAM

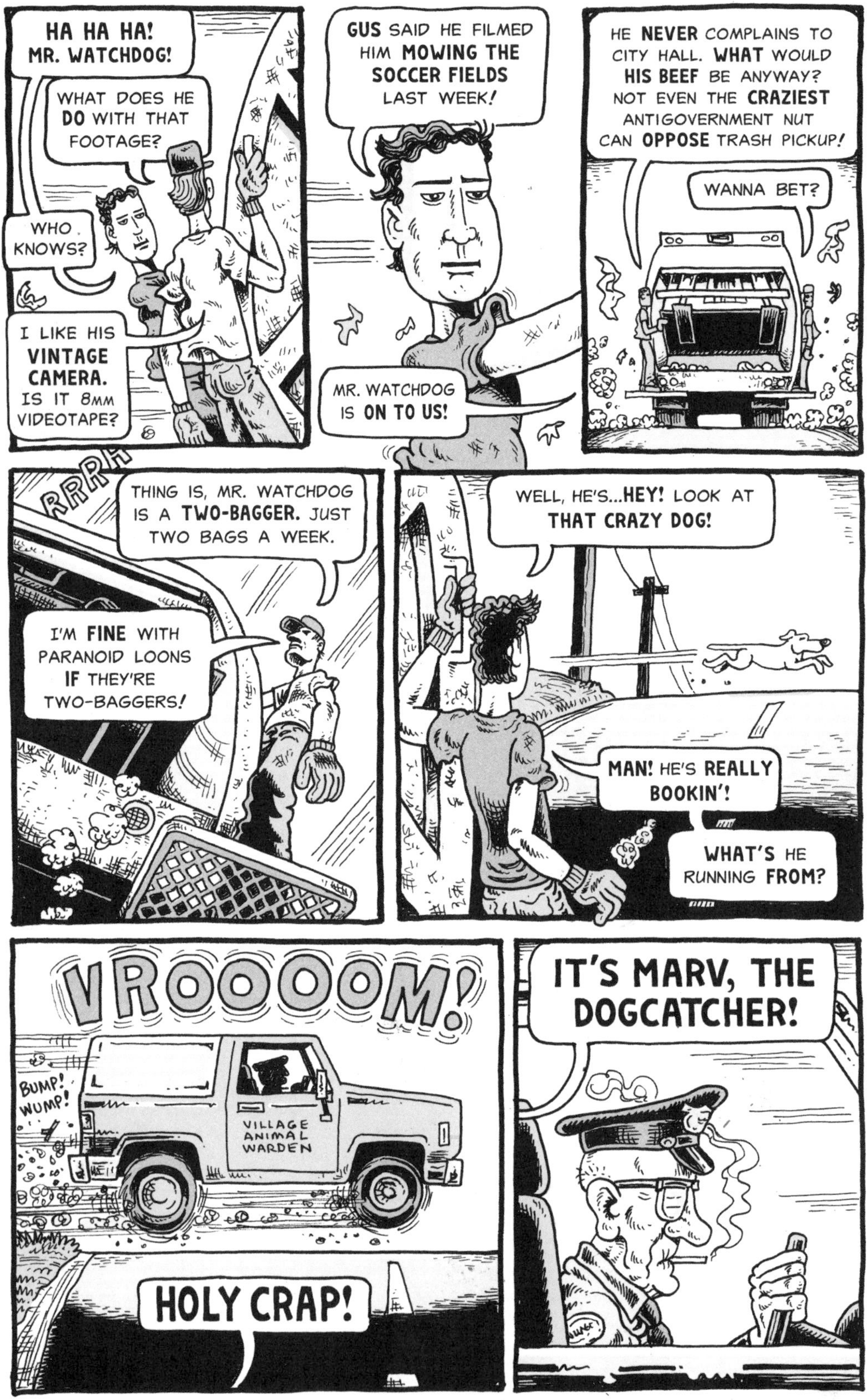
HA HA HA! MR. WATCHDOG!
WHAT DOES HE DO WITH THAT FOOTAGE?
WHO KNOWS?
I LIKE HIS VINTAGE CAMERA. IS IT 8MM VIDEOTAPE?
GUS SAID HE FILMED HIM MOWING THE SOCCER FIELDS LAST WEEK!
MR. WATCHDOG IS ON TO US!
HE NEVER COMPLAINS TO CITY HALL. WHAT WOULD HIS BEEF BE ANYWAY? NOT EVEN THE CRAZIEST ANTIGOVERNMENT NUT CAN OPPOSE TRASH PICKUP!
WANNA BET?
RRRR
THING IS, MR. WATCHDOG IS A TWO-BAGGER. JUST TWO BAGS A WEEK.
I'M FINE WITH PARANOID LOONS IF THEY'RE TWO-BAGGERS!
WELL, HE'S...HEY! LOOK AT THAT CRAZY DOG!
MAN! HE'S REALLY BOOKIN'!
WHAT'S HE RUNNING FROM?
VROOOOM!
BUMP! WUMP!
VILLAGE ANIMAL WARDEN
HOLY CRAP!
IT'S MARV, THE DOGCATCHER!

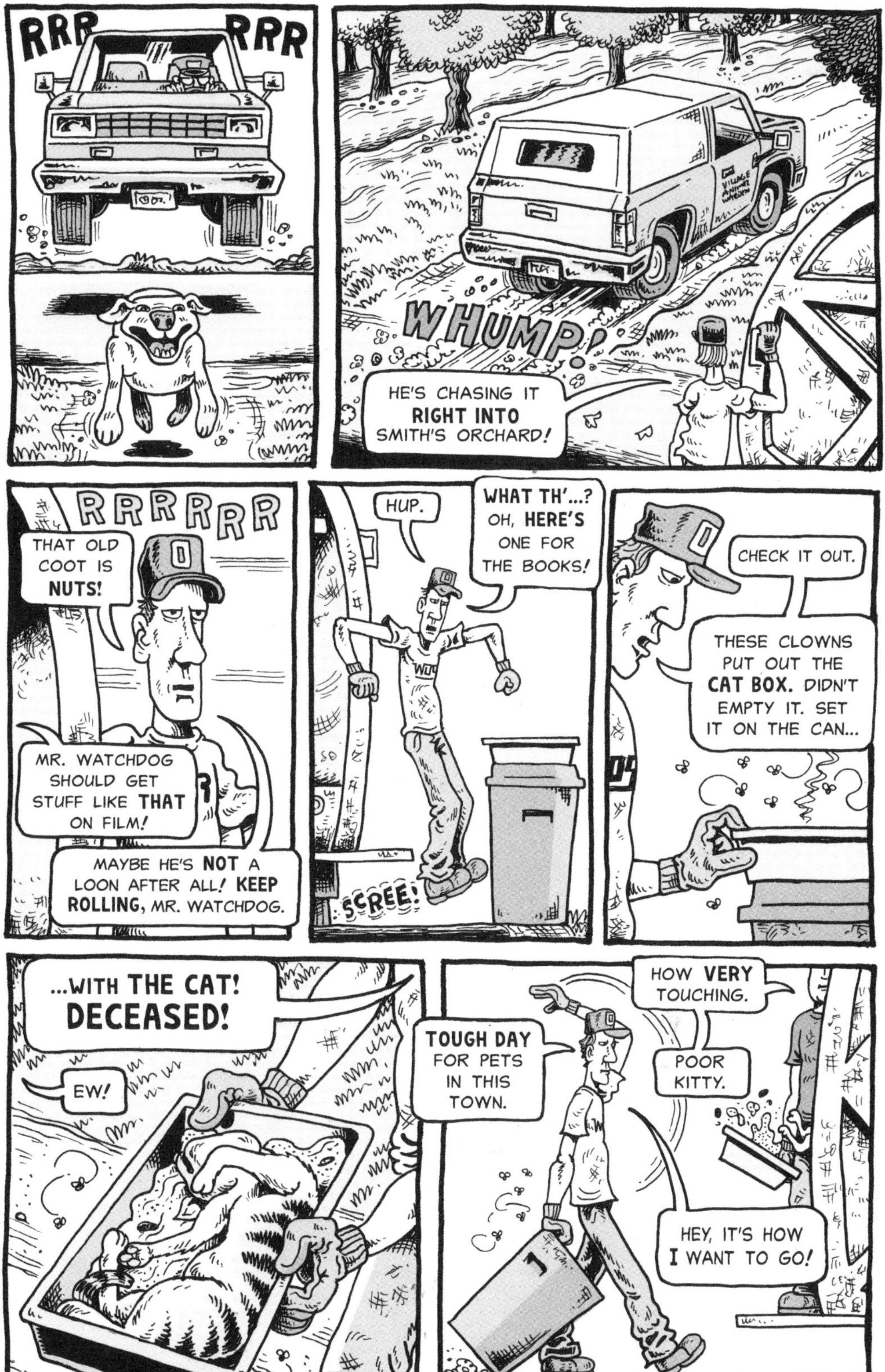
RRR RRR
WHUMP!
HE'S CHASING IT **RIGHT INTO** SMITH'S ORCHARD!
RRRRRRR
THAT OLD COOT IS **NUTS!**
MR. WATCHDOG SHOULD GET STUFF LIKE **THAT** ON FILM!
MAYBE HE'S **NOT** A LOON AFTER ALL! **KEEP ROLLING**, MR. WATCHDOG.
HUP.
SCREE!
WHAT TH'...? OH, **HERE'S** ONE FOR THE BOOKS!
CHECK IT OUT.
THESE CLOWNS PUT OUT THE **CAT BOX.** DIDN'T EMPTY IT. SET IT ON THE CAN...
...WITH **THE CAT! DECEASED!**
EW!
TOUGH DAY FOR PETS IN THIS TOWN.
HOW **VERY** TOUCHING.
POOR KITTY.
HEY, IT'S HOW I WANT TO GO!

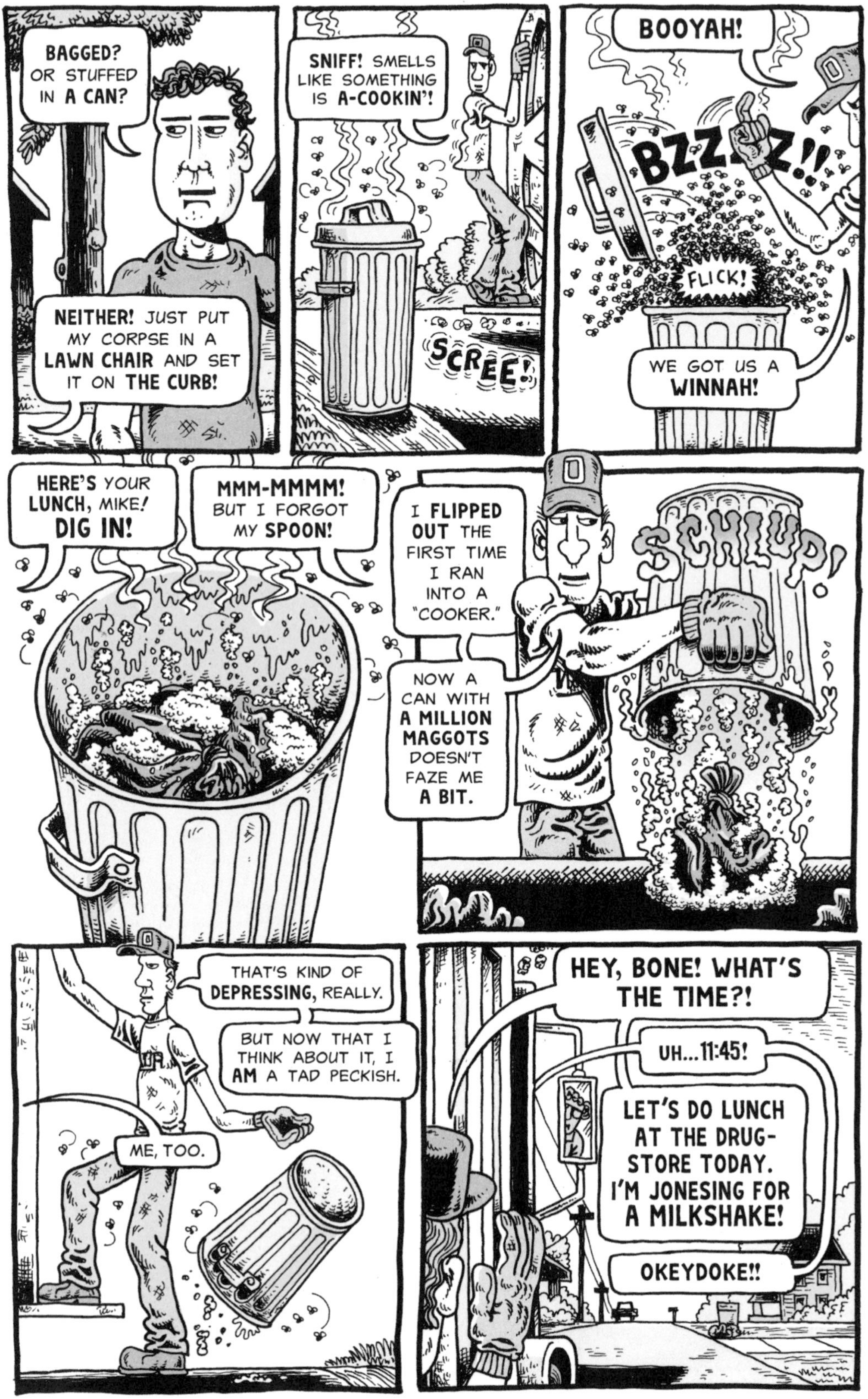
BAGGED? OR STUFFED IN A CAN?
NEITHER! JUST PUT MY CORPSE IN A LAWN CHAIR AND SET IT ON THE CURB!
SNIFF! SMELLS LIKE SOMETHING IS A-COOKIN'!
SCREE!
BOOYAH!
BZZZZ!!
FLICK!
WE GOT US A WINNAH!
HERE'S YOUR LUNCH, MIKE! DIG IN!
MMM-MMMM! BUT I FORGOT MY SPOON!
I FLIPPED OUT THE FIRST TIME I RAN INTO A "COOKER."
NOW A CAN WITH A MILLION MAGGOTS DOESN'T FAZE ME A BIT.
SCHLUP!
THAT'S KIND OF DEPRESSING, REALLY.
BUT NOW THAT I THINK ABOUT IT, I AM A TAD PECKISH.
ME, TOO.
HEY, BONE! WHAT'S THE TIME?!
UH...11:45!
LET'S DO LUNCH AT THE DRUG-STORE TODAY. I'M JONESING FOR A MILKSHAKE!
OKEYDOKE!!

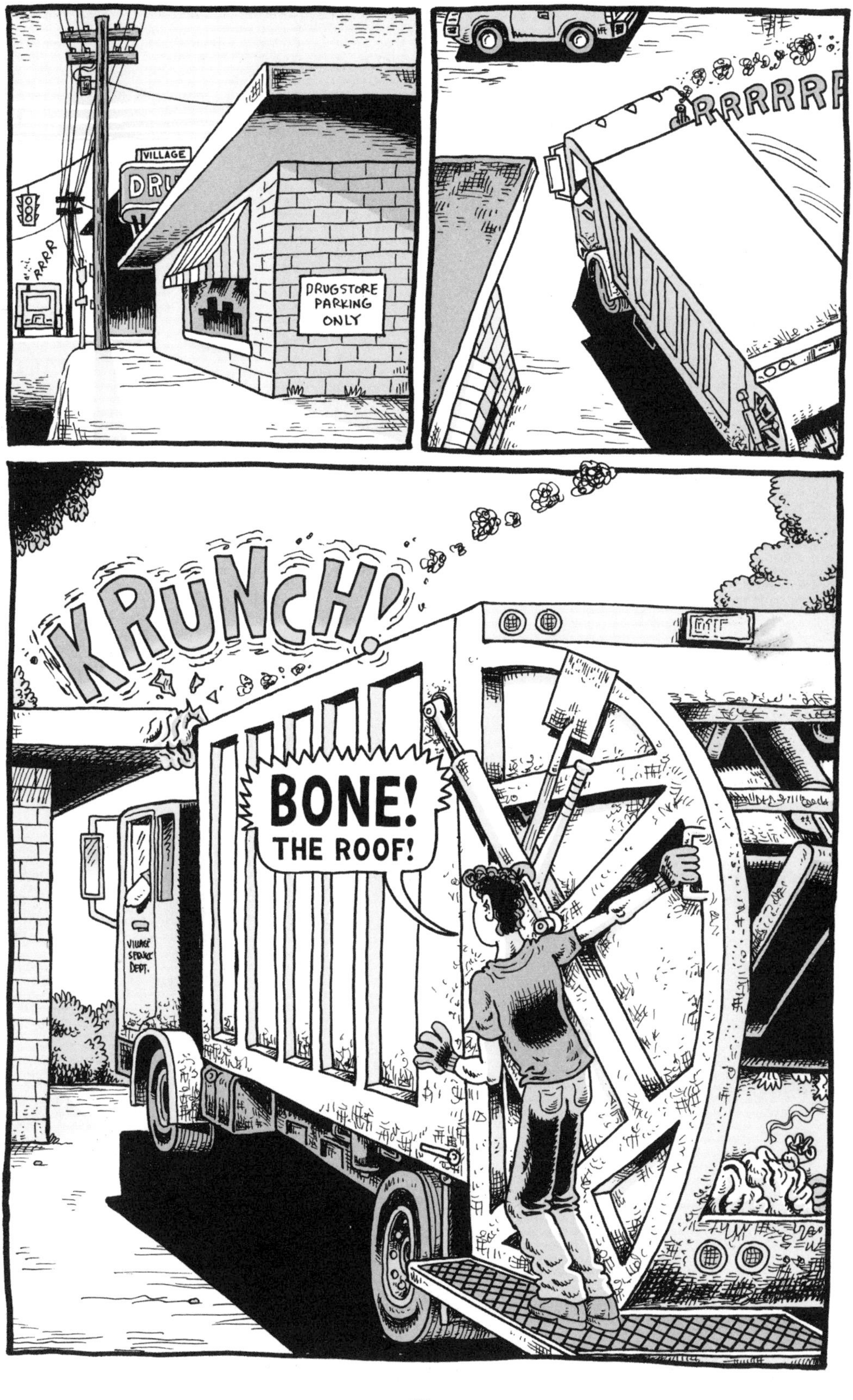
VILLAGE
DRU
RRRR
DRUGSTORE PARKING ONLY
RRRRRR
KRUNCH!
VILLAGE SERVICE DEPT.
BONE! THE ROOF!

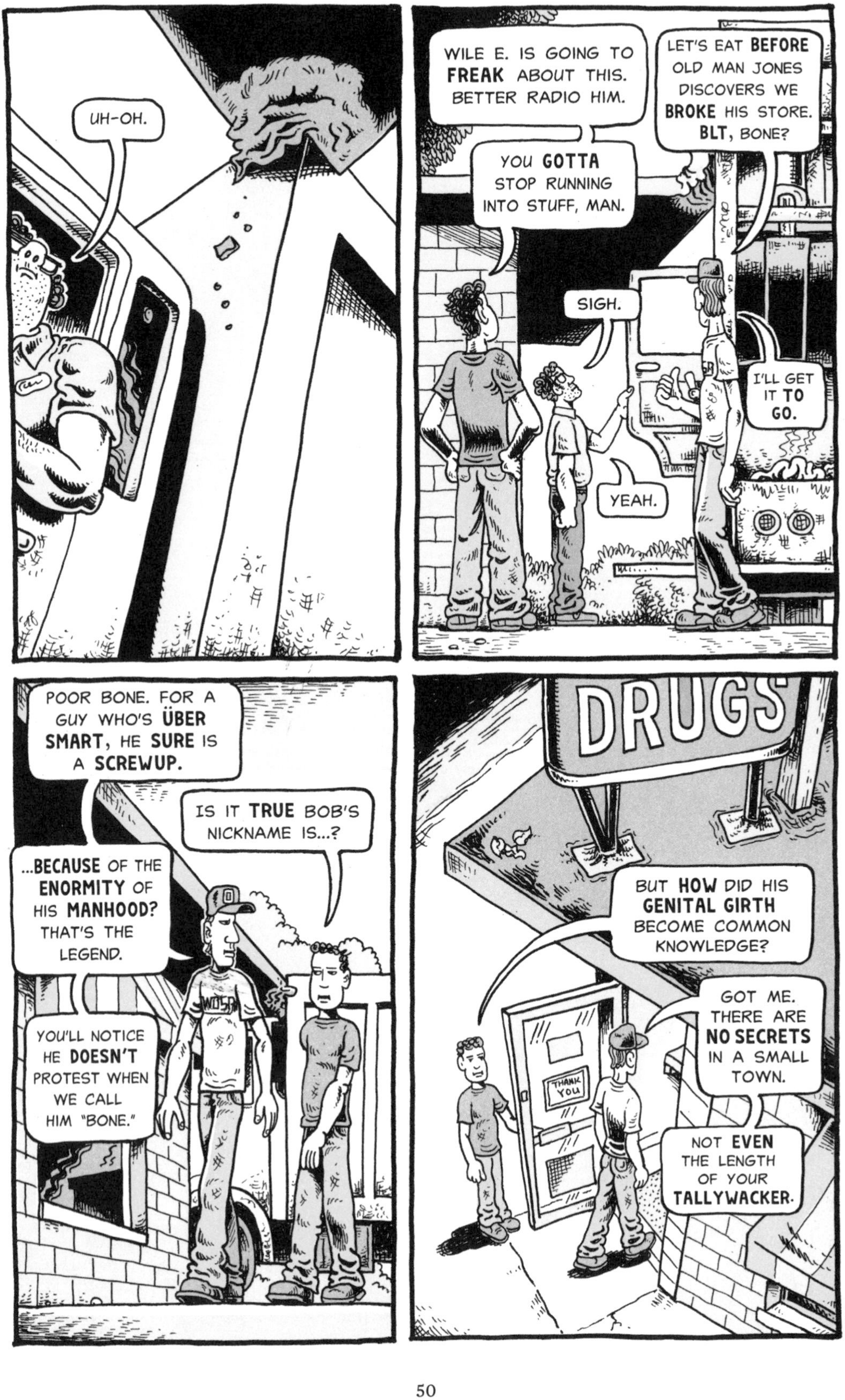
UH-OH.
WILE E. IS GOING TO FREAK ABOUT THIS. BETTER RADIO HIM.
YOU GOTTA STOP RUNNING INTO STUFF, MAN.
LET'S EAT BEFORE OLD MAN JONES DISCOVERS WE BROKE HIS STORE. BLT, BONE?
SIGH.
I'LL GET IT TO GO.
YEAH.
POOR BONE. FOR A GUY WHO'S ÜBER SMART, HE SURE IS A SCREWUP.
IS IT TRUE BOB'S NICKNAME IS...?
...BECAUSE OF THE ENORMITY OF HIS MANHOOD? THAT'S THE LEGEND.
YOU'LL NOTICE HE DOESN'T PROTEST WHEN WE CALL HIM "BONE."
WDSR
DRUGS
BUT HOW DID HIS GENITAL GIRTH BECOME COMMON KNOWLEDGE?
GOT ME. THERE ARE NO SECRETS IN A SMALL TOWN.
NOT EVEN THE LENGTH OF YOUR TALLYWACKER.
THANK YOU

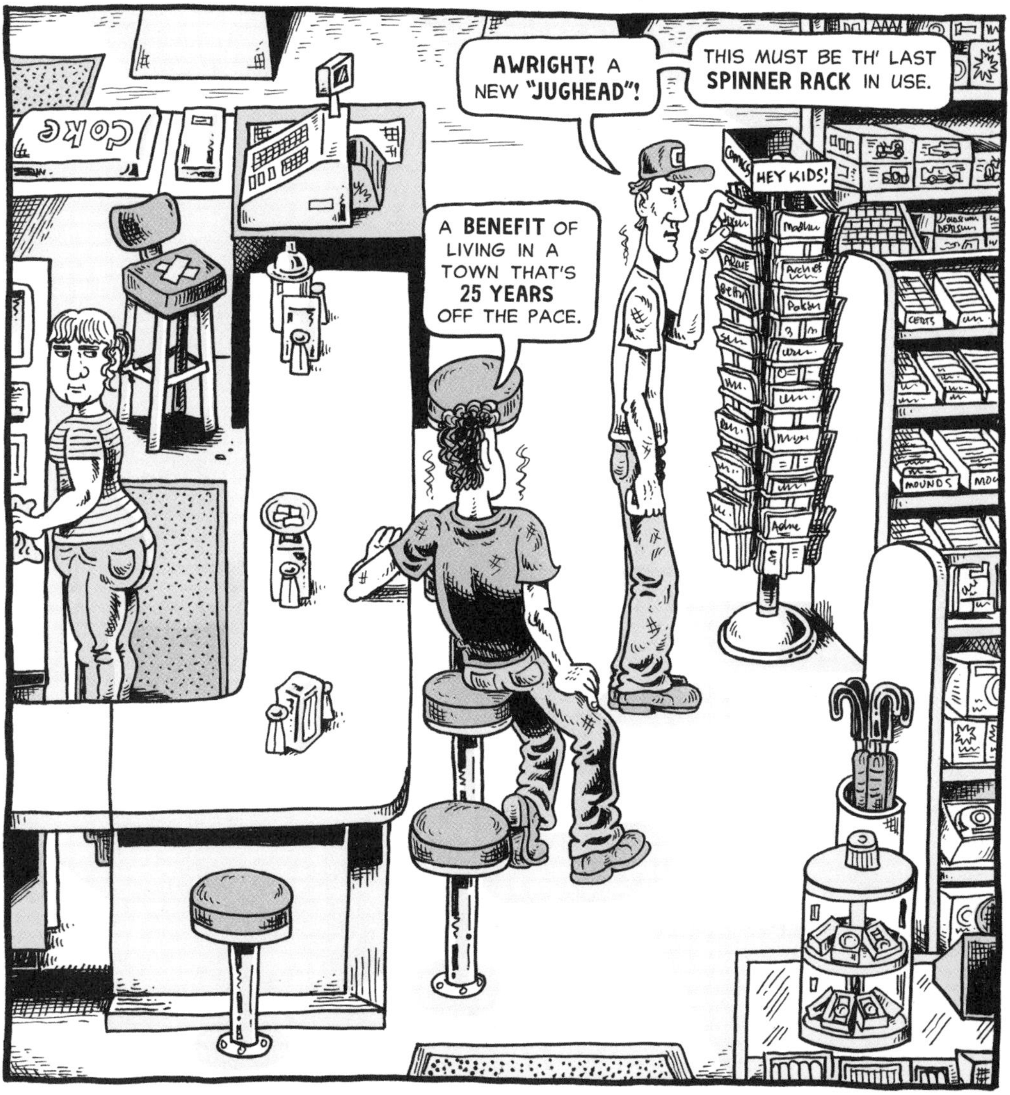
AWRIGHT! A NEW "JUGHEAD"!
THIS MUST BE TH' LAST SPINNER RACK IN USE.
A BENEFIT OF LIVING IN A TOWN THAT'S 25 YEARS OFF THE PACE.
COKE
COMICS
HEY KIDS!
Archie
Betty
Archet
MOUNDS
CERTS

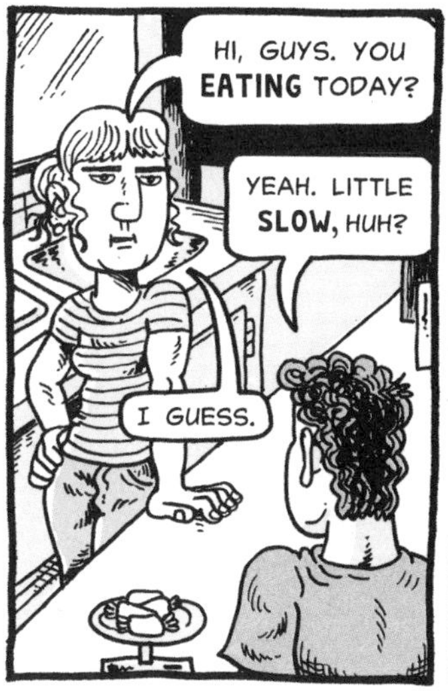
HI, GUYS. YOU EATING TODAY?
YEAH. LITTLE SLOW, HUH?
I GUESS.

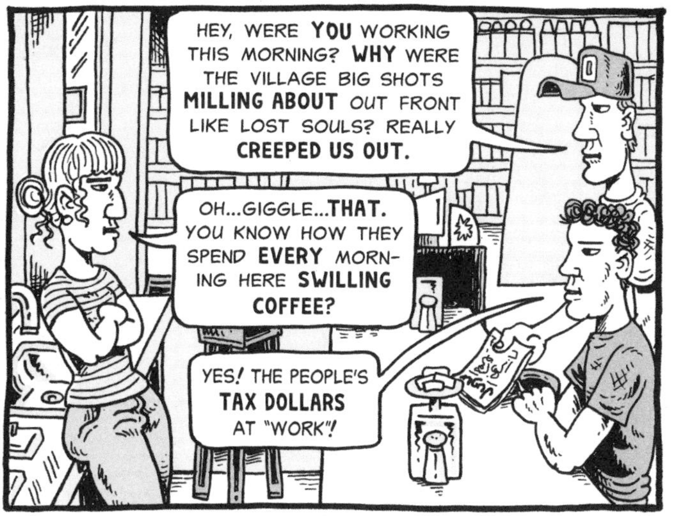
HEY, WERE YOU WORKING THIS MORNING? WHY WERE THE VILLAGE BIG SHOTS MILLING ABOUT OUT FRONT LIKE LOST SOULS? REALLY CREEPED US OUT.
OH...GIGGLE...THAT. YOU KNOW HOW THEY SPEND EVERY MORNING HERE SWILLING COFFEE?
YES! THE PEOPLE'S TAX DOLLARS AT "WORK"!

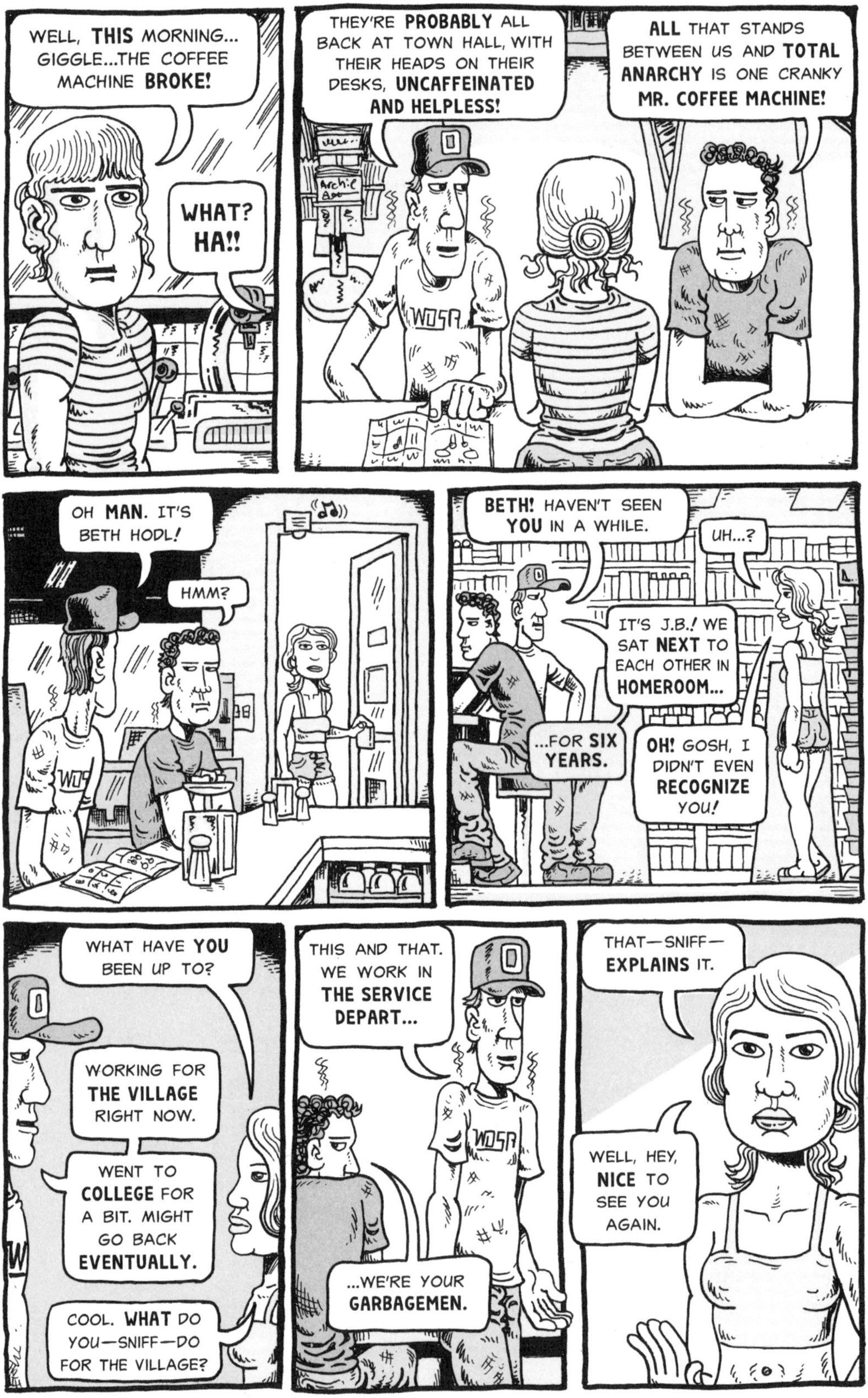
WELL, THIS MORNING... GIGGLE...THE COFFEE MACHINE BROKE!
WHAT? HA!!
THEY'RE PROBABLY ALL BACK AT TOWN HALL, WITH THEIR HEADS ON THEIR DESKS, UNCAFFEINATED AND HELPLESS!
ALL THAT STANDS BETWEEN US AND TOTAL ANARCHY IS ONE CRANKY MR. COFFEE MACHINE!
OH MAN. IT'S BETH HODL!
HMM?
BETH! HAVEN'T SEEN YOU IN A WHILE.
UH...?
IT'S J.B.! WE SAT NEXT TO EACH OTHER IN HOMEROOM...
...FOR SIX YEARS.
OH! GOSH, I DIDN'T EVEN RECOGNIZE YOU!
WHAT HAVE YOU BEEN UP TO?
WORKING FOR THE VILLAGE RIGHT NOW.
WENT TO COLLEGE FOR A BIT. MIGHT GO BACK EVENTUALLY.
COOL. WHAT DO YOU—SNIFF—DO FOR THE VILLAGE?
THIS AND THAT. WE WORK IN THE SERVICE DEPART...
...WE'RE YOUR GARBAGEMEN.
THAT—SNIFF—EXPLAINS IT.
WELL, HEY, NICE TO SEE YOU AGAIN.

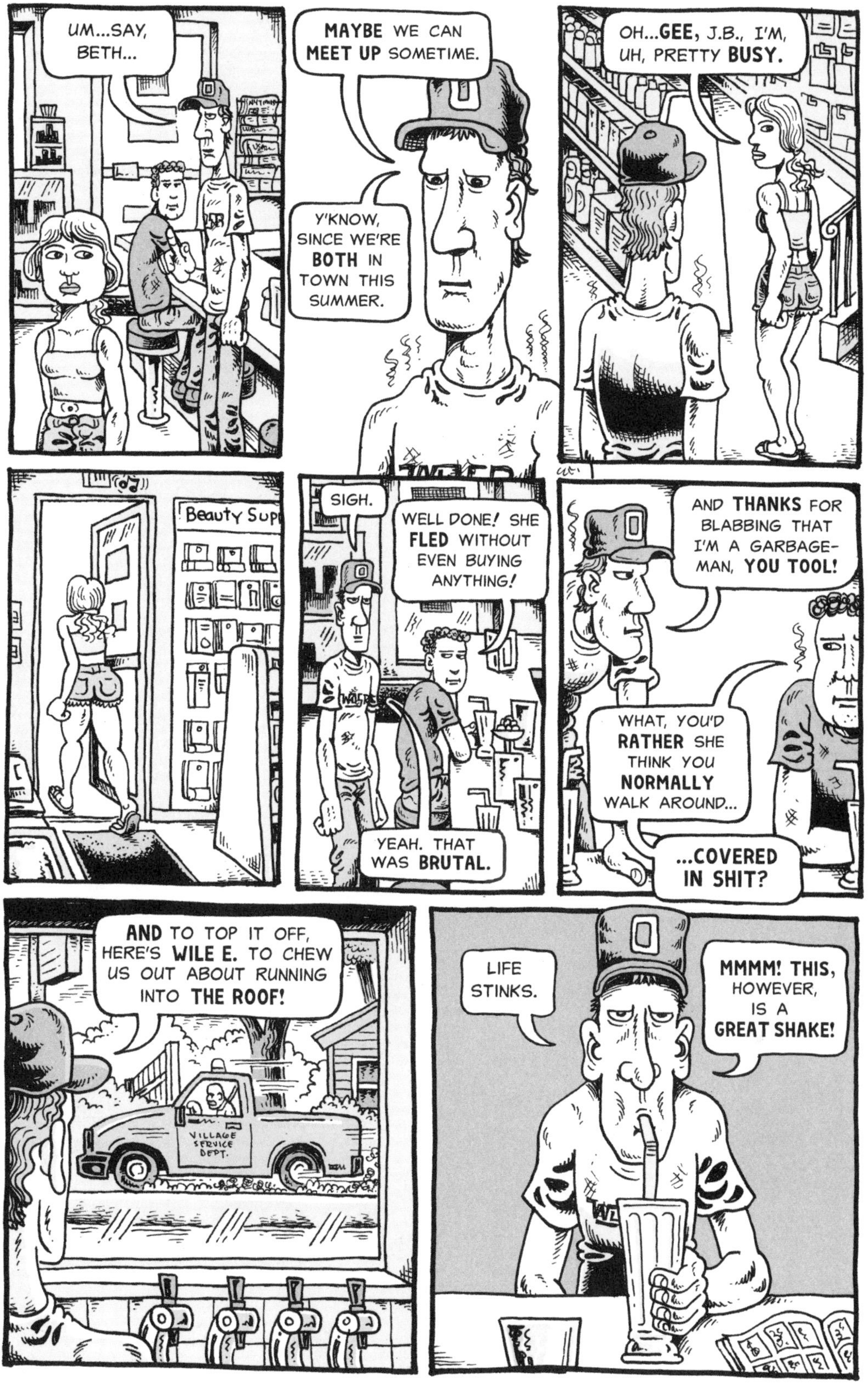
UM...SAY, BETH...
MAYBE WE CAN MEET UP SOMETIME.
Y'KNOW, SINCE WE'RE BOTH IN TOWN THIS SUMMER.
OH...GEE, J.B., I'M, UH, PRETTY BUSY.
Beauty Sup
SIGH.
WELL DONE! SHE FLED WITHOUT EVEN BUYING ANYTHING!
YEAH. THAT WAS BRUTAL.
AND THANKS FOR BLABBING THAT I'M A GARBAGE-MAN, YOU TOOL!
WHAT, YOU'D RATHER SHE THINK YOU NORMALLY WALK AROUND...
...COVERED IN SHIT?
AND TO TOP IT OFF, HERE'S WILE E. TO CHEW US OUT ABOUT RUNNING INTO THE ROOF!
VILLAGE SERVICE DEPT.
LIFE STINKS.
MMMM! THIS, HOWEVER, IS A GREAT SHAKE!

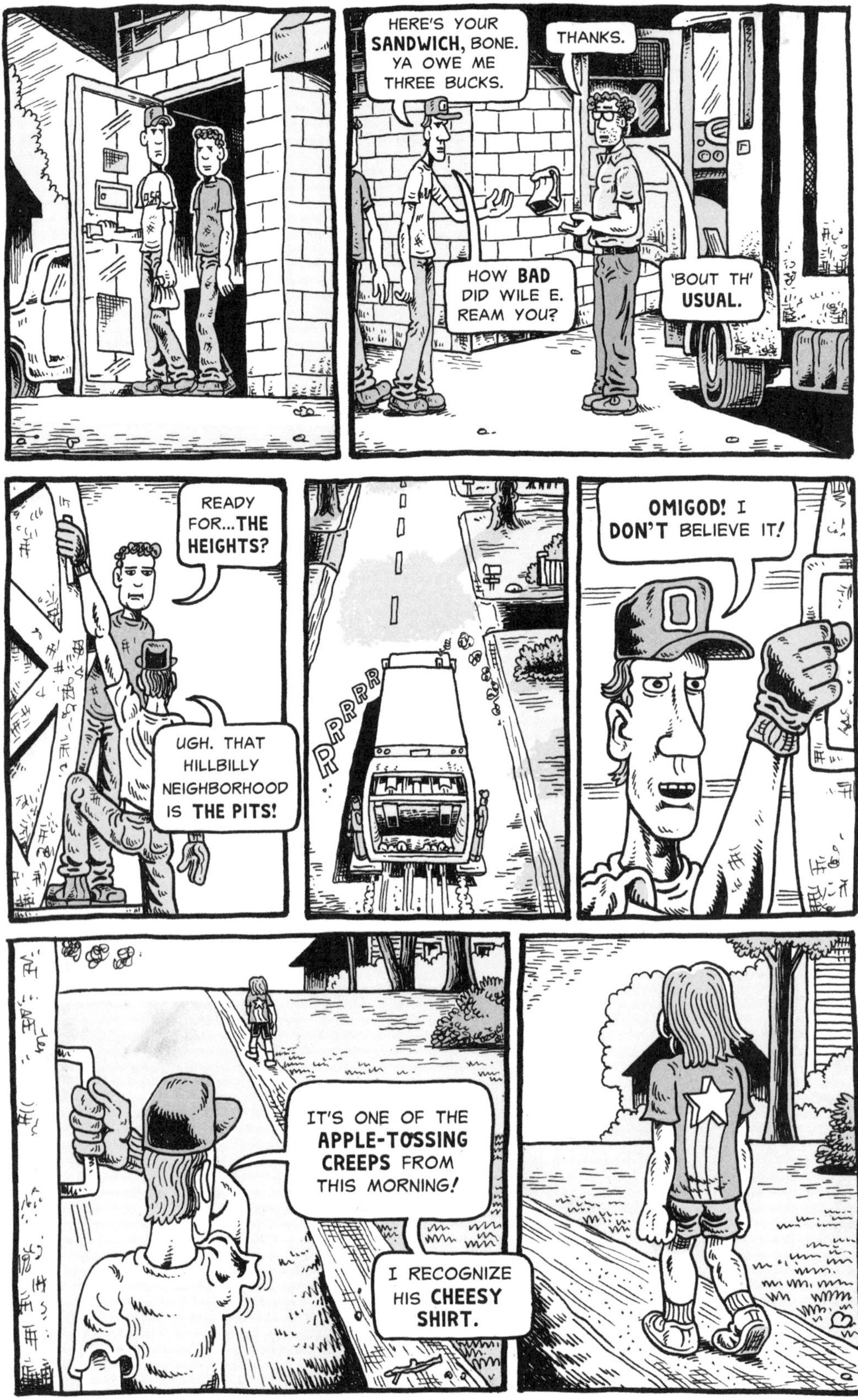
HERE'S YOUR SANDWICH, BONE. YA OWE ME THREE BUCKS.
THANKS.
HOW BAD DID WILE E. REAM YOU?
'BOUT TH' USUAL.
READY FOR...THE HEIGHTS?
UGH. THAT HILLBILLY NEIGHBORHOOD IS THE PITS!
RRRRRR
OMIGOD! I DON'T BELIEVE IT!
IT'S ONE OF THE APPLE-TOSSING CREEPS FROM THIS MORNING!
I RECOGNIZE HIS CHEESY SHIRT.

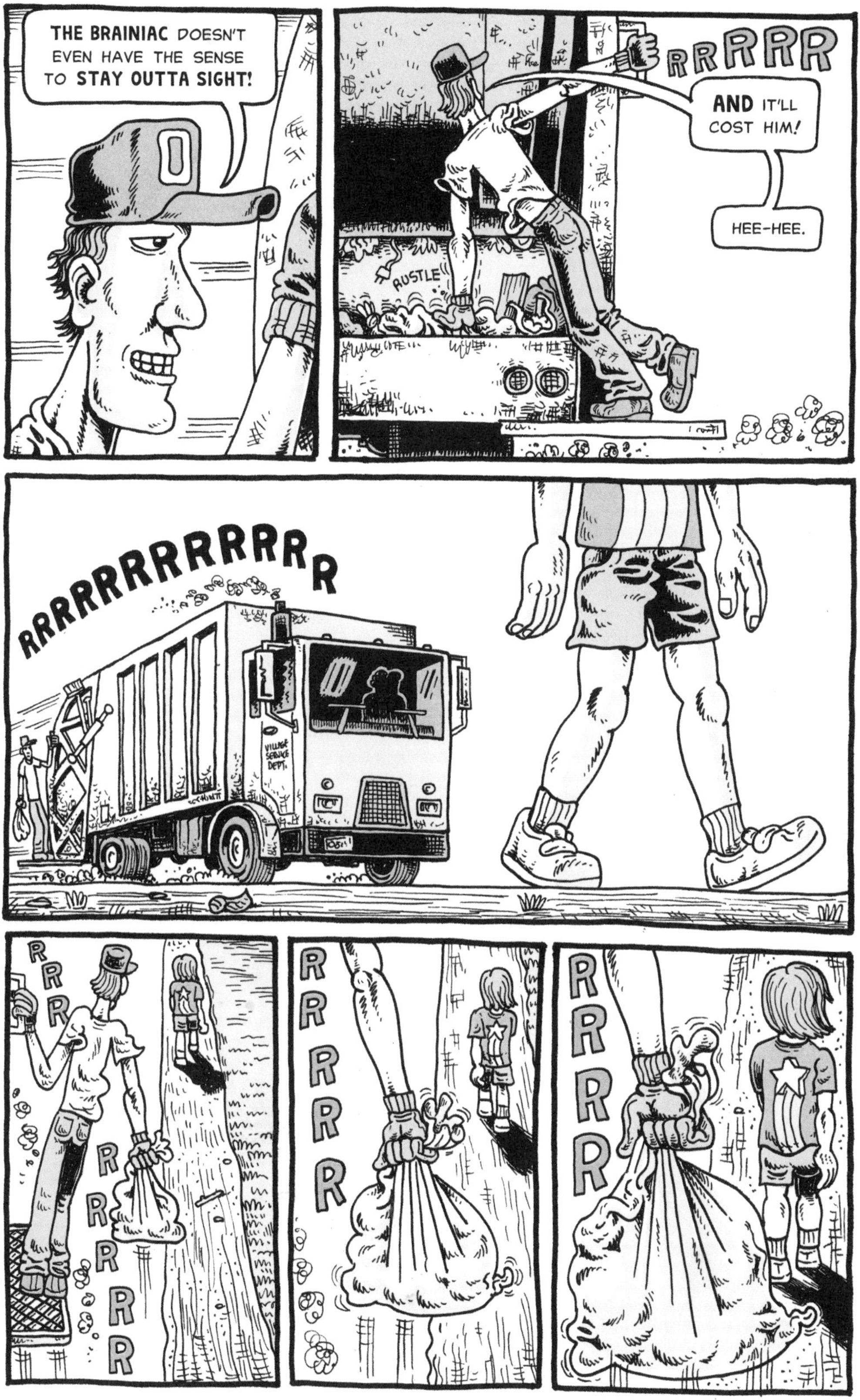
THE BRAINIAC DOESN'T EVEN HAVE THE SENSE TO STAY OUTTA SIGHT!
RRRRR
AND IT'LL COST HIM!
HEE-HEE.
RUSTLE
RRRRRRRRRRRR
VILLAGE SERVICE DEPT.
RRR
RRRRR
RRRRR
RRRR

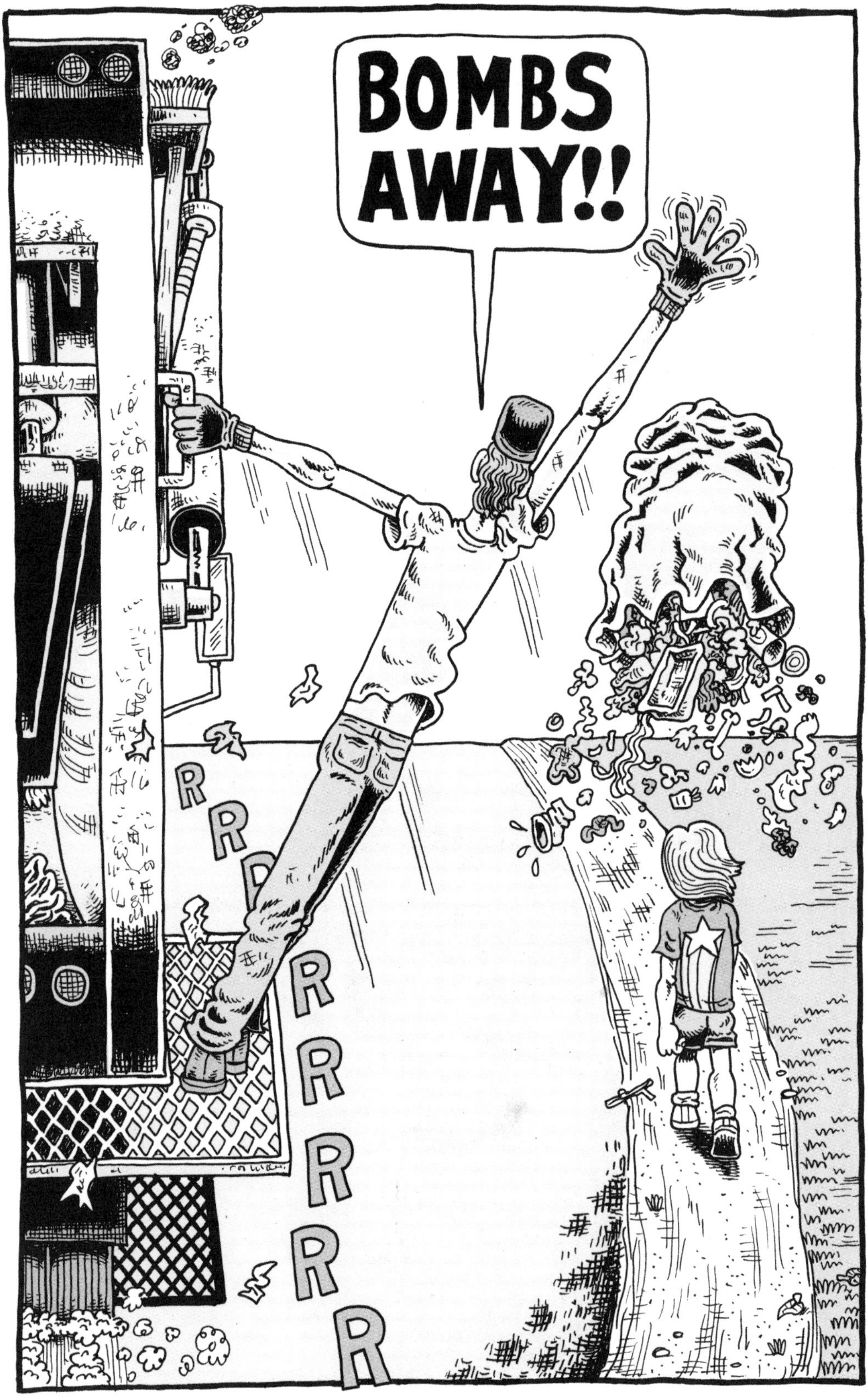
BOMBS AWAY!!
RRRRRRRRR

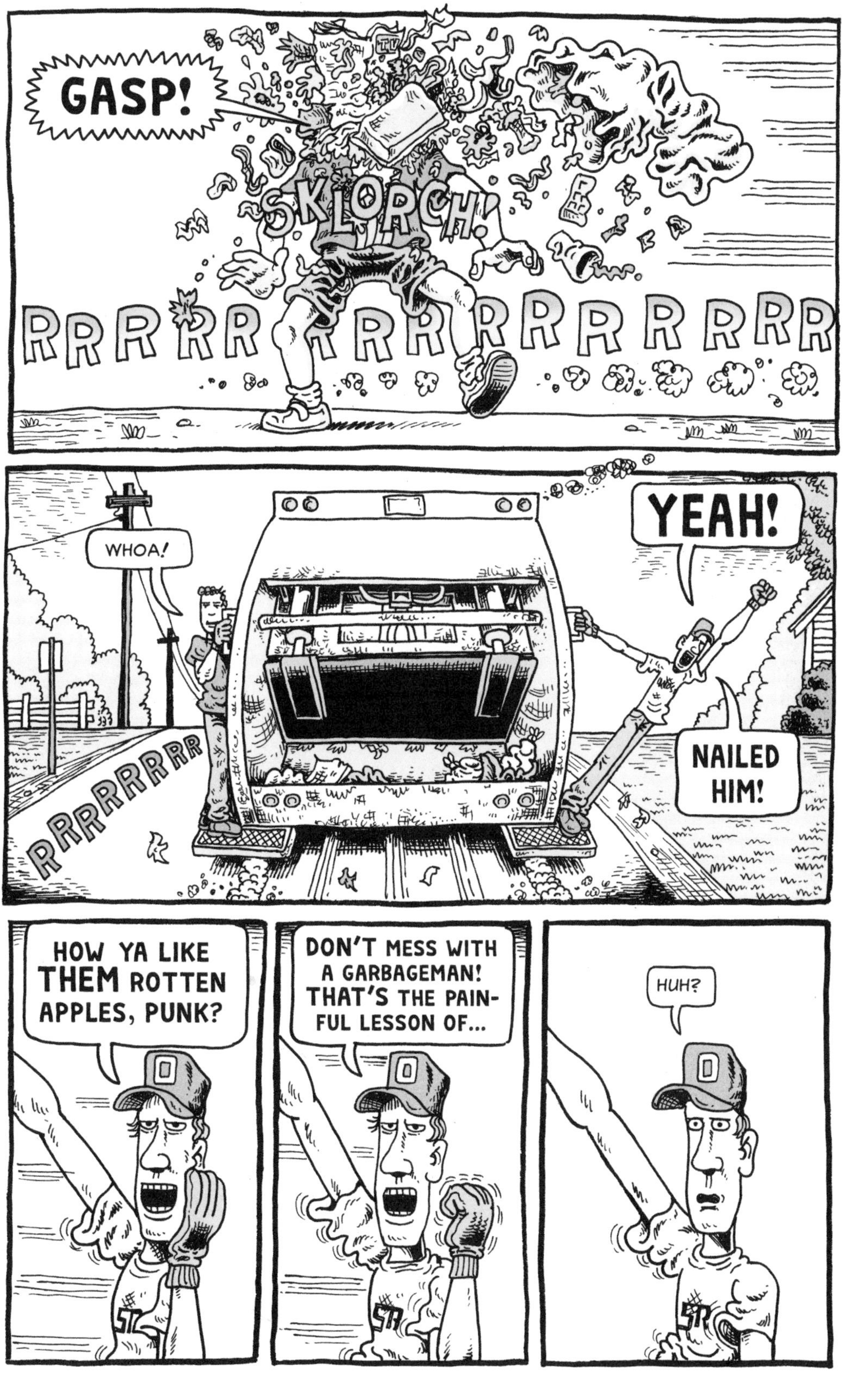
GASP!
SKLORCH!
RRRRRRRRRRRRR
WHOA!
YEAH!
RRRRRRRR
NAILED HIM!
HOW YA LIKE THEM ROTTEN APPLES, PUNK?
DON'T MESS WITH A GARBAGEMAN! THAT'S THE PAIN-FUL LESSON OF...
HUH?

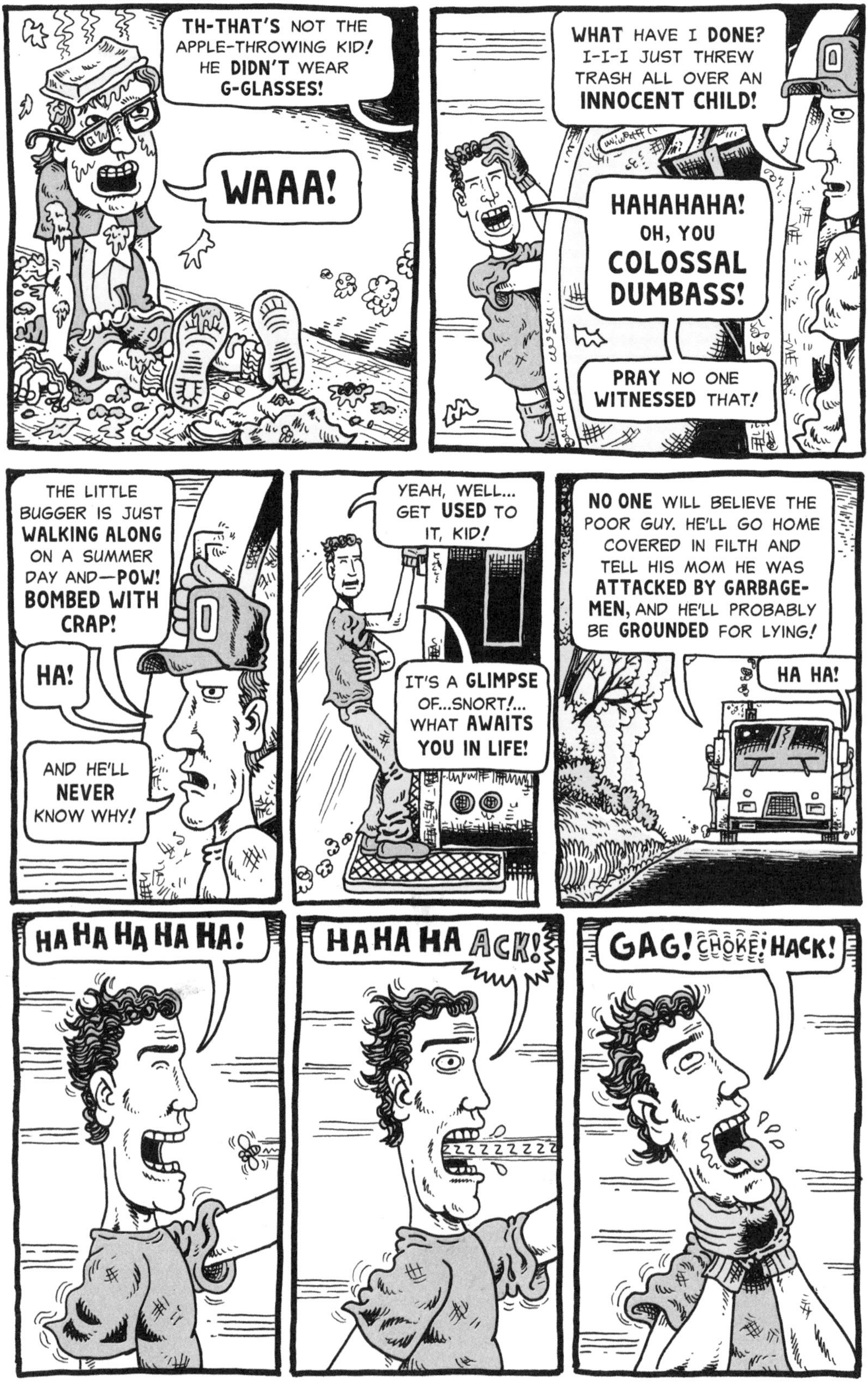
TH-THAT'S NOT THE APPLE-THROWING KID! HE DIDN'T WEAR G-GLASSES!
WAAA!
WHAT HAVE I DONE? I-I-I JUST THREW TRASH ALL OVER AN INNOCENT CHILD!
HAHAHAHA! OH, YOU COLOSSAL DUMBASS!
PRAY NO ONE WITNESSED THAT!
THE LITTLE BUGGER IS JUST WALKING ALONG ON A SUMMER DAY AND—POW! BOMBED WITH CRAP!
HA!
AND HE'LL NEVER KNOW WHY!
YEAH, WELL... GET USED TO IT, KID!
IT'S A GLIMPSE OF...SNORT!... WHAT AWAITS YOU IN LIFE!
NO ONE WILL BELIEVE THE POOR GUY. HE'LL GO HOME COVERED IN FILTH AND TELL HIS MOM HE WAS ATTACKED BY GARBAGE-MEN, AND HE'LL PROBABLY BE GROUNDED FOR LYING!
HA HA!
HA HA HA HA HA!
HA HA HA ACK!
ZZZZZZZZZ
GAG! CHOKE! HACK!

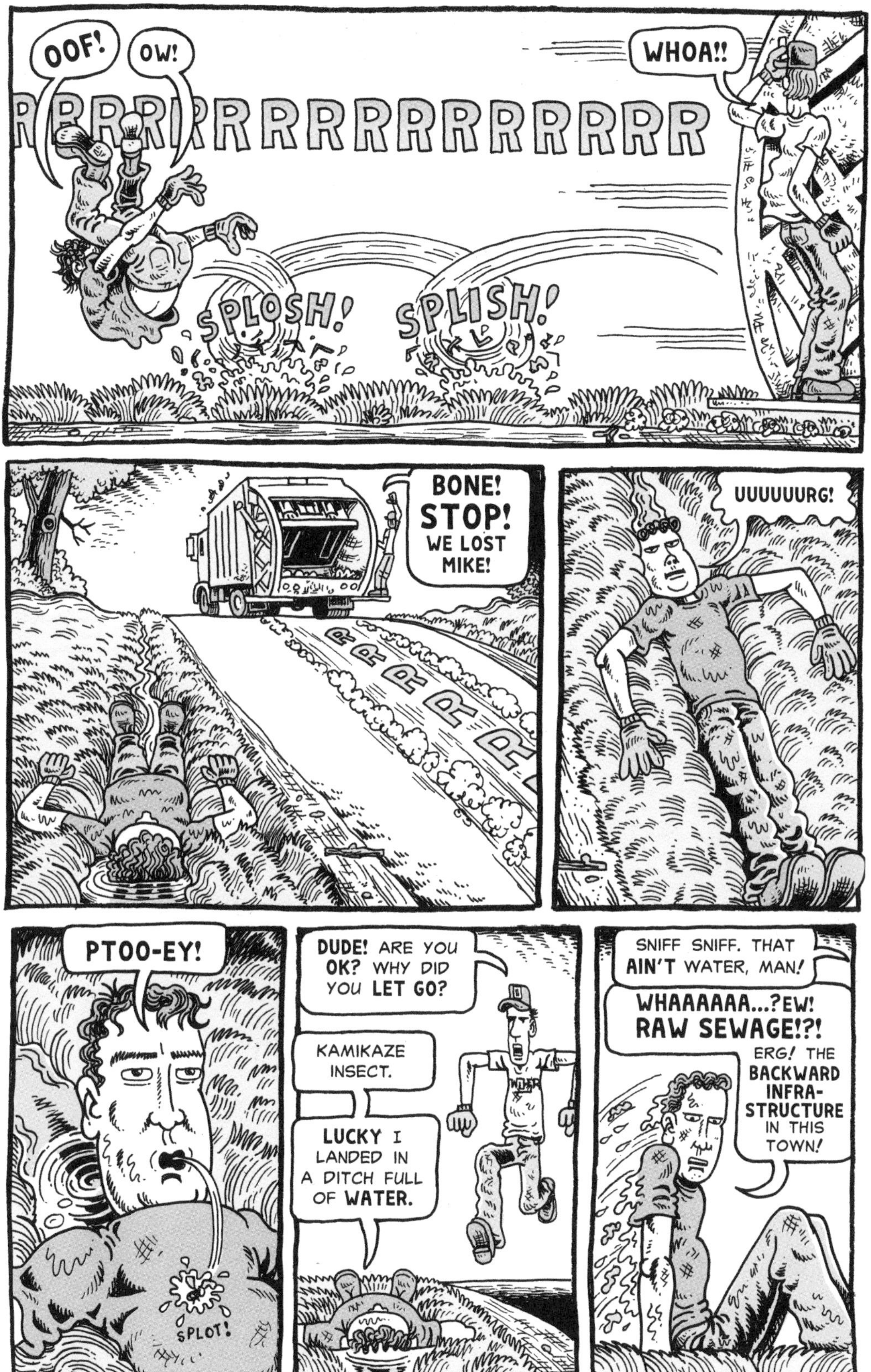
OOF!
OW!
WHOA!!
RRRRRRRRRRRRRRR
SPLOSH!
SPLISH!
BONE! STOP! WE LOST MIKE!
UUUUUURG!
PTOO-EY!
SPLOT!
DUDE! ARE YOU OK? WHY DID YOU LET GO?
KAMIKAZE INSECT.
LUCKY I LANDED IN A DITCH FULL OF WATER.
SNIFF SNIFF. THAT AIN'T WATER, MAN!
WHAAAAAA...?EW! RAW SEWAGE!?!
ERG! THE BACKWARD INFRA-STRUCTURE IN THIS TOWN!

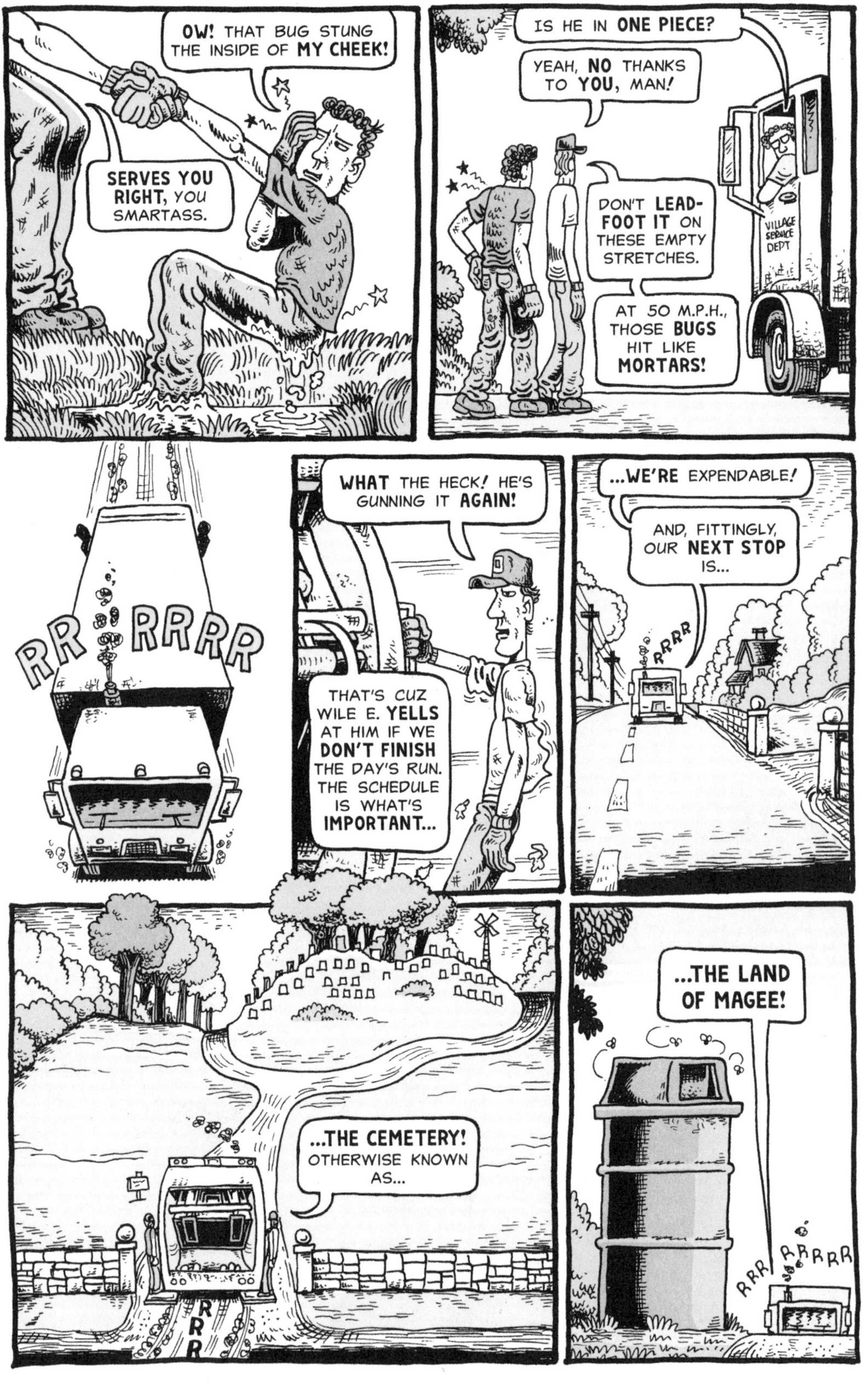
OW! THAT BUG STUNG THE INSIDE OF MY CHEEK!
SERVES YOU RIGHT, YOU SMARTASS.
IS HE IN ONE PIECE?
YEAH, NO THANKS TO YOU, MAN!
DON'T LEAD-FOOT IT ON THESE EMPTY STRETCHES.
AT 50 M.P.H., THOSE BUGS HIT LIKE MORTARS!
VILLAGE SERVICE DEPT
RR RRRRR
WHAT THE HECK! HE'S GUNNING IT AGAIN!
THAT'S CUZ WILE E. YELLS AT HIM IF WE DON'T FINISH THE DAY'S RUN. THE SCHEDULE IS WHAT'S IMPORTANT...
...WE'RE EXPENDABLE!
AND, FITTINGLY, OUR NEXT STOP IS...
RRRR
...THE CEMETERY! OTHERWISE KNOWN AS...
RRR
...THE LAND OF MAGEE!
RRR RRRRRR

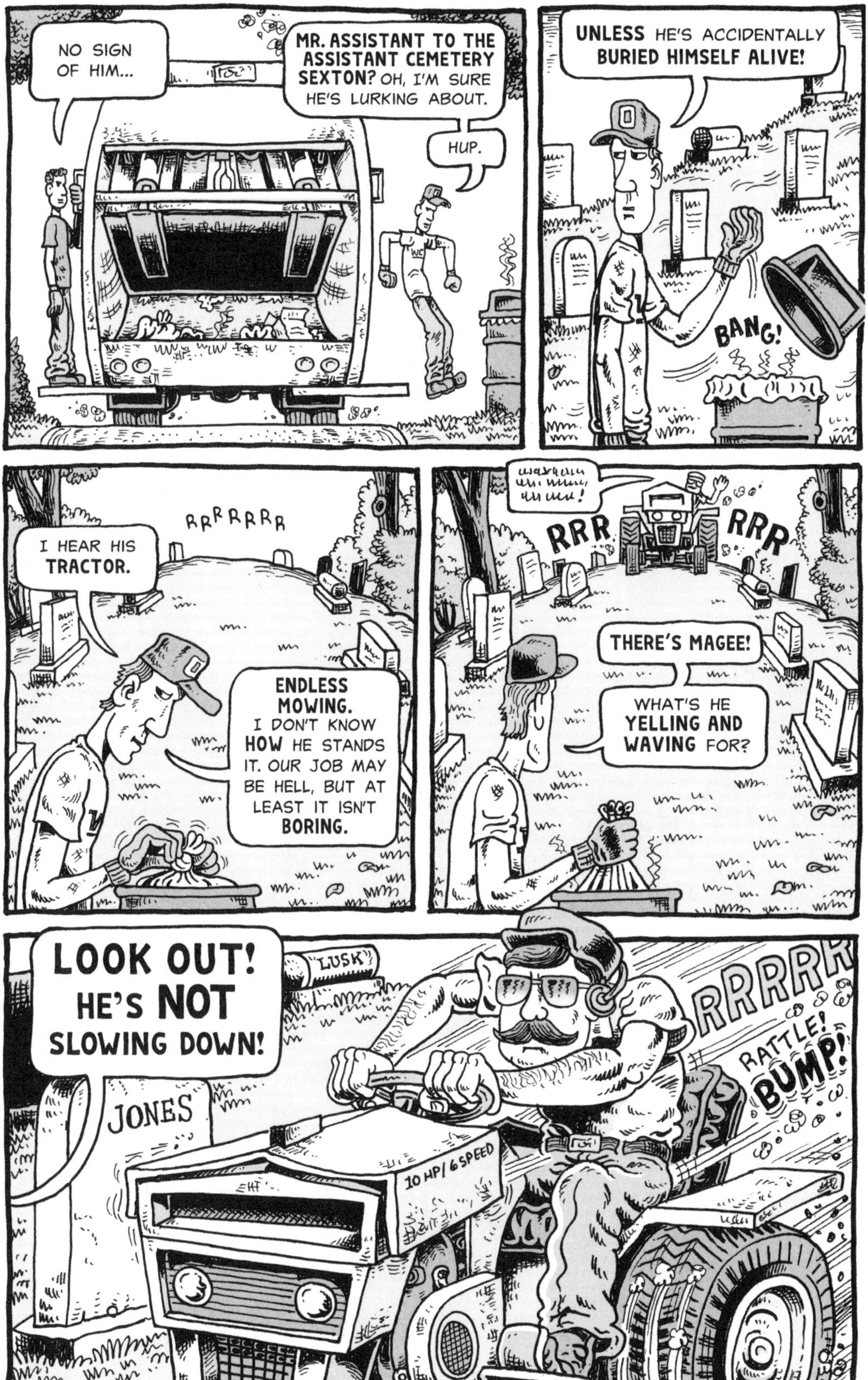
NO SIGN OF HIM...
MR. ASSISTANT TO THE ASSISTANT CEMETERY SEXTON? OH, I'M SURE HE'S LURKING ABOUT.
HUP.
UNLESS HE'S ACCIDENTALLY BURIED HIMSELF ALIVE!
BANG!
RRRRRRRR
I HEAR HIS TRACTOR.
ENDLESS MOWING. I DON'T KNOW HOW HE STANDS IT. OUR JOB MAY BE HELL, BUT AT LEAST IT ISN'T BORING.
RRR
RRR
THERE'S MAGEE!
WHAT'S HE YELLING AND WAVING FOR?
LOOK OUT! HE'S NOT SLOWING DOWN!
LUSK
JONES
10 HP/6 SPEED
RRRRR
RATTLE!
BUMP!

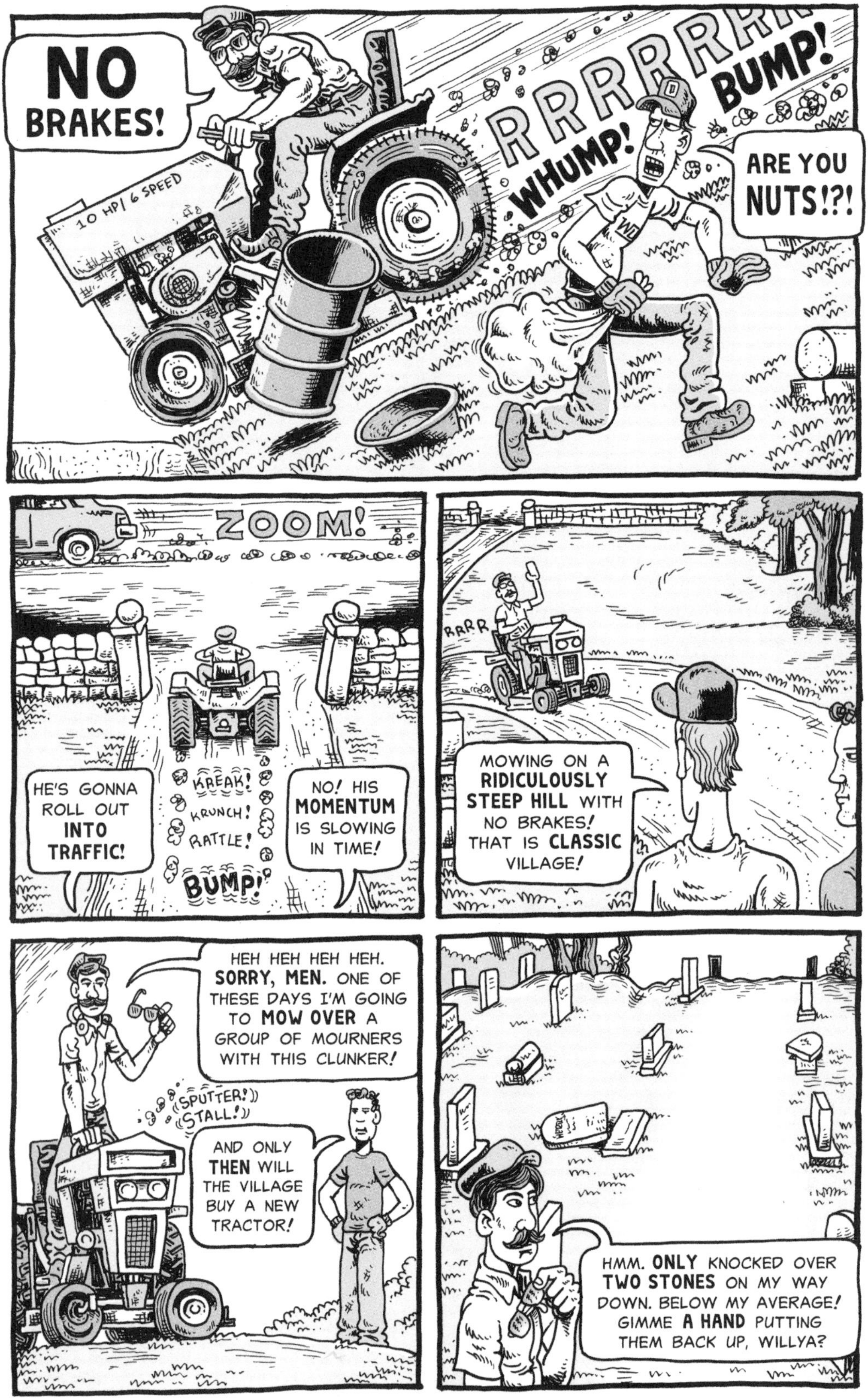

NO BRAKES!
RRRRRRRR
BUMP!
WHUMP!
ARE YOU NUTS!?!
10 HP/6 SPEED
ZOOM!
HE'S GONNA ROLL OUT INTO TRAFFIC!
KREAK! KRUNCH! RATTLE!
BUMP!
NO! HIS MOMENTUM IS SLOWING IN TIME!
RRRR
MOWING ON A RIDICULOUSLY STEEP HILL WITH NO BRAKES! THAT IS CLASSIC VILLAGE!
HEH HEH HEH HEH. SORRY, MEN. ONE OF THESE DAYS I'M GOING TO MOW OVER A GROUP OF MOURNERS WITH THIS CLUNKER!
SPUTTER! STALL!
AND ONLY THEN WILL THE VILLAGE BUY A NEW TRACTOR!
HMM. ONLY KNOCKED OVER TWO STONES ON MY WAY DOWN. BELOW MY AVERAGE! GIMME A HAND PUTTING THEM BACK UP, WILLYA?

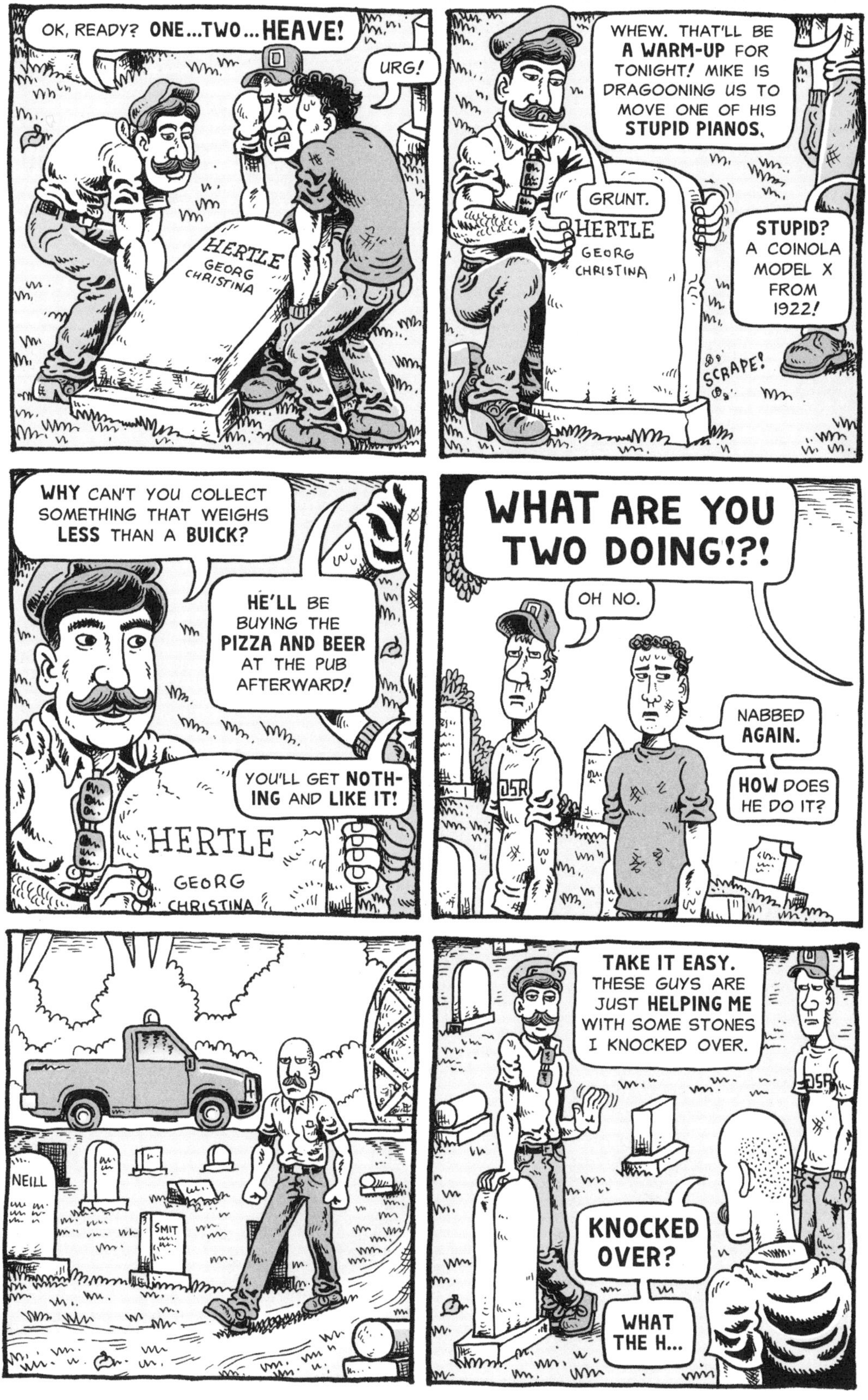

OK, READY? ONE...TWO...HEAVE!
URG!
HERTLE
GEORG
CHRISTINA
WHEW. THAT'LL BE A WARM-UP FOR TONIGHT! MIKE IS DRAGOONING US TO MOVE ONE OF HIS STUPID PIANOS.
GRUNT.
STUPID? A COINOLA MODEL X FROM 1922!
SCRAPE!
WHY CAN'T YOU COLLECT SOMETHING THAT WEIGHS LESS THAN A BUICK?
HE'LL BE BUYING THE PIZZA AND BEER AT THE PUB AFTERWARD!
YOU'LL GET NOTHING AND LIKE IT!
WHAT ARE YOU TWO DOING!?!
OH NO.
NABBED AGAIN.
HOW DOES HE DO IT?
NEILL
SMIT
TAKE IT EASY. THESE GUYS ARE JUST HELPING ME WITH SOME STONES I KNOCKED OVER.
KNOCKED OVER?
WHAT THE H...

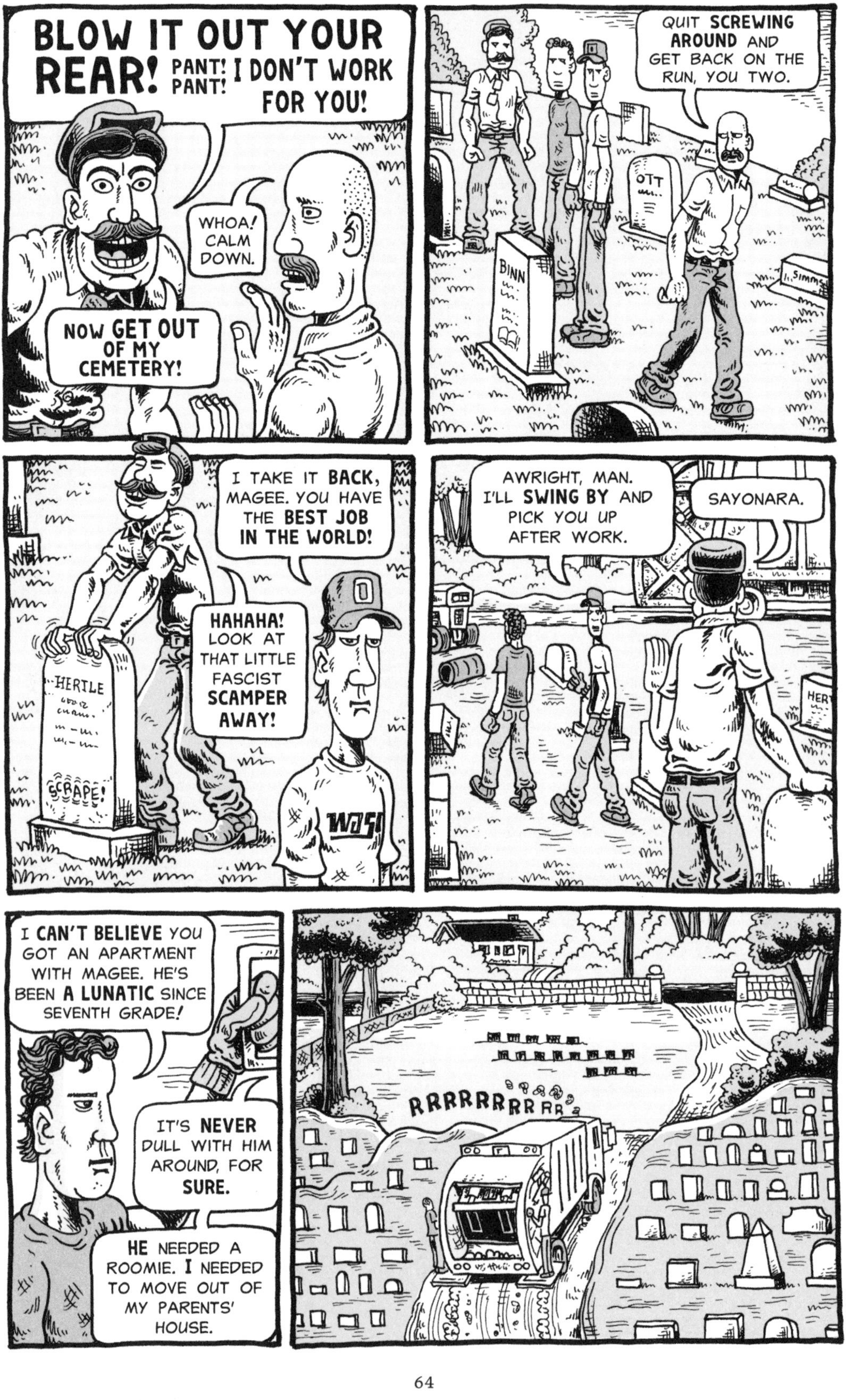
BLOW IT OUT YOUR REAR! PANT! PANT! I DON'T WORK FOR YOU!
WHOA! CALM DOWN.
NOW GET OUT OF MY CEMETERY!
QUIT SCREWING AROUND AND GET BACK ON THE RUN, YOU TWO.
BINN
OTT
SIMMS
I TAKE IT BACK, MAGEE. YOU HAVE THE BEST JOB IN THE WORLD!
HAHAHA! LOOK AT THAT LITTLE FASCIST SCAMPER AWAY!
HERTLE
SCRAPE!
AWRIGHT, MAN. I'LL SWING BY AND PICK YOU UP AFTER WORK.
SAYONARA.
I CAN'T BELIEVE YOU GOT AN APARTMENT WITH MAGEE. HE'S BEEN A LUNATIC SINCE SEVENTH GRADE!
IT'S NEVER DULL WITH HIM AROUND, FOR SURE.
HE NEEDED A ROOMIE. I NEEDED TO MOVE OUT OF MY PARENTS' HOUSE.
RRRRRRRRRRR

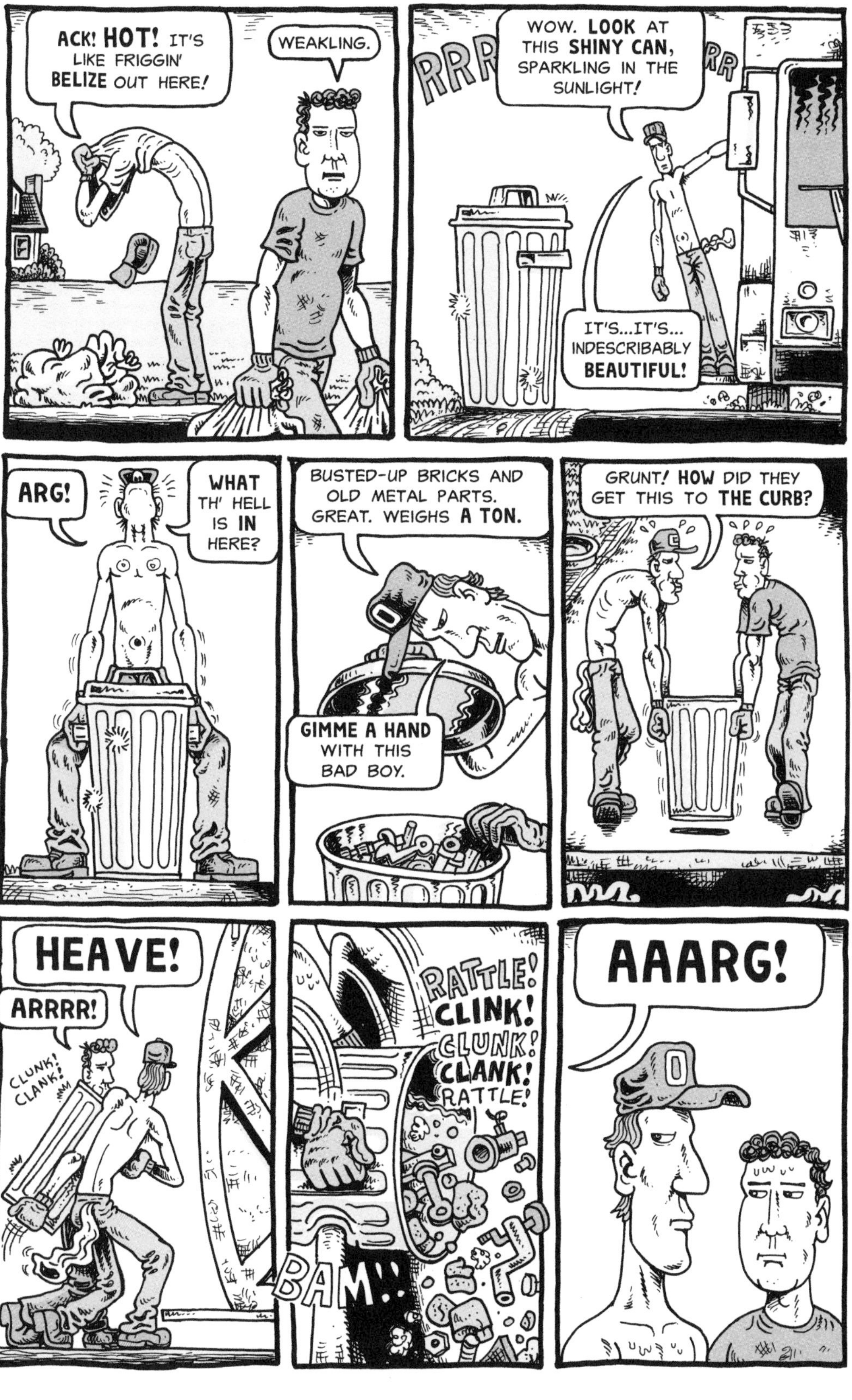

ACK! HOT! IT'S LIKE FRIGGIN' BELIZE OUT HERE!
WEAKLING.
RRR
RR
WOW. LOOK AT THIS SHINY CAN, SPARKLING IN THE SUNLIGHT!
IT'S...IT'S... INDESCRIBABLY BEAUTIFUL!
ARG!
WHAT TH' HELL IS IN HERE?
BUSTED-UP BRICKS AND OLD METAL PARTS. GREAT. WEIGHS A TON.
GIMME A HAND WITH THIS BAD BOY.
GRUNT! HOW DID THEY GET THIS TO THE CURB?
HEAVE!
ARRRR!
CLUNK! CLANK!
RATTLE! CLINK! CLUNK! CLANK! RATTLE!
BAM!!
AAARG!

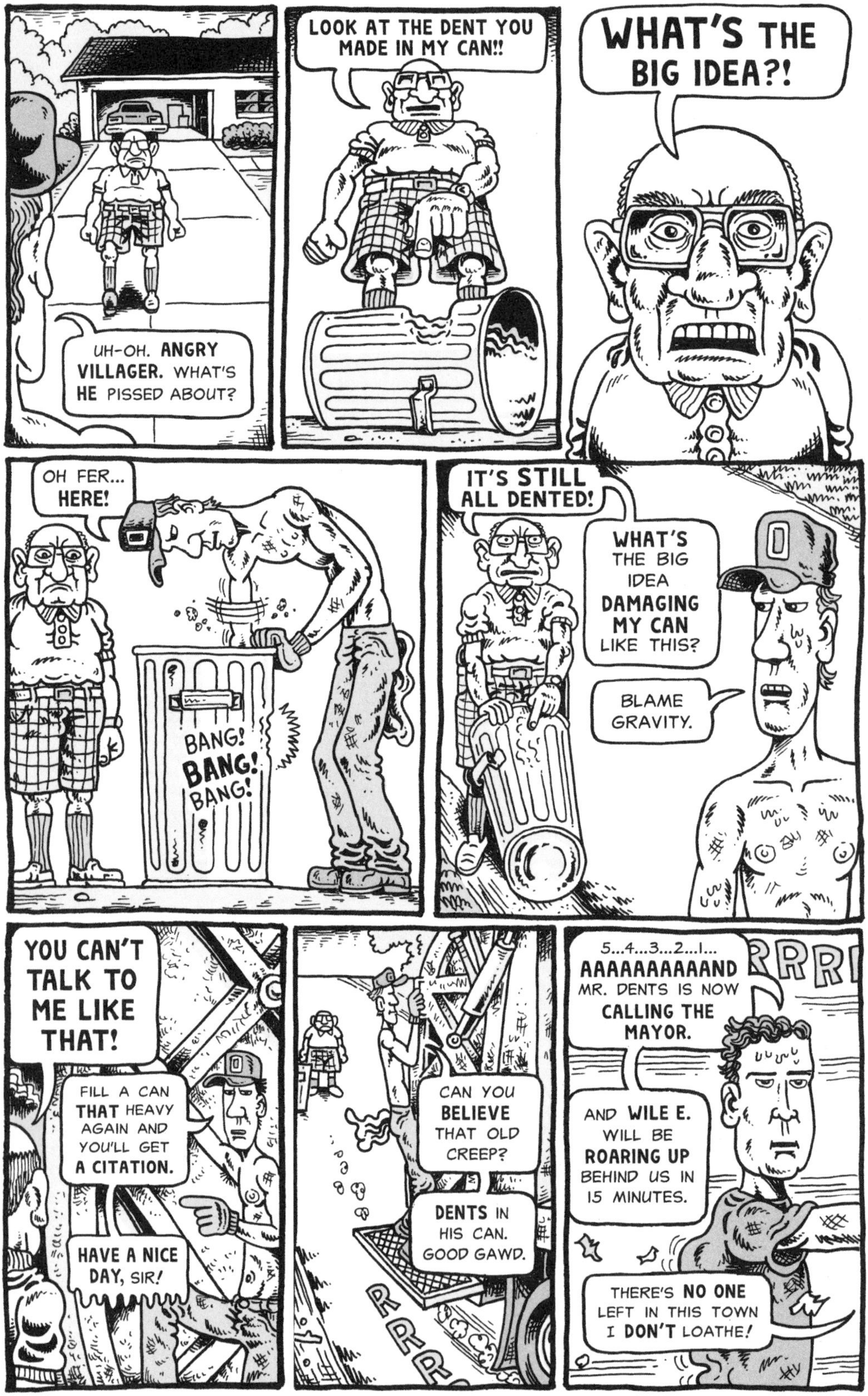
UH-OH. ANGRY VILLAGER. WHAT'S HE PISSED ABOUT?
LOOK AT THE DENT YOU MADE IN MY CAN!!
WHAT'S THE BIG IDEA?!
OH FER... HERE!
BANG! BANG! BANG!
IT'S STILL ALL DENTED!
WHAT'S THE BIG IDEA DAMAGING MY CAN LIKE THIS?
BLAME GRAVITY.
YOU CAN'T TALK TO ME LIKE THAT!
FILL A CAN THAT HEAVY AGAIN AND YOU'LL GET A CITATION.
HAVE A NICE DAY, SIR!
CAN YOU BELIEVE THAT OLD CREEP?
DENTS IN HIS CAN. GOOD GAWD.
RRR
5...4...3...2...1... AAAAAAAAAAND MR. DENTS IS NOW CALLING THE MAYOR.
AND WILE E. WILL BE ROARING UP BEHIND US IN 15 MINUTES.
THERE'S NO ONE LEFT IN THIS TOWN I DON'T LOATHE!
RRRI

THE LAST STOP! THE RUN, SHE IS **FINITO!**

WHEW.

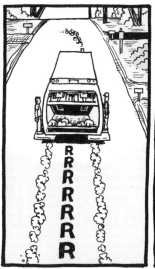

RRRRRR

SCREE!

WHAT'S BONE **STOPPING** FOR? WE **ALREADY DID** THIS STREET.

HA.

OUR **UNHAPPY VILLAGER** WITH **THE STOVE**...HE TOOK OFF **THE DOORS,** AS INSTRUCTED.

HEY, LOOK.

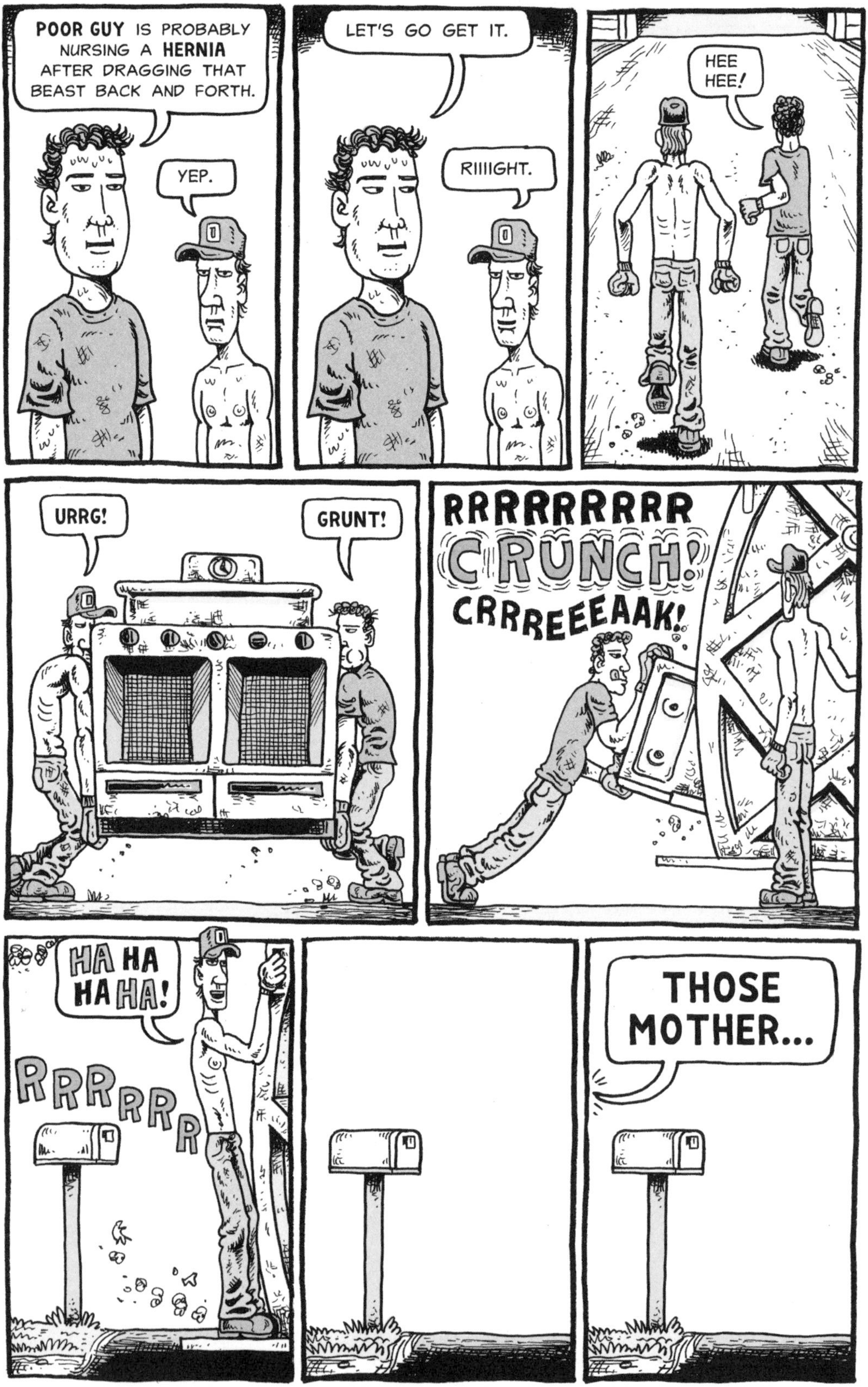
POOR GUY IS PROBABLY NURSING A HERNIA AFTER DRAGGING THAT BEAST BACK AND FORTH.
YEP.
LET'S GO GET IT.
RIIIIGHT.
HEE HEE!
URRG!
GRUNT!
RRRRRRRRR
CRUNCH!
CRRREEEAAK!
HA HA HA HA!
RRRRRRR
THOSE MOTHER...

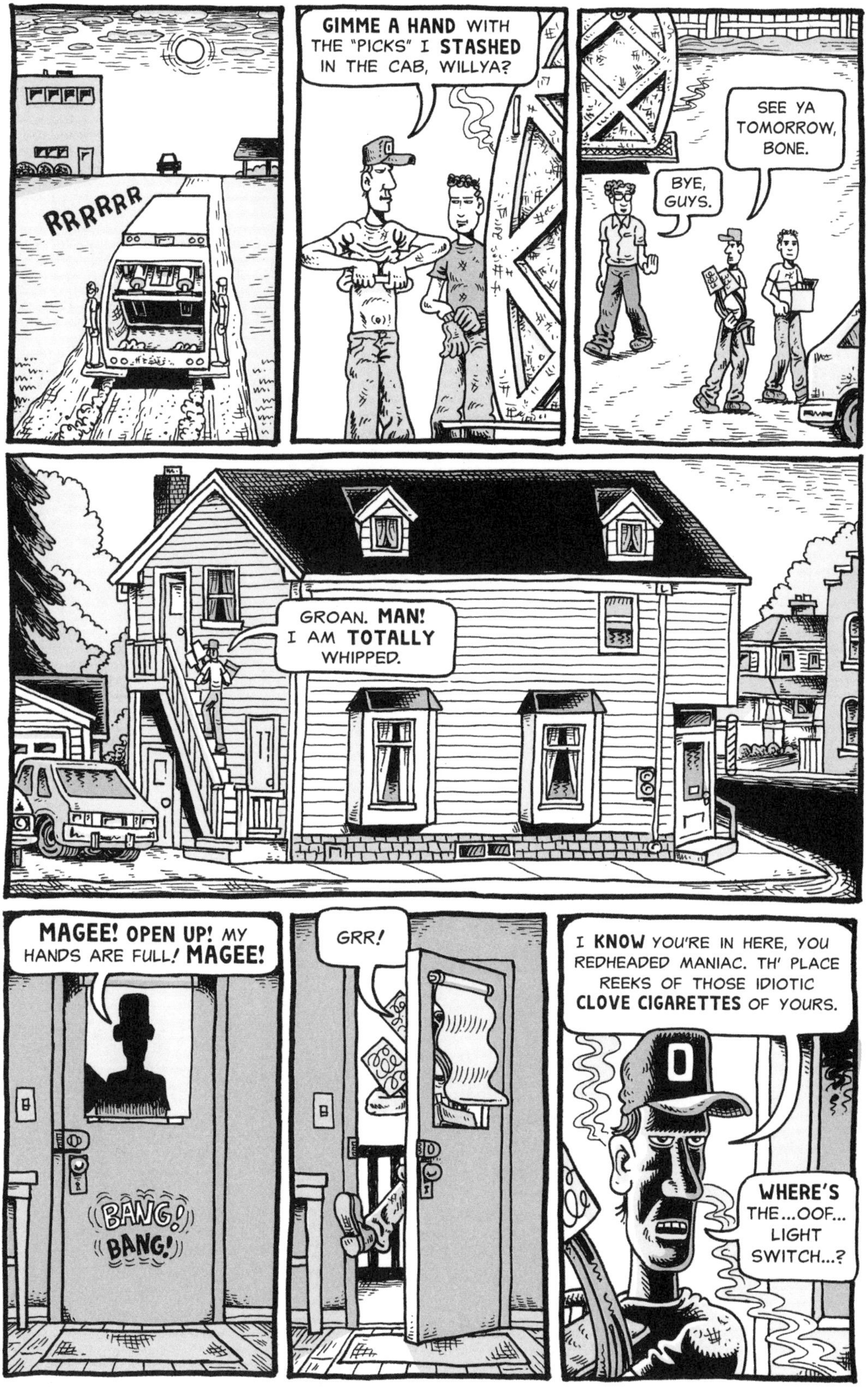
RRRRRR
GIMME A HAND WITH THE "PICKS" I STASHED IN THE CAB, WILLYA?
BYE, GUYS.
SEE YA TOMORROW, BONE.
GROAN. MAN! I AM TOTALLY WHIPPED.
MAGEE! OPEN UP! MY HANDS ARE FULL! MAGEE!
BANG! BANG!
GRR!
I KNOW YOU'RE IN HERE, YOU REDHEADED MANIAC. TH' PLACE REEKS OF THOSE IDIOTIC CLOVE CIGARETTES OF YOURS.
WHERE'S THE...OOF... LIGHT SWITCH...?

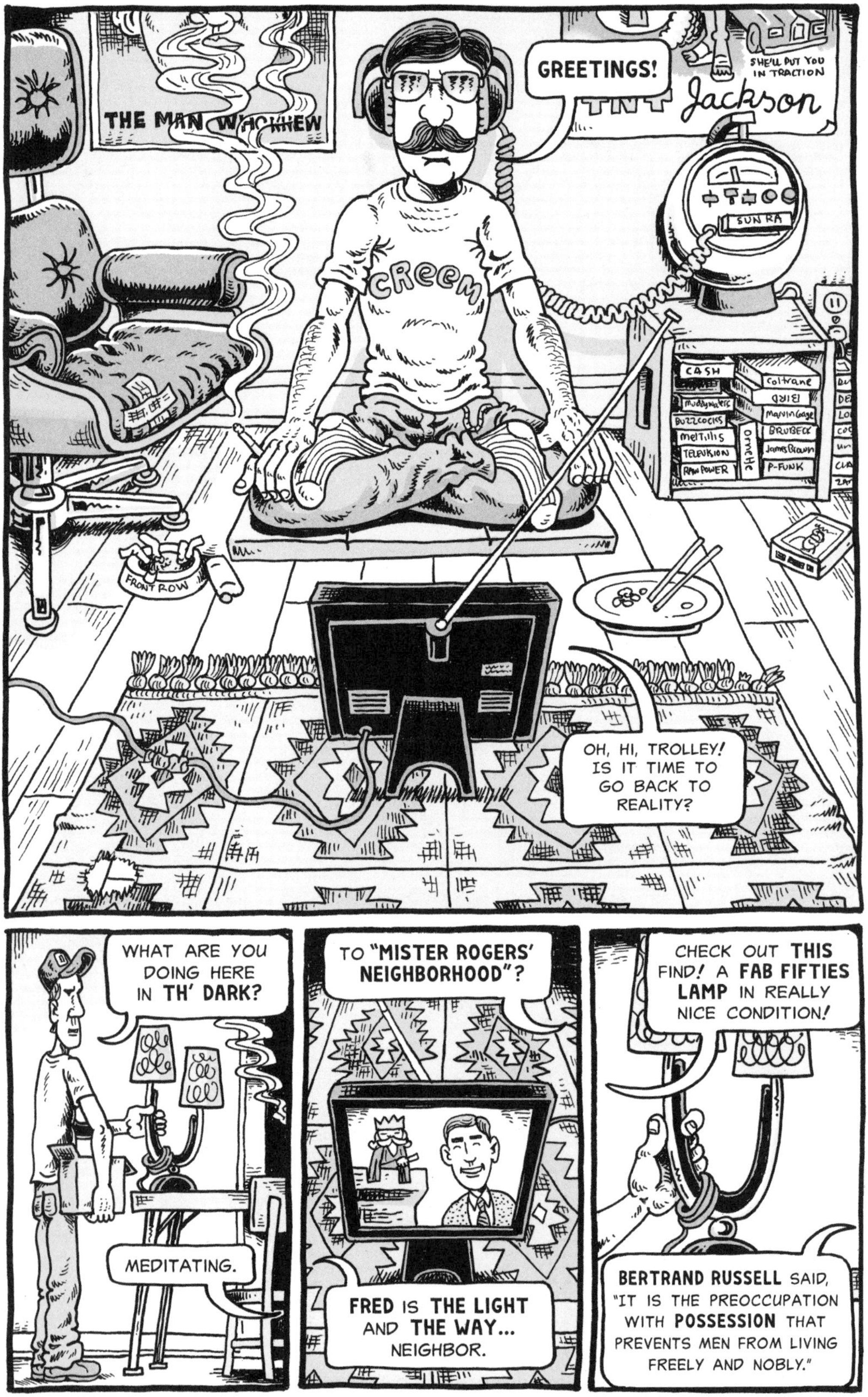
GREETINGS!
THE MAN WHO KNEW
SHE'LL PUT YOU IN TRACTION
TNT Jackson
CREEM
SUN RA
CASH
Coltrane
BIRD
Muddy Waters
Marvin Gaye
BUZZCOCKS
BRUBECK
MELTILLS
Ornette
TELEVISION
James Brown
RAW POWER
P-FUNK
FRONT ROW
OH, HI, TROLLEY! IS IT TIME TO GO BACK TO REALITY?
WHAT ARE YOU DOING HERE IN TH' DARK?
MEDITATING.
TO "MISTER ROGERS' NEIGHBORHOOD"?
FRED IS THE LIGHT AND THE WAY... NEIGHBOR.
CHECK OUT THIS FIND! A FAB FIFTIES LAMP IN REALLY NICE CONDITION!
BERTRAND RUSSELL SAID, "IT IS THE PREOCCUPATION WITH POSSESSION THAT PREVENTS MEN FROM LIVING FREELY AND NOBLY."

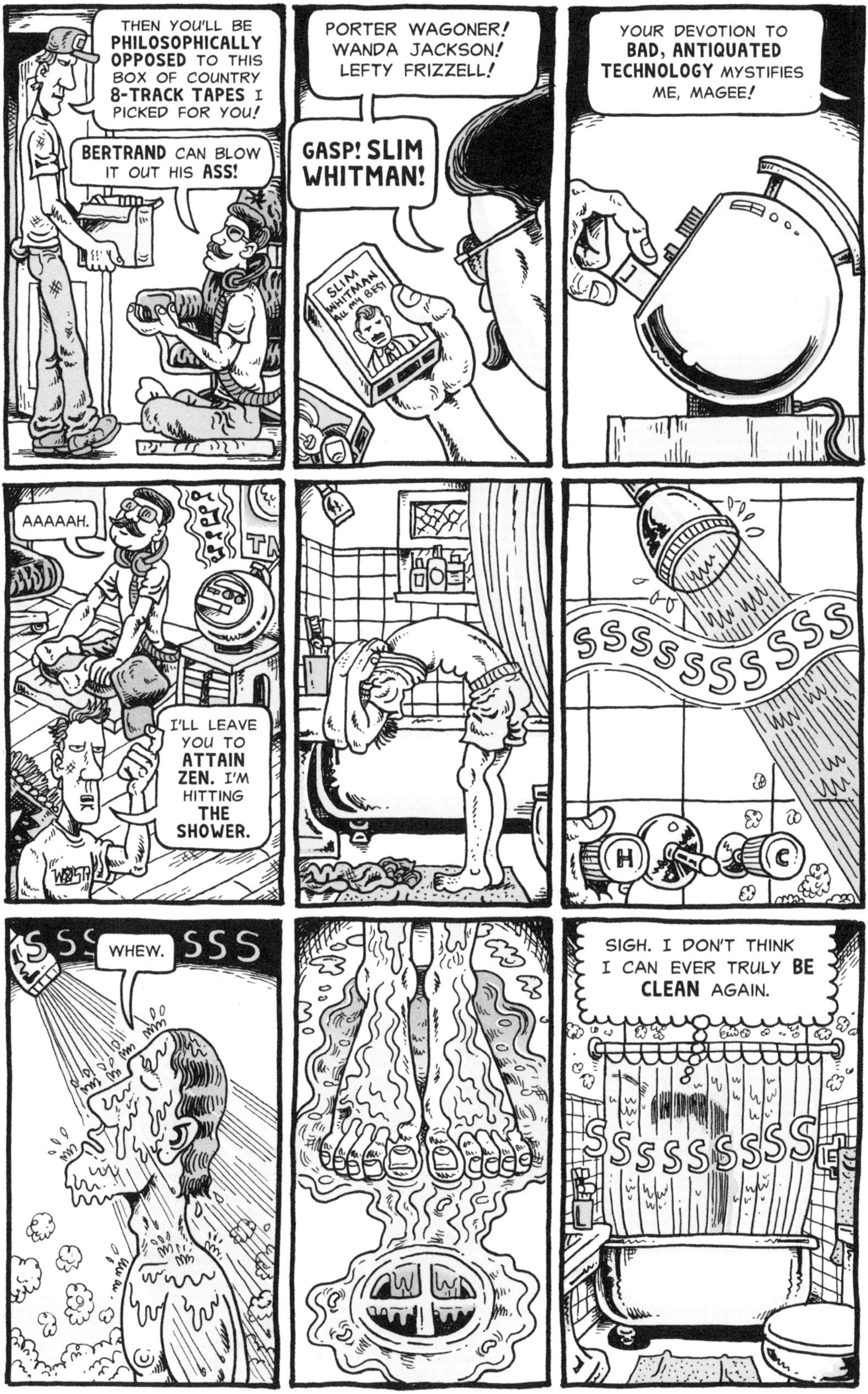

THEN YOU'LL BE PHILOSOPHICALLY OPPOSED TO THIS BOX OF COUNTRY 8-TRACK TAPES I PICKED FOR YOU!
BERTRAND CAN BLOW IT OUT HIS ASS!
PORTER WAGONER! WANDA JACKSON! LEFTY FRIZZELL!
GASP! SLIM WHITMAN!
SLIM WHITMAN ALL MY BEST
YOUR DEVOTION TO BAD, ANTIQUATED TECHNOLOGY MYSTIFIES ME, MAGEE!
AAAAAH.
I'LL LEAVE YOU TO ATTAIN ZEN. I'M HITTING THE SHOWER.
SSSSSSSSSSSS
H
C
SSS
WHEW.
SSS
SIGH. I DON'T THINK I CAN EVER TRULY BE CLEAN AGAIN.
SSSSSSSSSS

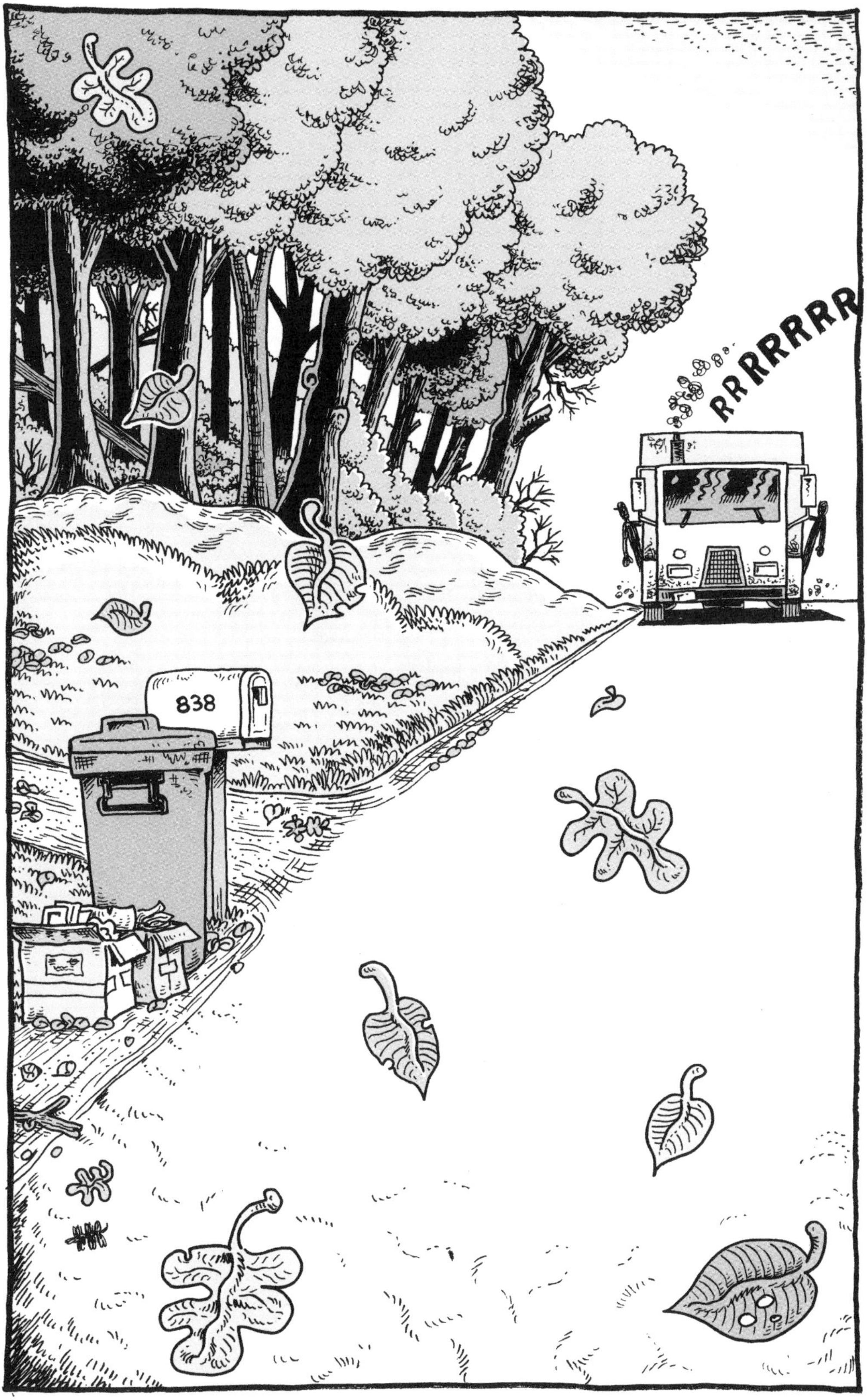
838
RRRRRRR

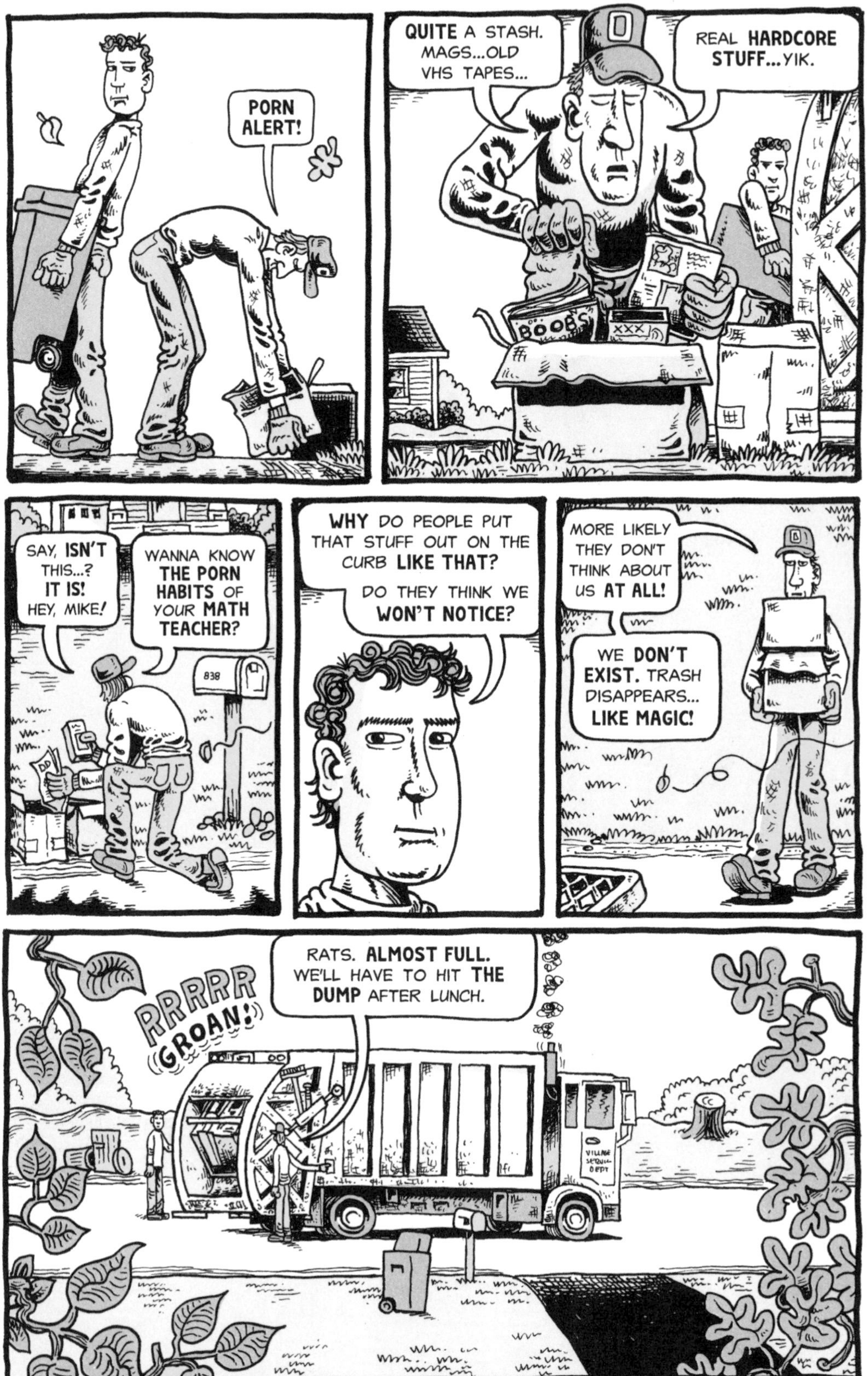
PORN ALERT!
QUITE A STASH. MAGS...OLD VHS TAPES...
REAL HARDCORE STUFF...YIK.
BOOBS
XXX
SAY, ISN'T THIS...? IT IS! HEY, MIKE!
WANNA KNOW THE PORN HABITS OF YOUR MATH TEACHER?
838
WHY DO PEOPLE PUT THAT STUFF OUT ON THE CURB LIKE THAT?
DO THEY THINK WE WON'T NOTICE?
MORE LIKELY THEY DON'T THINK ABOUT US AT ALL!
WE DON'T EXIST. TRASH DISAPPEARS... LIKE MAGIC!
RRRRR GROAN!
RATS. ALMOST FULL. WE'LL HAVE TO HIT THE DUMP AFTER LUNCH.

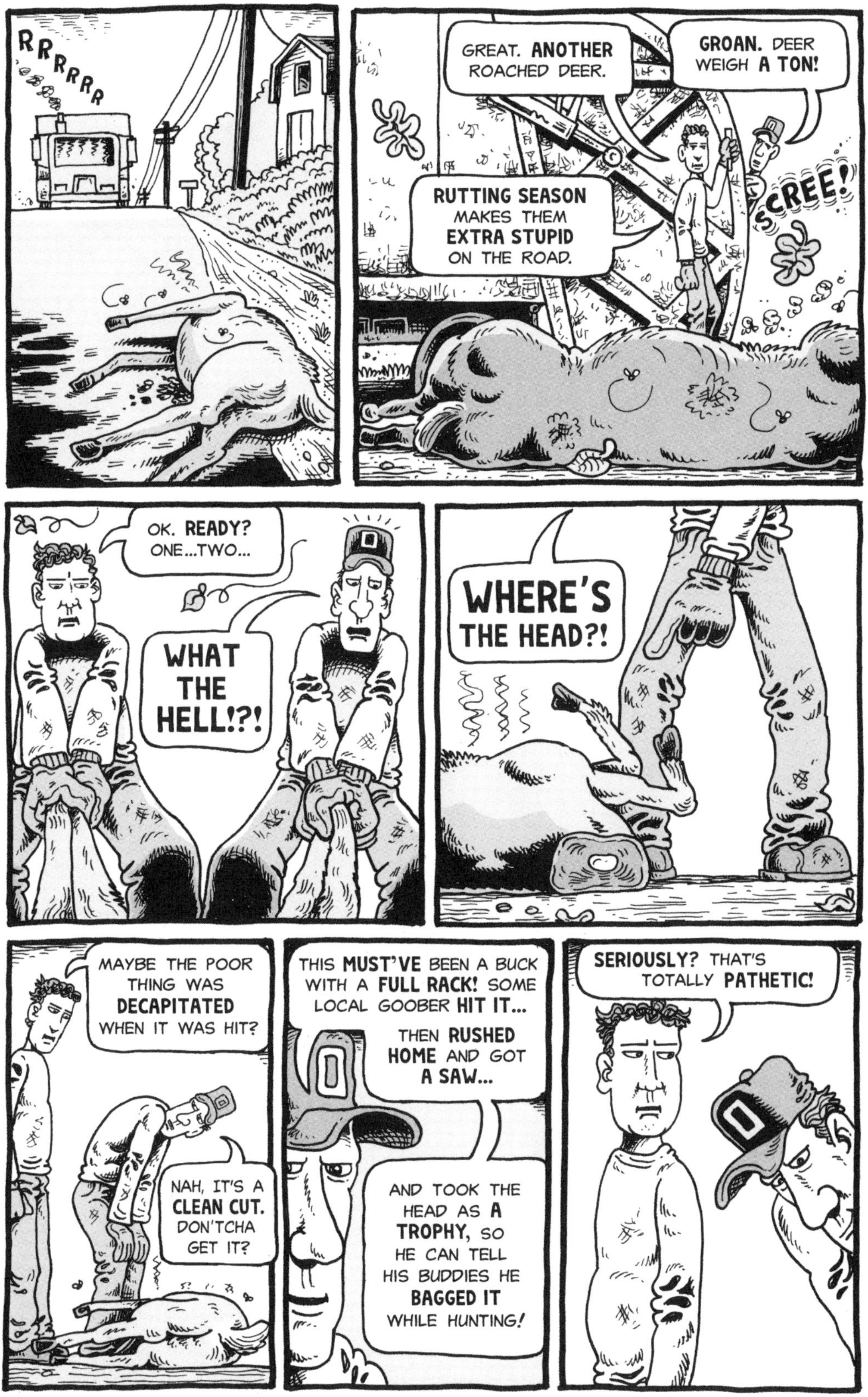
RRRRRRR
GREAT. ANOTHER ROACHED DEER.
GROAN. DEER WEIGH A TON!
RUTTING SEASON MAKES THEM EXTRA STUPID ON THE ROAD.
SCREE!
OK. READY? ONE...TWO...
WHAT THE HELL!?!
WHERE'S THE HEAD?!
MAYBE THE POOR THING WAS DECAPITATED WHEN IT WAS HIT?
NAH, IT'S A CLEAN CUT. DON'TCHA GET IT?
THIS MUST'VE BEEN A BUCK WITH A FULL RACK! SOME LOCAL GOOBER HIT IT...
THEN RUSHED HOME AND GOT A SAW...
AND TOOK THE HEAD AS A TROPHY, SO HE CAN TELL HIS BUDDIES HE BAGGED IT WHILE HUNTING!
SERIOUSLY? THAT'S TOTALLY PATHETIC!

YEAH! HE PRETENDS HE'S THE **GREAT WHITE HUNTER**...

...WHEN, IN REALITY, HE'S THE **GREAT WHITE DISTRACTED DRIVER!**

WHUMP!

WAIT UNTIL THE HOPPER IS **FULL** BEFORE CYCLING. IT'S **GROSS** WHEN A DEER BODY GETS **CHEWED UP.**

UGH. ROTTEN **JACK-O'-LANTERNS.**

SKY HOOK!

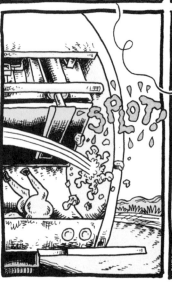

SPLOT!

HOLIDAYS ARE A **DRAG.** AND NEXT UP IS THE WORST ONE: **THANKSGIVING!** THE RACCOONS GO **MEDIEVAL** ON EVERY CAN WITH A **TURKEY CARCASS!**

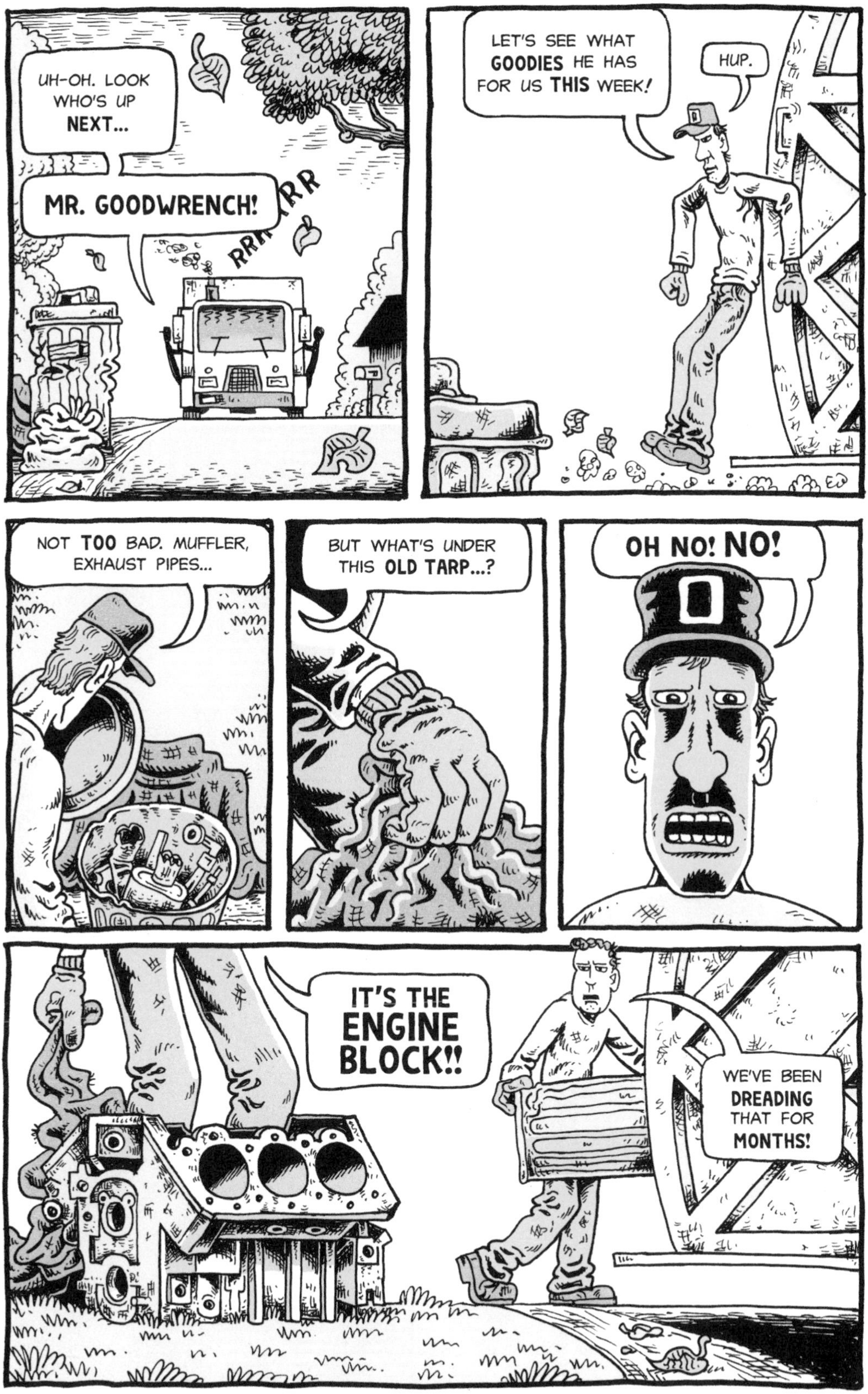
UH-OH. LOOK WHO'S UP NEXT...
MR. GOODWRENCH!
RRRRRR
LET'S SEE WHAT GOODIES HE HAS FOR US THIS WEEK!
HUP.
NOT TOO BAD. MUFFLER, EXHAUST PIPES...
BUT WHAT'S UNDER THIS OLD TARP...?
OH NO! NO!
IT'S THE ENGINE BLOCK!!
WE'VE BEEN DREADING THAT FOR MONTHS!

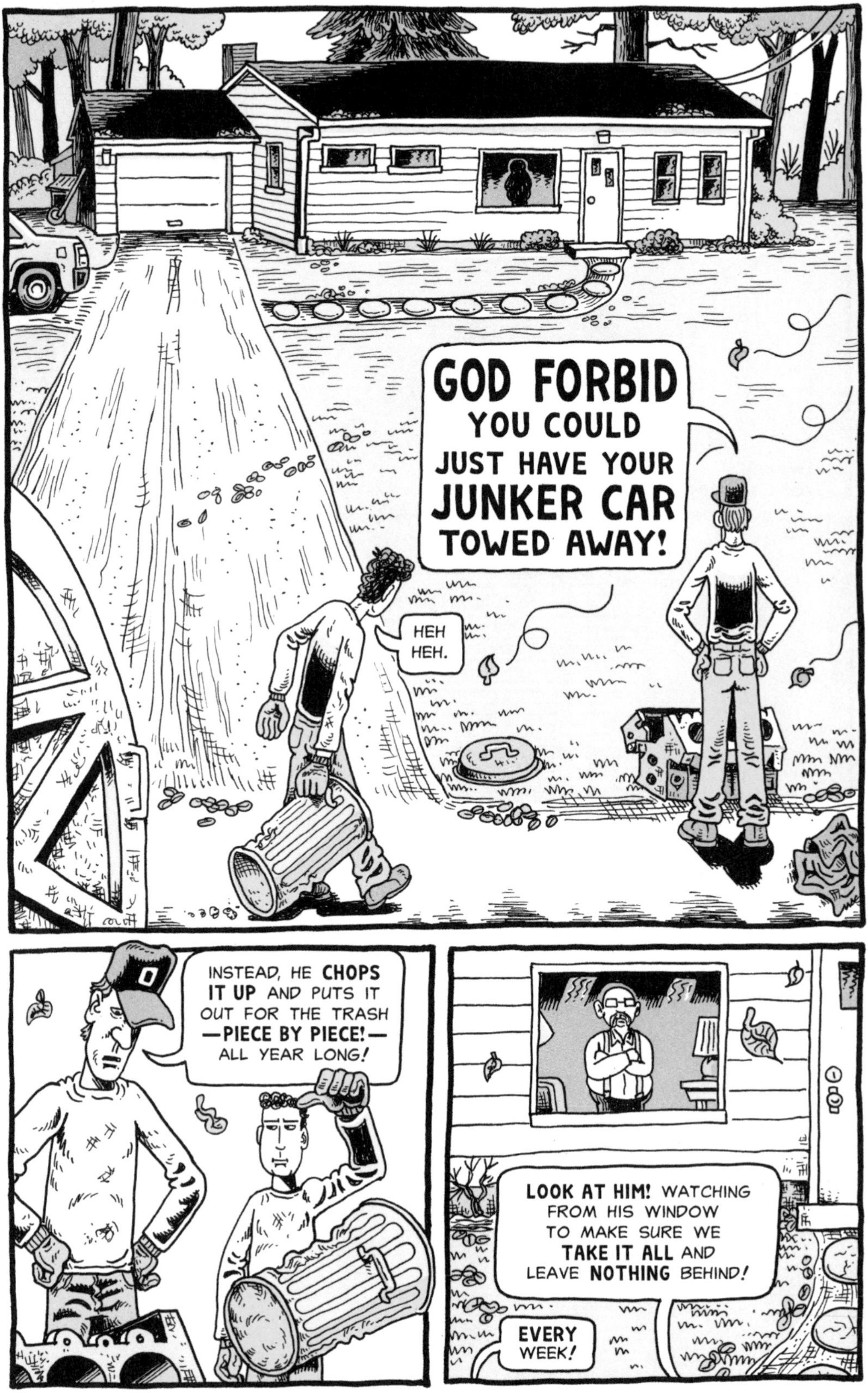
GOD FORBID YOU COULD JUST HAVE YOUR JUNKER CAR TOWED AWAY!
HEH HEH.
INSTEAD, HE CHOPS IT UP AND PUTS IT OUT FOR THE TRASH —PIECE BY PIECE!— ALL YEAR LONG!
LOOK AT HIM! WATCHING FROM HIS WINDOW TO MAKE SURE WE TAKE IT ALL AND LEAVE NOTHING BEHIND!
EVERY WEEK!

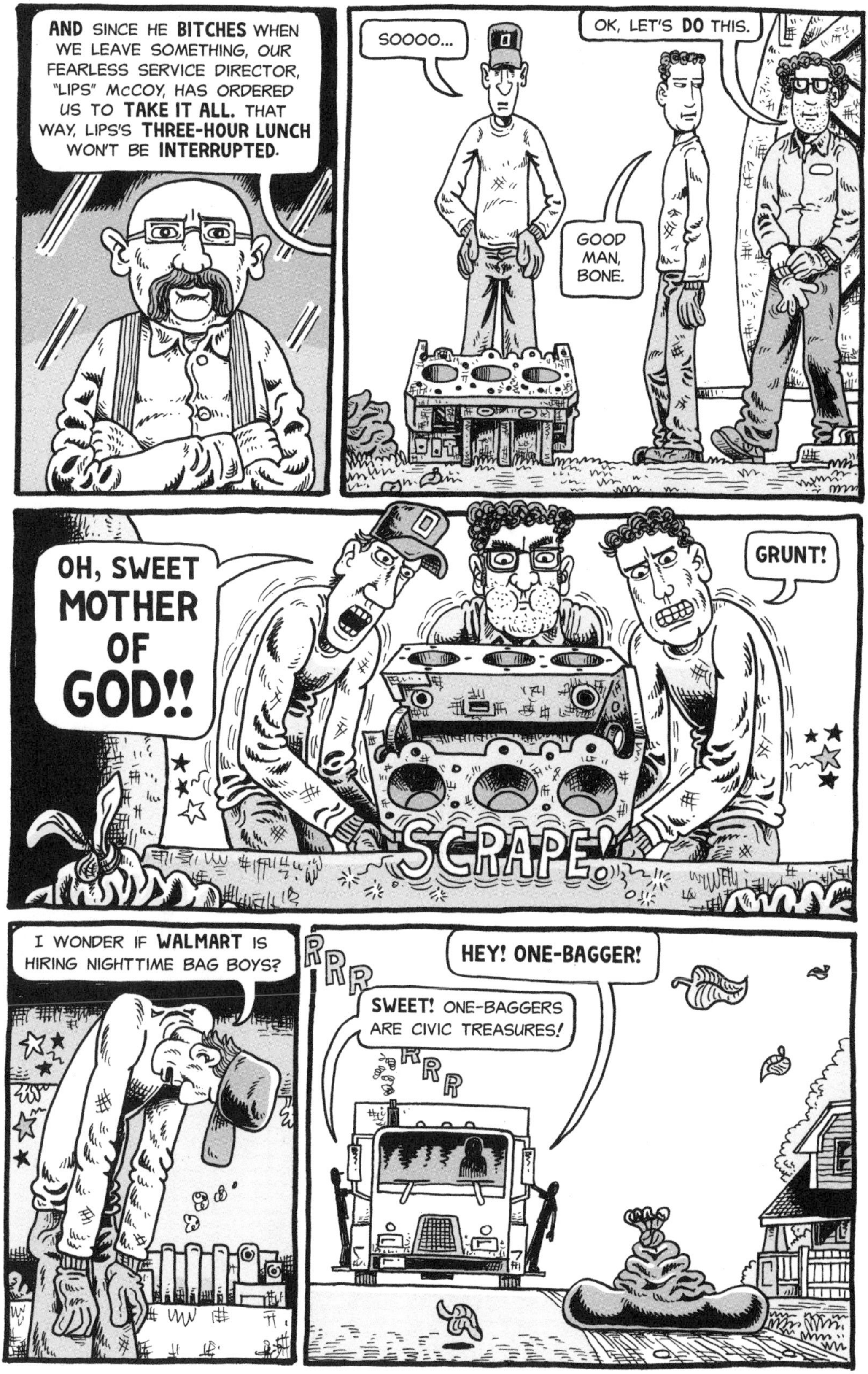
AND SINCE HE BITCHES WHEN WE LEAVE SOMETHING, OUR FEARLESS SERVICE DIRECTOR, "LIPS" McCOY, HAS ORDERED US TO TAKE IT ALL. THAT WAY, LIPS'S THREE-HOUR LUNCH WON'T BE INTERRUPTED.
SOOOO...
OK, LET'S DO THIS.
GOOD MAN, BONE.
OH, SWEET MOTHER OF GOD!!
GRUNT!
SCRAPE!
I WONDER IF WALMART IS HIRING NIGHTTIME BAG BOYS?
RRR
HEY! ONE-BAGGER!
SWEET! ONE-BAGGERS ARE CIVIC TREASURES!
RRR

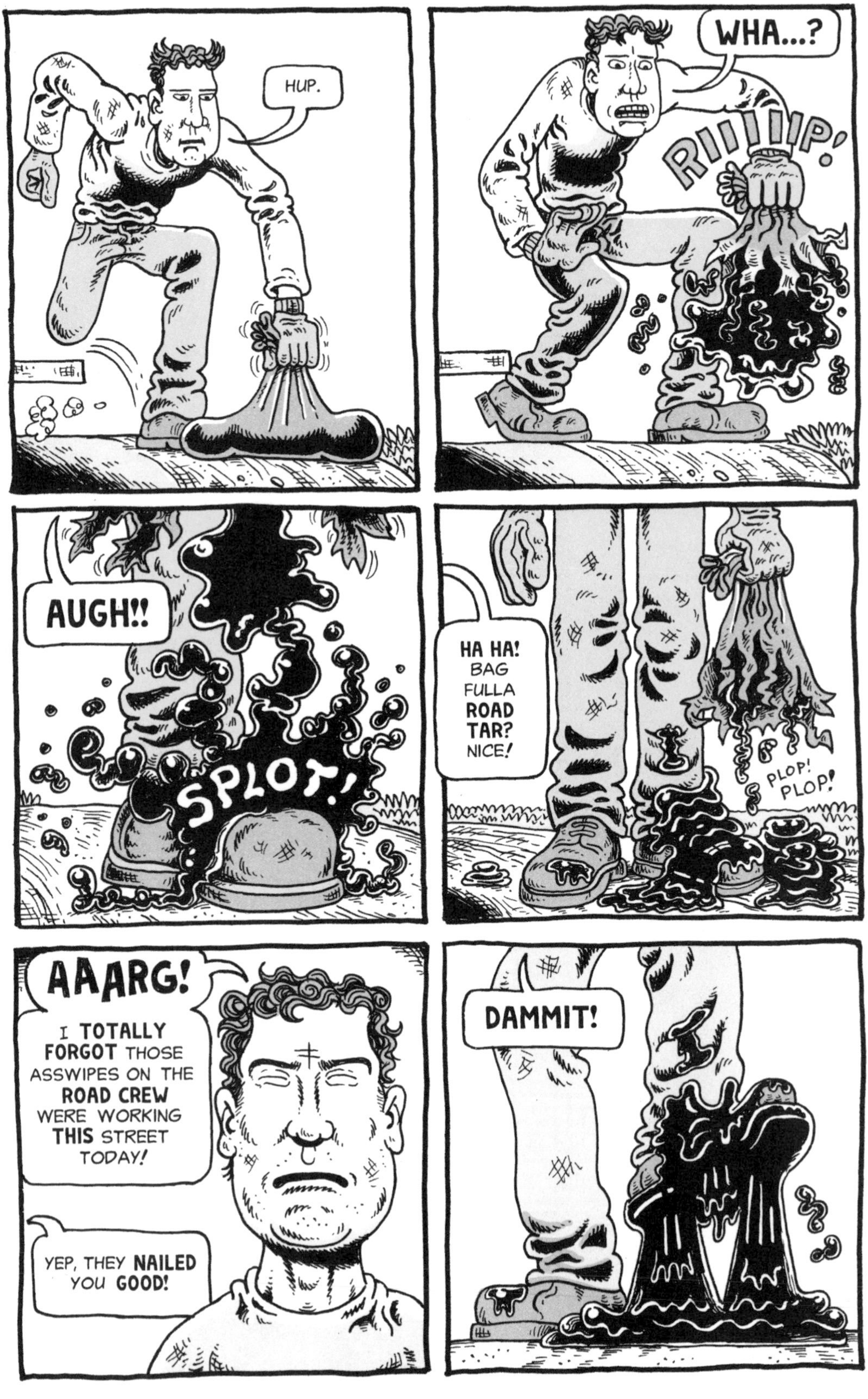
HUP.
WHA...?
RIIIIIP!
AUGH!!
SPLOT!
HA HA! BAG FULLA ROAD TAR? NICE!
PLOP! PLOP!
AAARG!
I TOTALLY FORGOT THOSE ASSWIPES ON THE ROAD CREW WERE WORKING THIS STREET TODAY!
YEP, THEY NAILED YOU GOOD!
DAMMIT!

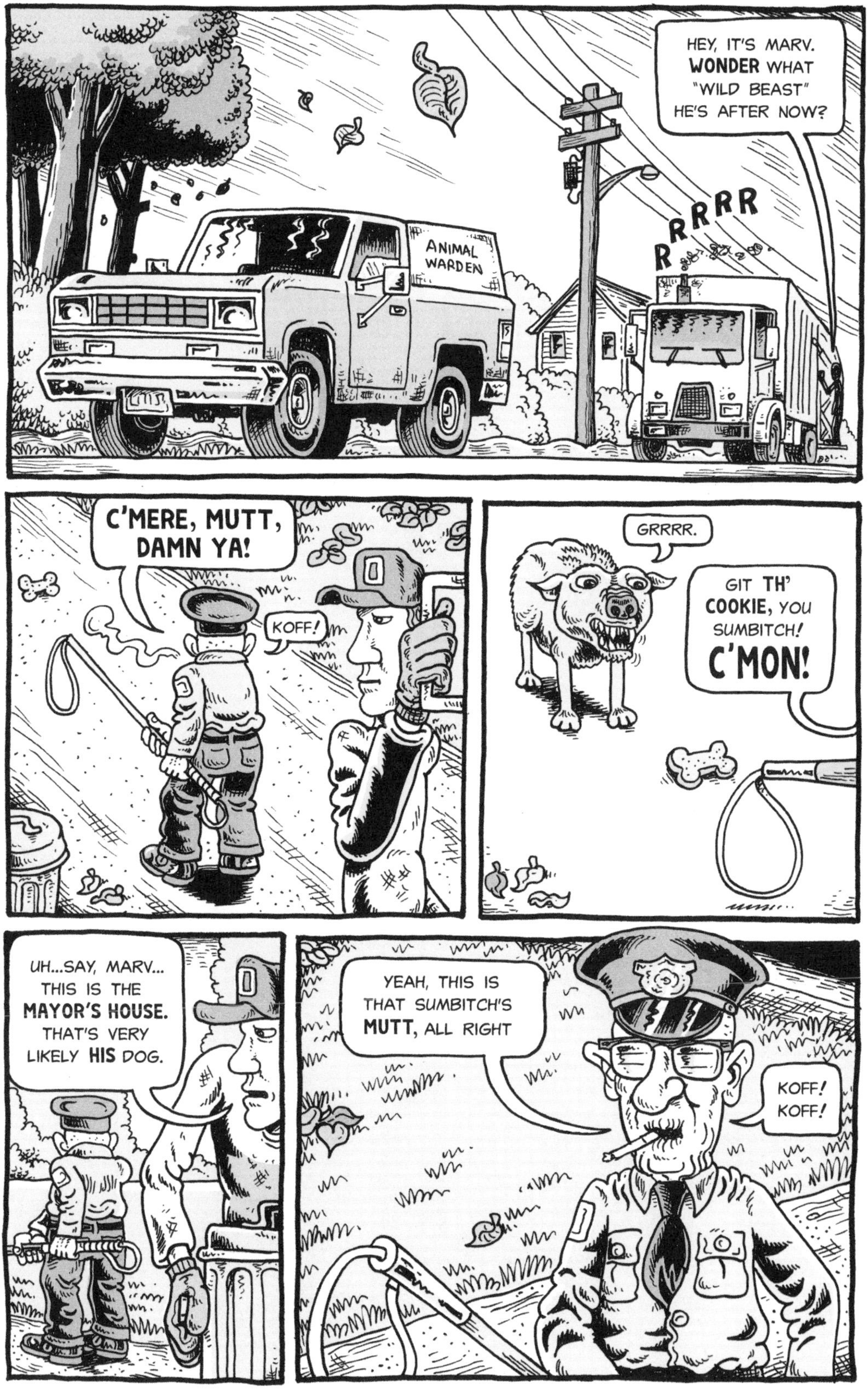

HEY, IT'S MARV. WONDER WHAT "WILD BEAST" HE'S AFTER NOW?
ANIMAL WARDEN
RRRRRR
C'MERE, MUTT, DAMN YA!
KOFF!
GRRRR.
GIT TH' COOKIE, YOU SUMBITCH! C'MON!
UH...SAY, MARV... THIS IS THE MAYOR'S HOUSE. THAT'S VERY LIKELY HIS DOG.
YEAH, THIS IS THAT SUMBITCH'S MUTT, ALL RIGHT
KOFF! KOFF!

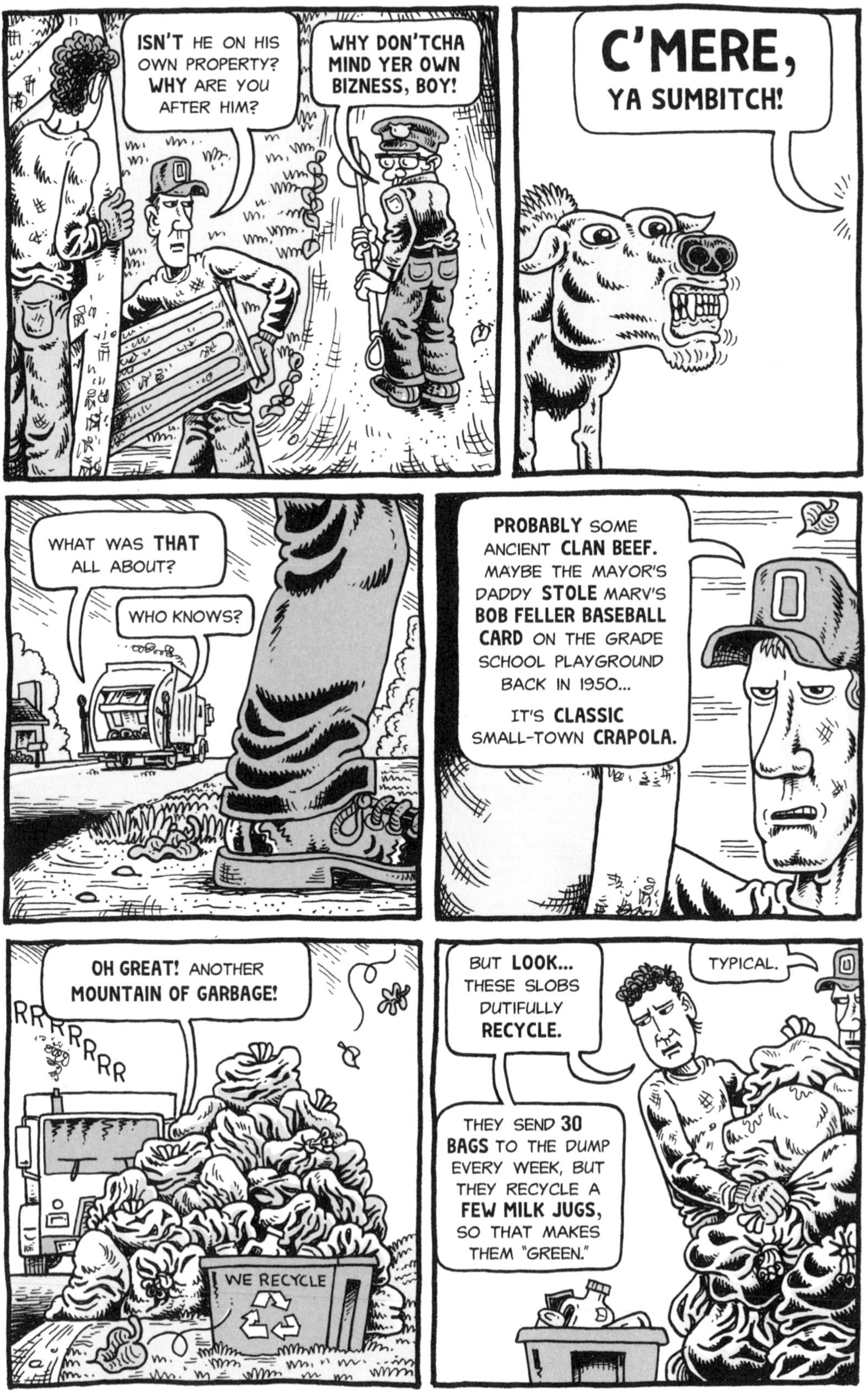
ISN'T HE ON HIS OWN PROPERTY? WHY ARE YOU AFTER HIM?
WHY DON'TCHA MIND YER OWN BIZNESS, BOY!
C'MERE, YA SUMBITCH!
WHAT WAS THAT ALL ABOUT?
WHO KNOWS?
PROBABLY SOME ANCIENT CLAN BEEF. MAYBE THE MAYOR'S DADDY STOLE MARV'S BOB FELLER BASEBALL CARD ON THE GRADE SCHOOL PLAYGROUND BACK IN 1950...
IT'S CLASSIC SMALL-TOWN CRAPOLA.
OH GREAT! ANOTHER MOUNTAIN OF GARBAGE!
RRRRRRR
WE RECYCLE
BUT LOOK... THESE SLOBS DUTIFULLY RECYCLE.
TYPICAL.
THEY SEND 30 BAGS TO THE DUMP EVERY WEEK, BUT THEY RECYCLE A FEW MILK JUGS, SO THAT MAKES THEM "GREEN."

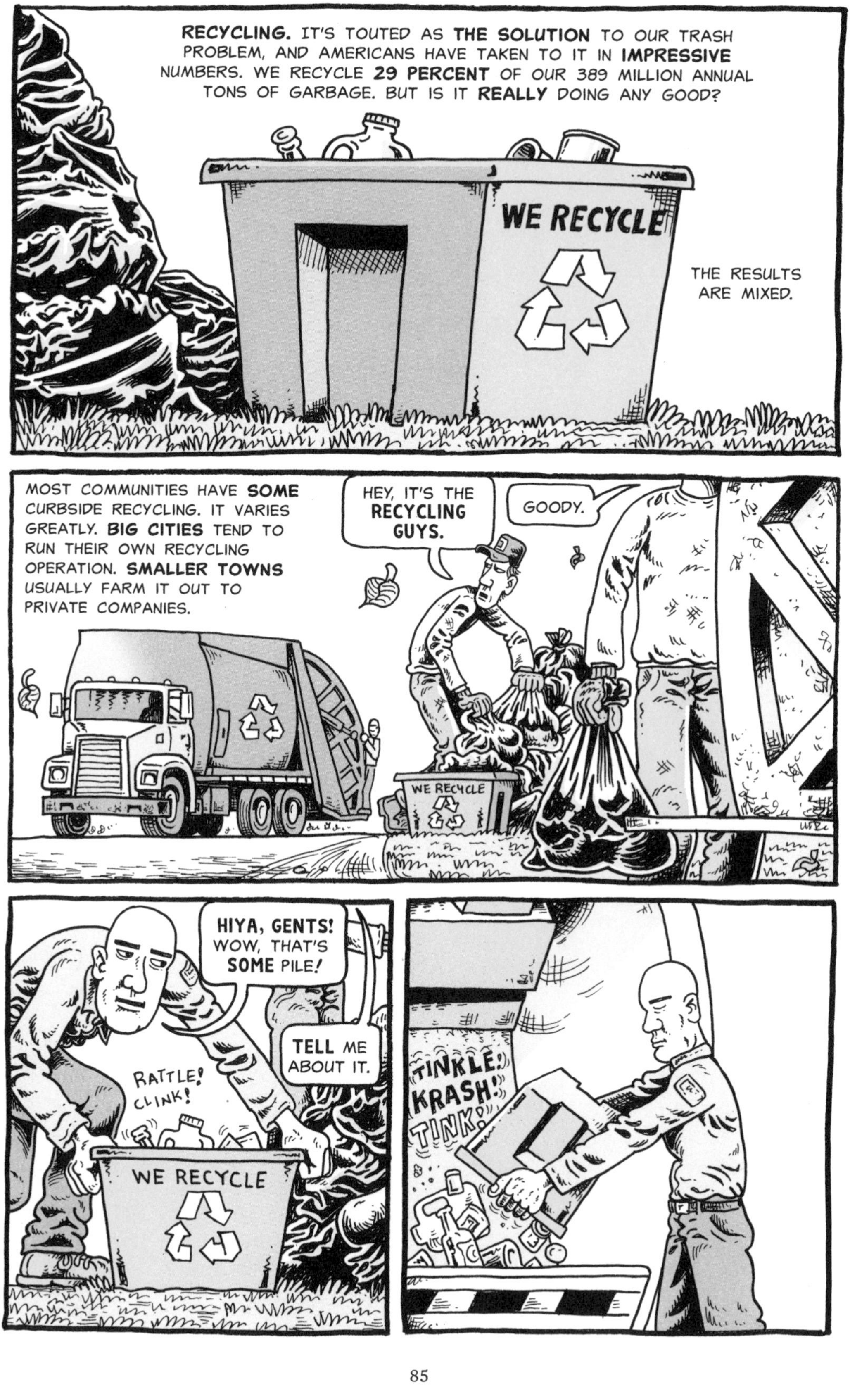
RECYCLING. IT'S TOUTED AS **THE SOLUTION** TO OUR TRASH PROBLEM, AND AMERICANS HAVE TAKEN TO IT IN **IMPRESSIVE** NUMBERS. WE RECYCLE **29 PERCENT** OF OUR 389 MILLION ANNUAL TONS OF GARBAGE. BUT IS IT **REALLY** DOING ANY GOOD?
WE RECYCLE
THE RESULTS ARE MIXED.
MOST COMMUNITIES HAVE **SOME** CURBSIDE RECYCLING. IT VARIES GREATLY. **BIG CITIES** TEND TO RUN THEIR OWN RECYCLING OPERATION. **SMALLER TOWNS** USUALLY FARM IT OUT TO PRIVATE COMPANIES.
HEY, IT'S THE **RECYCLING GUYS.**
GOODY.
WE RECYCLE
HIYA, GENTS! WOW, THAT'S **SOME** PILE!
TELL ME ABOUT IT.
RATTLE! CLINK!
WE RECYCLE
TINKLE! KRASH! TINK!

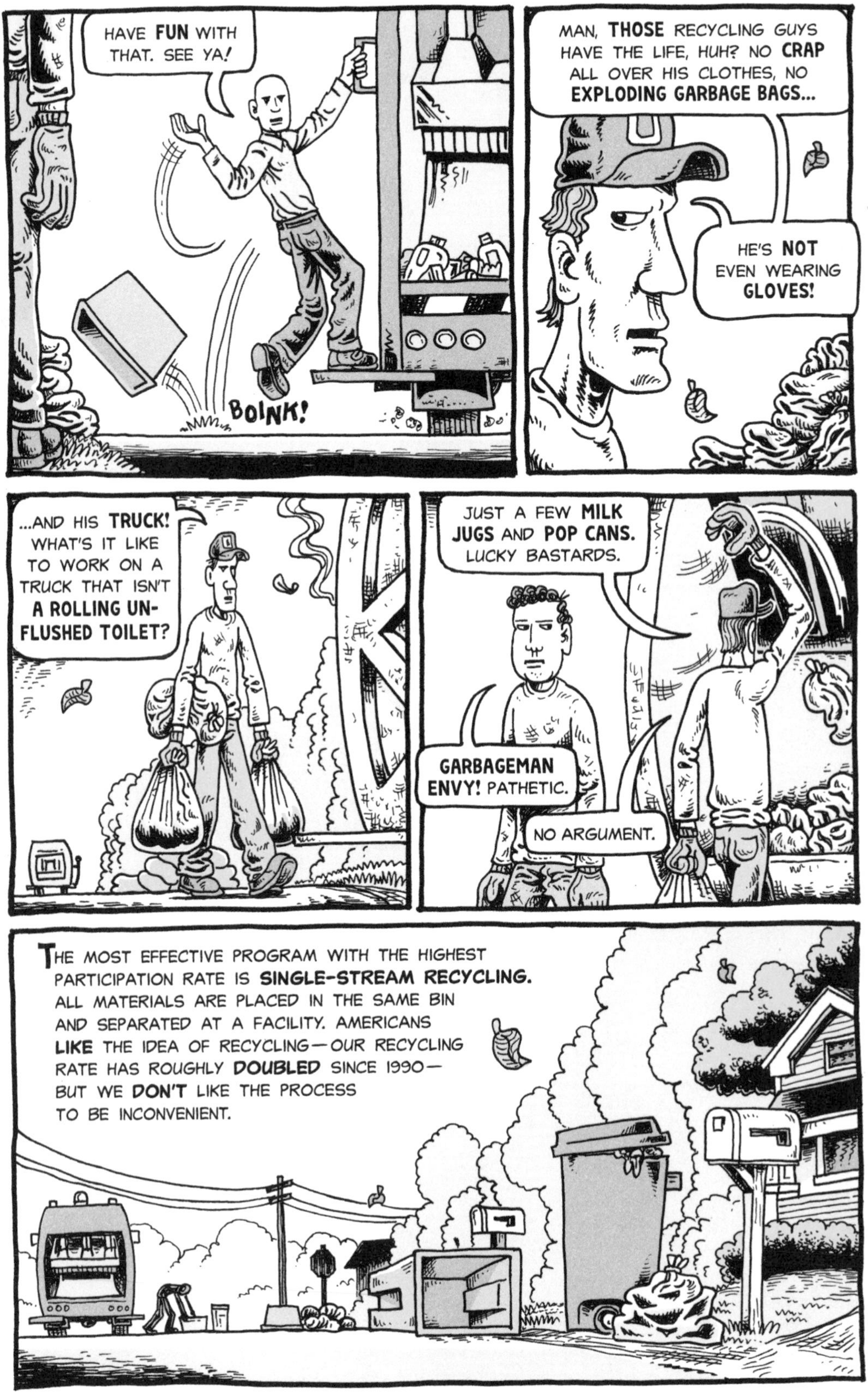
HAVE FUN WITH THAT. SEE YA!
BOINK!
MAN, THOSE RECYCLING GUYS HAVE THE LIFE, HUH? NO CRAP ALL OVER HIS CLOTHES, NO EXPLODING GARBAGE BAGS...
HE'S NOT EVEN WEARING GLOVES!
...AND HIS TRUCK! WHAT'S IT LIKE TO WORK ON A TRUCK THAT ISN'T A ROLLING UN-FLUSHED TOILET?
JUST A FEW MILK JUGS AND POP CANS. LUCKY BASTARDS.
GARBAGEMAN ENVY! PATHETIC.
NO ARGUMENT.
THE MOST EFFECTIVE PROGRAM WITH THE HIGHEST PARTICIPATION RATE IS SINGLE-STREAM RECYCLING. ALL MATERIALS ARE PLACED IN THE SAME BIN AND SEPARATED AT A FACILITY. AMERICANS LIKE THE IDEA OF RECYCLING—OUR RECYCLING RATE HAS ROUGHLY DOUBLED SINCE 1990—BUT WE DON'T LIKE THE PROCESS TO BE INCONVENIENT.

BUT EVEN WITH A **DRAMATIC RISE** IN RECYCLING, IT'S ONLY MADE A SLIGHT DENT OVER THE SAME PERIOD OF TIME IN THE STAGGERING AMOUNT OF GARBAGE WE SEND TO THE LANDFILL EACH YEAR.

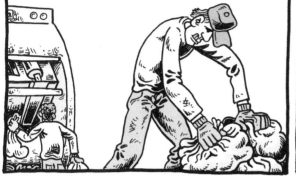

BUT THERE'S **BIG MONEY** IN RECYCLING, RIGHT? NOT REALLY. IT DEPENDS ON **THE PRICE** RECYCLED MATERIALS FETCH ON THE COMMODITIES MARKET, AND THOSE **HAVEN'T** BEEN HIGH ENOUGH TO MAKE RECYCLING PROFITABLE. AT LEAST TOWNS GET **SOMETHING** TO OFFSET THE COST OF COLLECTION, AS OPPOSED TO TRASH PICKUP, WHICH IS JUST AN OUTRIGHT EXPENS...AND **A GROWING ONE.**

SURE, RECYCLING ALL WE CAN IS THE **COMMONSENSE THING** TO DO—WHY ENTOMB RESOURCES IN A LANDFILL?—BUT IT'S **NOT** THE ANSWER TO OUR GARBAGE PROBLEM. THAT SOLUTION IS: USE **LESS STUFF** IN THE FIRST PLACE!

WHEW! WORKING UP **A SWEAT** HERE!

AND **THAT'S** THE UNSOLVABLE QUANDARY. ARE WE GOING TO **GIVE UP** OUR PLASTIC DRINK BOTTLES, DISPOSABLE DIAPERS, AND PLASTIC SHOPPING BAGS?

YOU **KNOW** THE ANSWER.

BOINK!

OK. THIS IS **THE LAST** OF IT.

FINALLY!

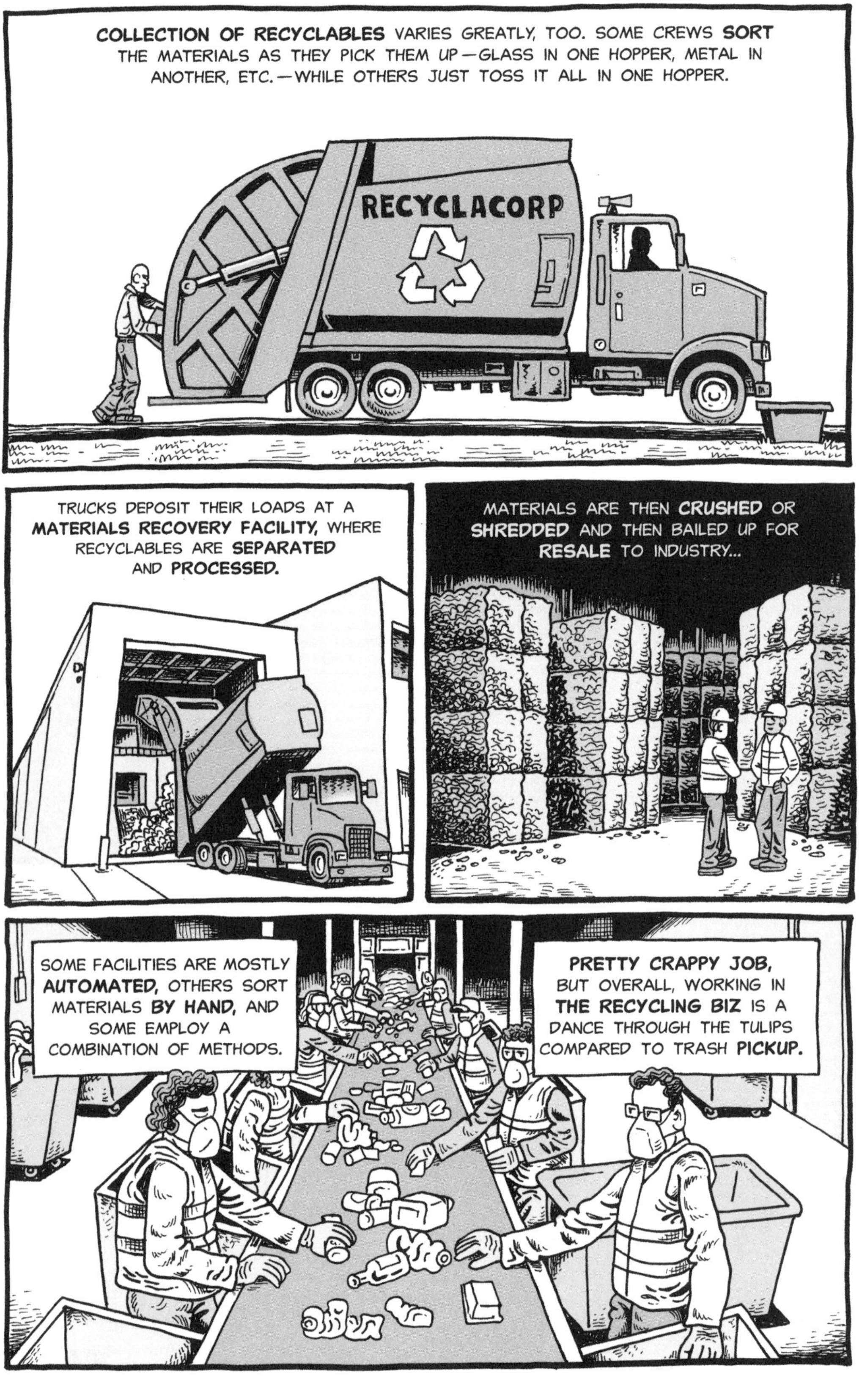
COLLECTION OF RECYCLABLES VARIES GREATLY, TOO. SOME CREWS SORT THE MATERIALS AS THEY PICK THEM UP—GLASS IN ONE HOPPER, METAL IN ANOTHER, ETC.—WHILE OTHERS JUST TOSS IT ALL IN ONE HOPPER.
RECYCLACORP
TRUCKS DEPOSIT THEIR LOADS AT A MATERIALS RECOVERY FACILITY, WHERE RECYCLABLES ARE SEPARATED AND PROCESSED.
MATERIALS ARE THEN CRUSHED OR SHREDDED AND THEN BAILED UP FOR RESALE TO INDUSTRY...
SOME FACILITIES ARE MOSTLY AUTOMATED, OTHERS SORT MATERIALS BY HAND, AND SOME EMPLOY A COMBINATION OF METHODS.
PRETTY CRAPPY JOB, BUT OVERALL, WORKING IN THE RECYCLING BIZ IS A DANCE THROUGH THE TULIPS COMPARED TO TRASH PICKUP.

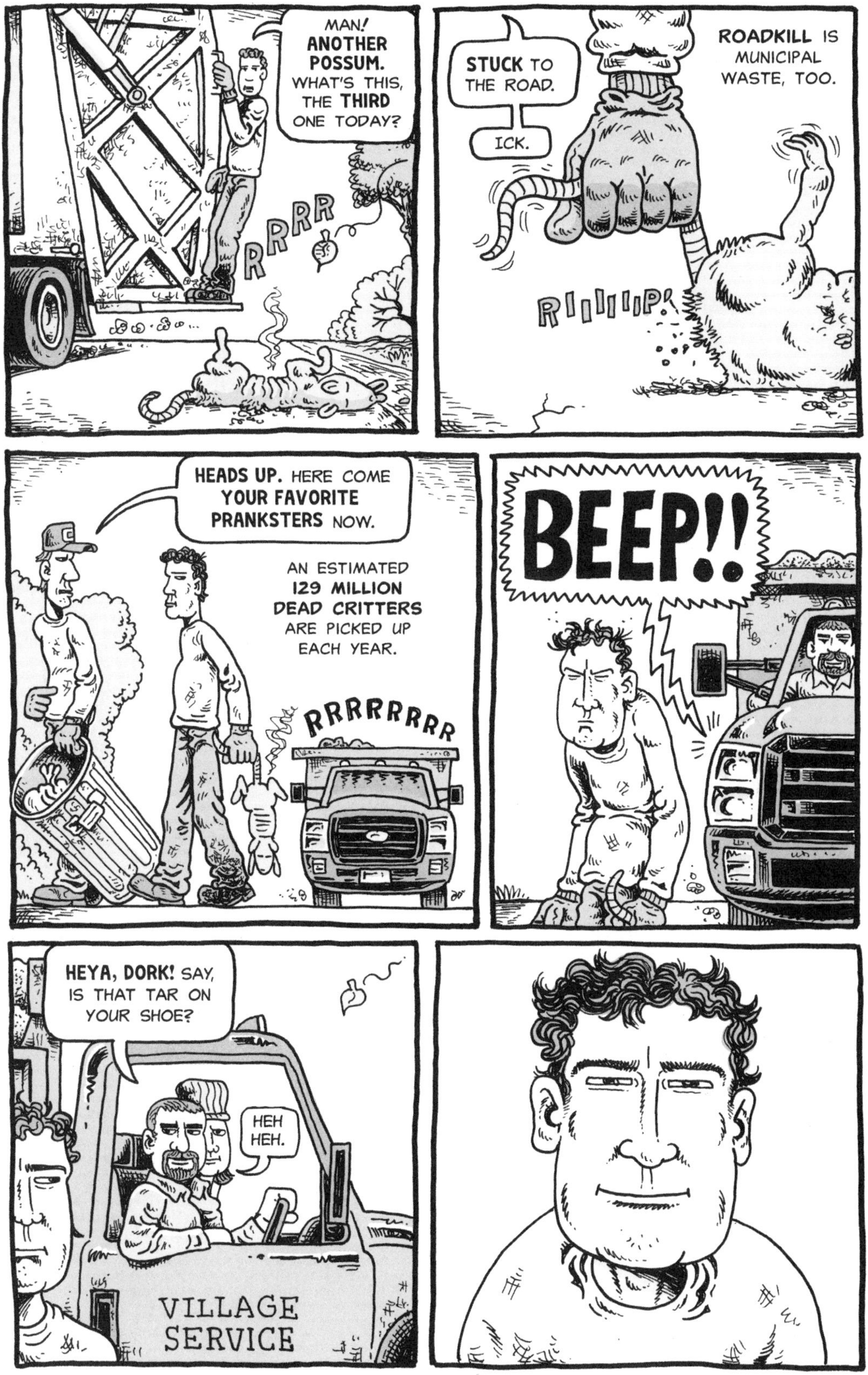
MAN! ANOTHER POSSUM. WHAT'S THIS, THE THIRD ONE TODAY?
RRRR
STUCK TO THE ROAD.
ICK.
ROADKILL IS MUNICIPAL WASTE, TOO.
RIIIIIIP!
HEADS UP. HERE COME YOUR FAVORITE PRANKSTERS NOW.
AN ESTIMATED 129 MILLION DEAD CRITTERS ARE PICKED UP EACH YEAR.
RRRRRRRR
BEEP!!
HEYA, DORK! SAY, IS THAT TAR ON YOUR SHOE?
HEH HEH.
VILLAGE SERVICE

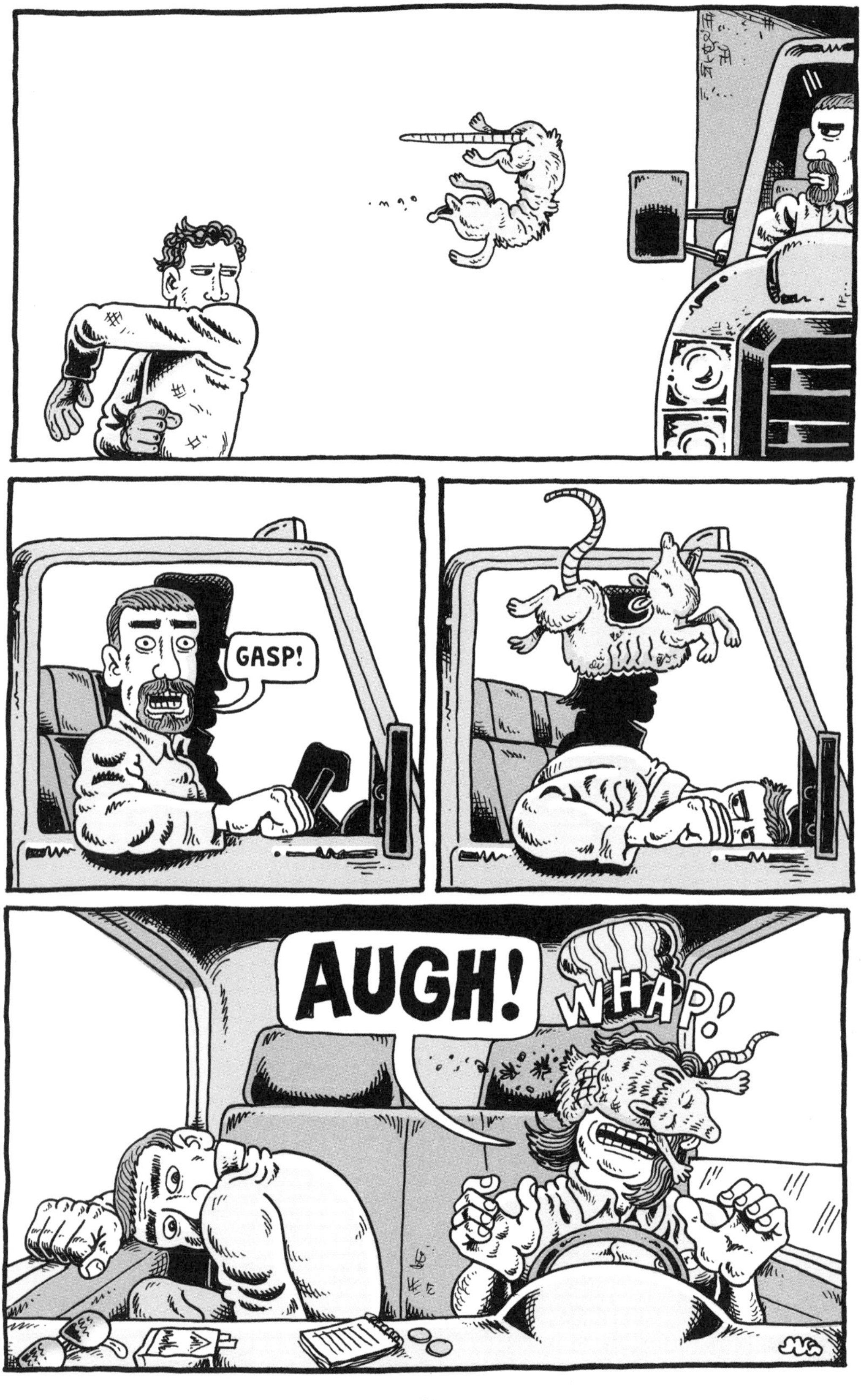
GASP!
AUGH!
WHAP!

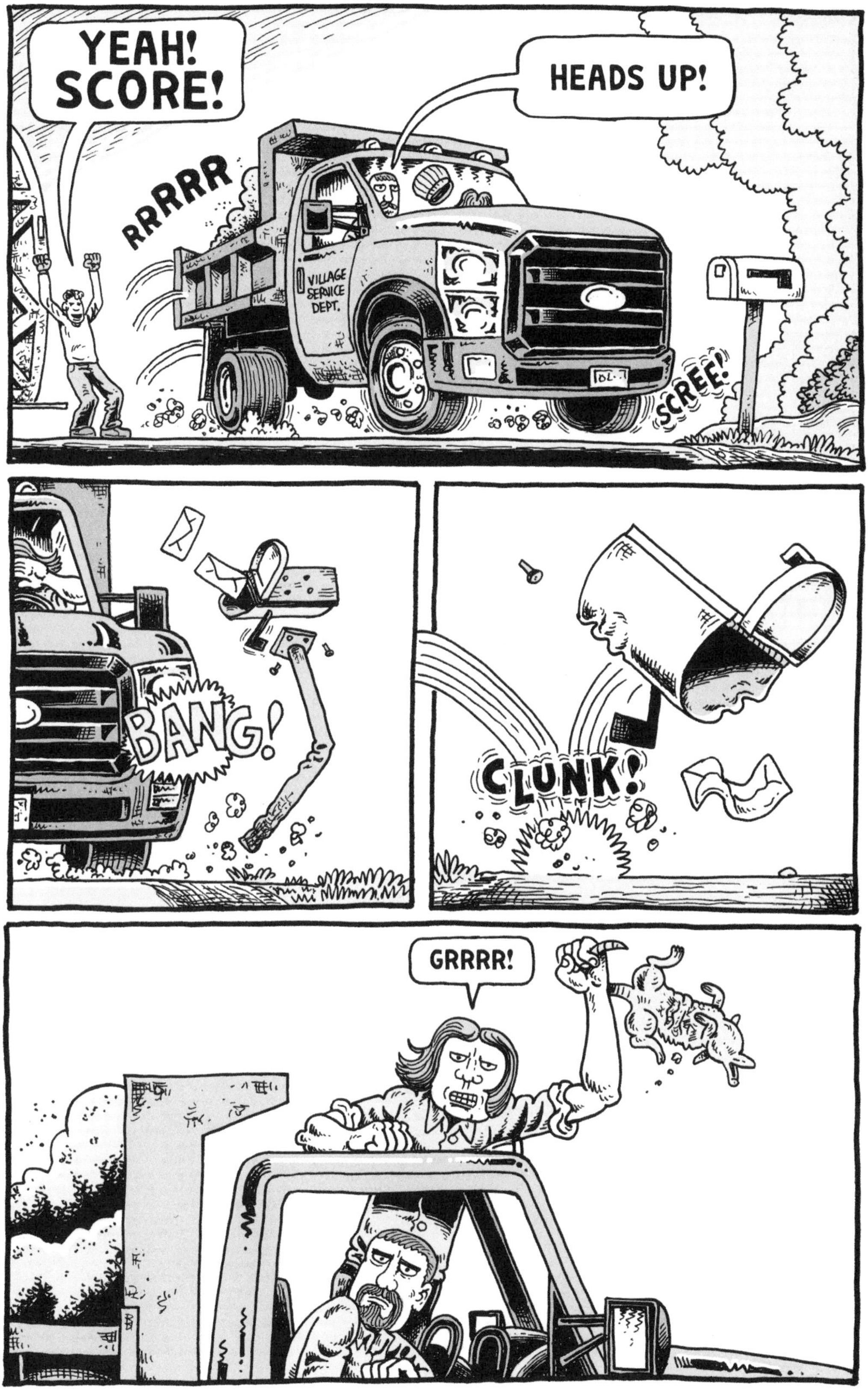
YEAH! SCORE!
HEADS UP!
RRRRR
VILLAGE SERVICE DEPT.
SCREE!
BANG!
CLUNK!
GRRRR!

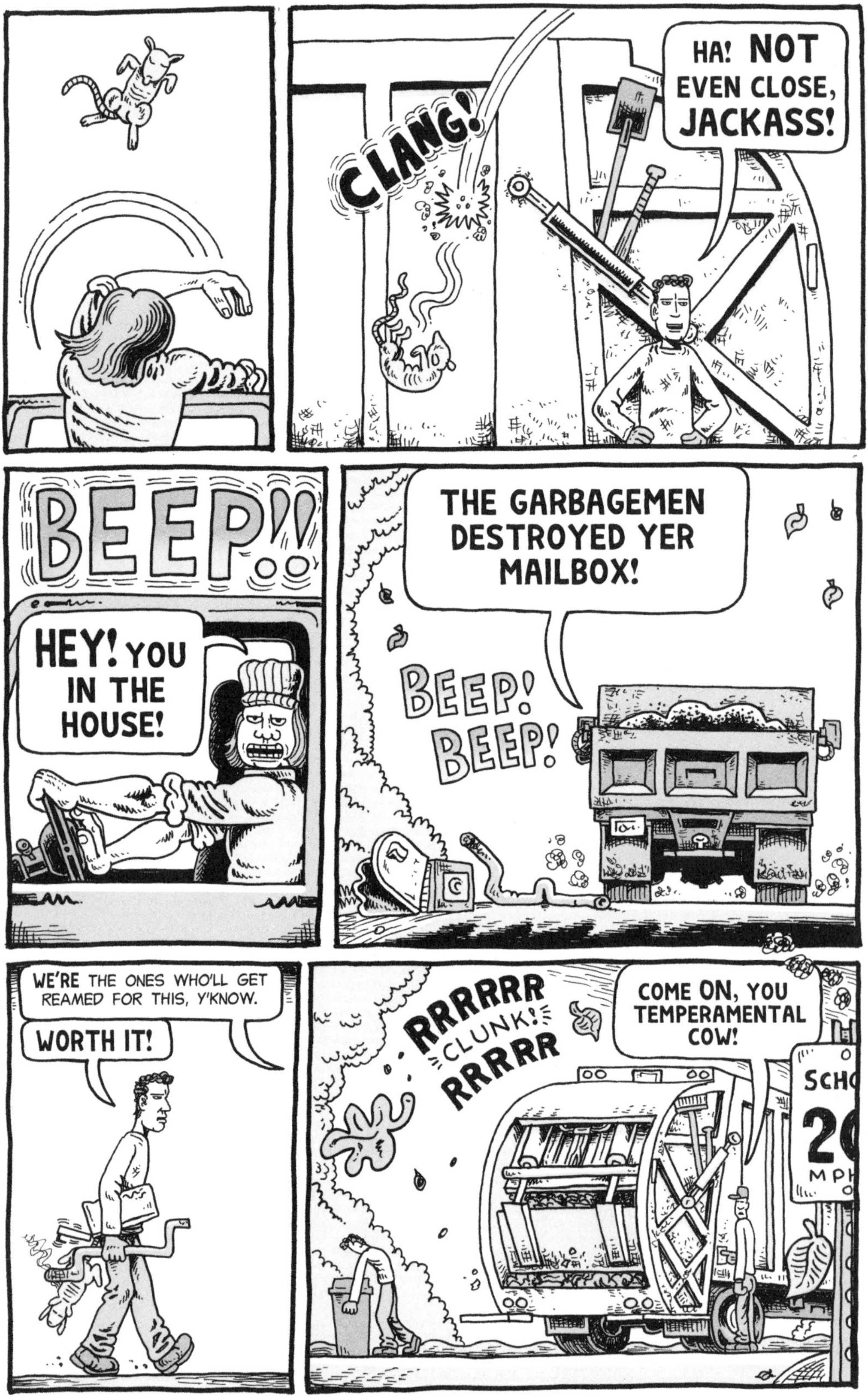
CLANG!
HA! NOT EVEN CLOSE, JACKASS!
BEEP!!
HEY! YOU IN THE HOUSE!
THE GARBAGEMEN DESTROYED YER MAILBOX!
BEEP! BEEP!
WE'RE THE ONES WHO'LL GET REAMED FOR THIS, Y'KNOW.
WORTH IT!
RRRRRR
CLUNK!
RRRRR
COME ON, YOU TEMPERAMENTAL COW!
SCHO
20
MPH

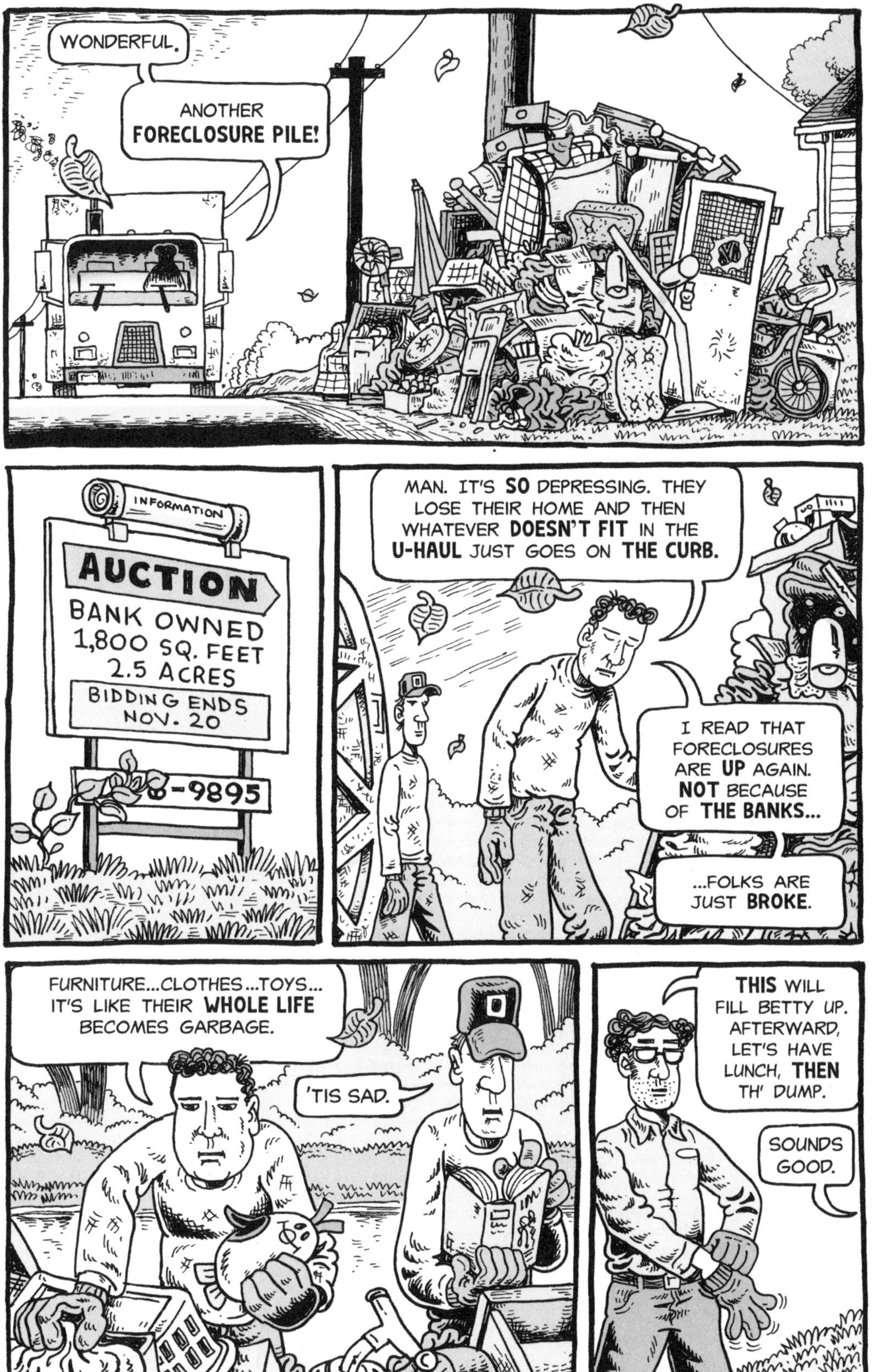
WONDERFUL.
ANOTHER FORECLOSURE PILE!
INFORMATION
AUCTION
BANK OWNED 1,800 SQ. FEET 2.5 ACRES
BIDDING ENDS NOV. 20
-9895
MAN. IT'S SO DEPRESSING. THEY LOSE THEIR HOME AND THEN WHATEVER DOESN'T FIT IN THE U-HAUL JUST GOES ON THE CURB.
I READ THAT FORECLOSURES ARE UP AGAIN. NOT BECAUSE OF THE BANKS...
...FOLKS ARE JUST BROKE.
FURNITURE...CLOTHES...TOYS... IT'S LIKE THEIR WHOLE LIFE BECOMES GARBAGE.
'TIS SAD.
THIS WILL FILL BETTY UP. AFTERWARD, LET'S HAVE LUNCH, THEN TH' DUMP.
SOUNDS GOOD.

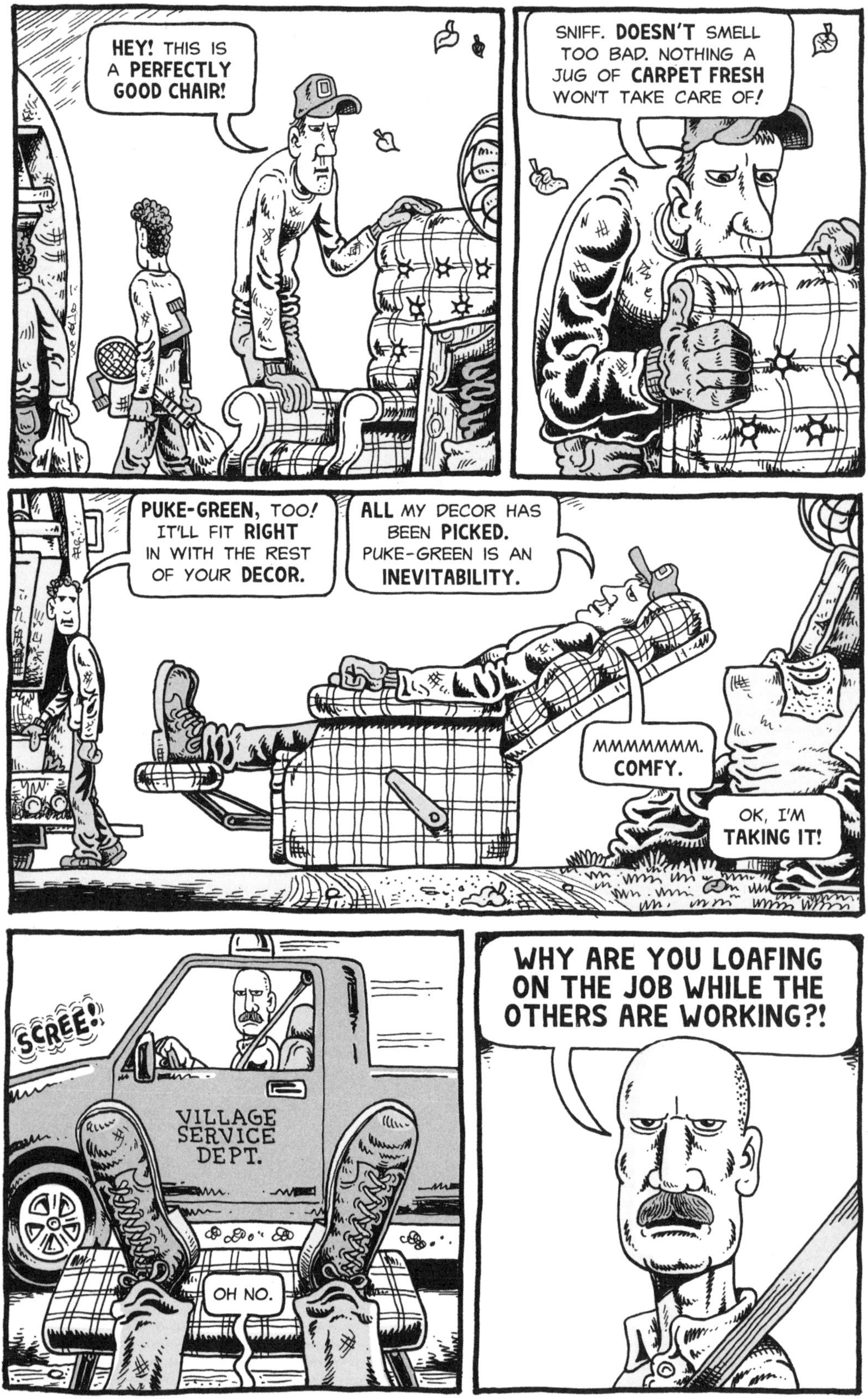
HEY! THIS IS A PERFECTLY GOOD CHAIR!
SNIFF. DOESN'T SMELL TOO BAD. NOTHING A JUG OF CARPET FRESH WON'T TAKE CARE OF!
PUKE-GREEN, TOO! IT'LL FIT RIGHT IN WITH THE REST OF YOUR DECOR.
ALL MY DECOR HAS BEEN PICKED. PUKE-GREEN IS AN INEVITABILITY.
MMMMMMM. COMFY.
OK, I'M TAKING IT!
SCREE!
VILLAGE SERVICE DEPT.
OH NO.
WHY ARE YOU LOAFING ON THE JOB WHILE THE OTHERS ARE WORKING?!

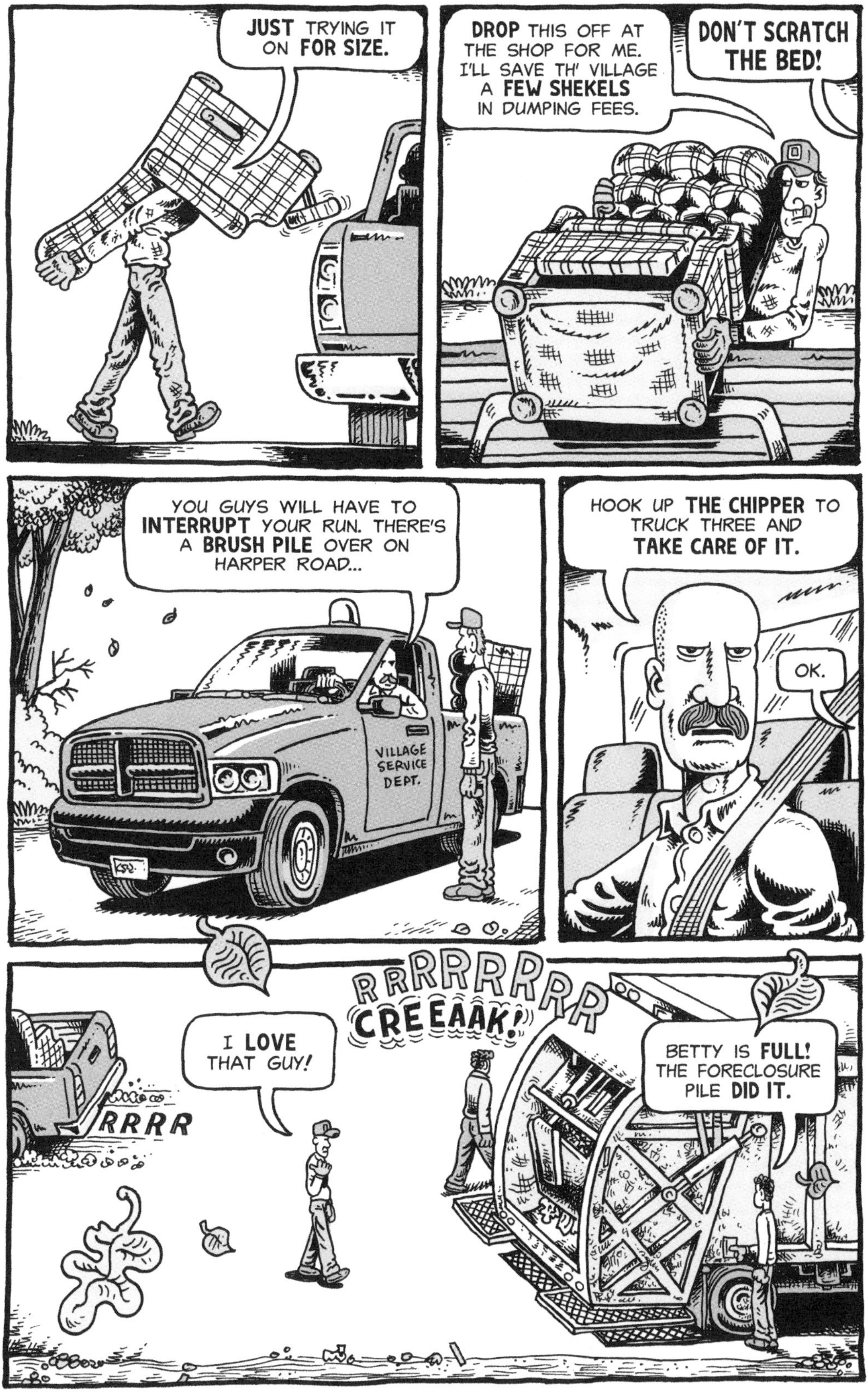
JUST TRYING IT ON FOR SIZE.
DROP THIS OFF AT THE SHOP FOR ME. I'LL SAVE TH' VILLAGE A FEW SHEKELS IN DUMPING FEES.
DON'T SCRATCH THE BED!
YOU GUYS WILL HAVE TO INTERRUPT YOUR RUN. THERE'S A BRUSH PILE OVER ON HARPER ROAD...
VILLAGE SERVICE DEPT.
HOOK UP THE CHIPPER TO TRUCK THREE AND TAKE CARE OF IT.
OK.
RRRR
I LOVE THAT GUY!
RRRRRRRRR
CREEAAK!
BETTY IS FULL! THE FORECLOSURE PILE DID IT.

THE GARBAGE TRUCK HAS BEEN ON THE STREETS FOR ALMOST 100 YEARS, BUT ITS BASIC DESIGN HAS CHANGED VERY LITTLE.

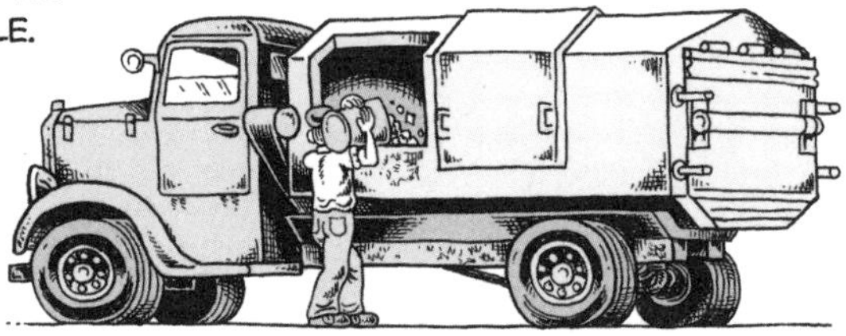

THE FIRST TRUCKS WERE **MODIFIED MODEL Ts.** IT WAS SIMPLY A MOTORIZED VERSION OF THE HORSE-DRAWN GARBAGE WAGON THAT HAD BEEN IN USE FOR CENTURIES. THE MODEL Ts DIDN'T HOLD MUCH OF A LOAD, AND GARBAGEMEN HAD TO HOIST HEAVY CANS OVER THEIR HEADS.

1934 REFUSE COMPRESSOR *from* **MARION METAL PRODUCTS**

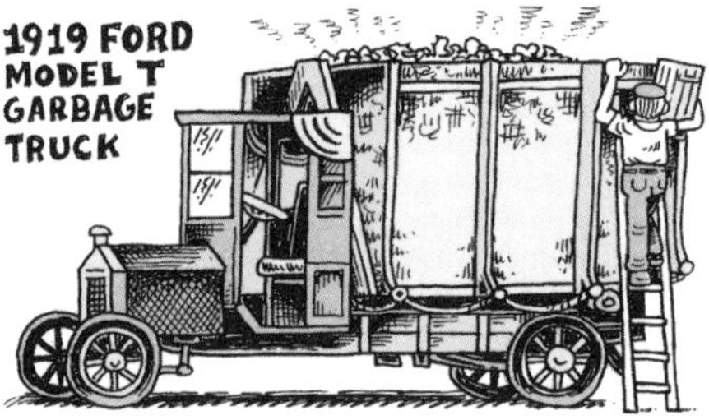

1919 FORD MODEL T GARBAGE TRUCK

THE FIRST BIG ADVANCE IN TRUCK DESIGN CAME WHEN MARION METAL INVENTED **THE PACKER.** AN INTERIOR METAL PLATE, POWERED BY HYDRAULICS, PACKED TRASH IN THE HOLD. BY COMPRESSING THE LOAD, A TRUCK COULD HOLD THREE TIMES MORE GARBAGE BEFORE HEADING TO THE DUMP. EFFICIENCY WAS SIGNIFICANTLY INCREASED.

HOW A TRUCK LIKE BETTY WORKS

IT'S ACTUALLY QUITE SIMPLE...

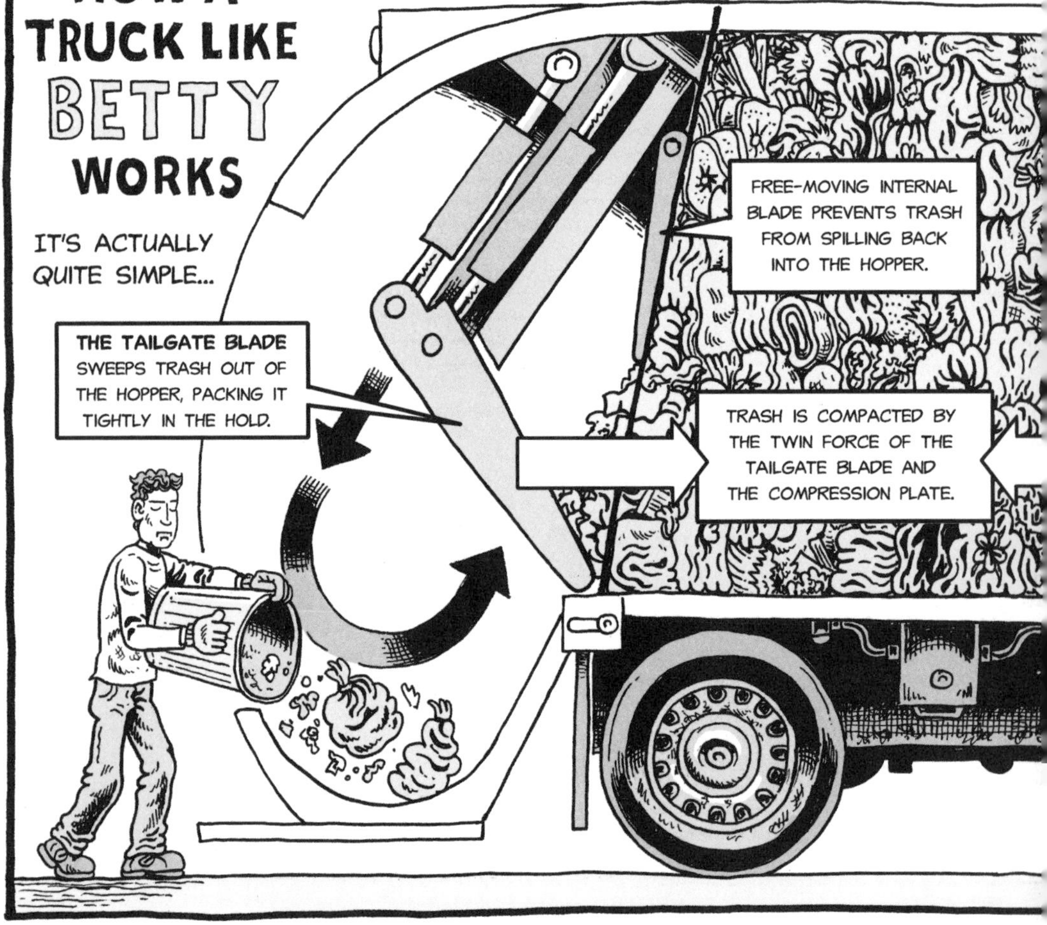
FREE-MOVING INTERNAL BLADE PREVENTS TRASH FROM SPILLING BACK INTO THE HOPPER.
THE TAILGATE BLADE SWEEPS TRASH OUT OF THE HOPPER, PACKING IT TIGHTLY IN THE HOLD.
TRASH IS COMPACTED BY THE TWIN FORCE OF THE TAILGATE BLADE AND THE COMPRESSION PLATE.

THE PACKER ALSO INTRODUCED THE ENCLOSED HOLD. THIS GREATLY REDUCED SMELLS AND SPILLS. AT THE DUMP, THE COMPRESSION PLATE ALSO EJECTED THE GARBAGE LOAD OUT A BACK HATCH. BOTH OF THESE FEATURES ARE STILL THE NORM TODAY.

1938 GARWOOD LOAD PACKER

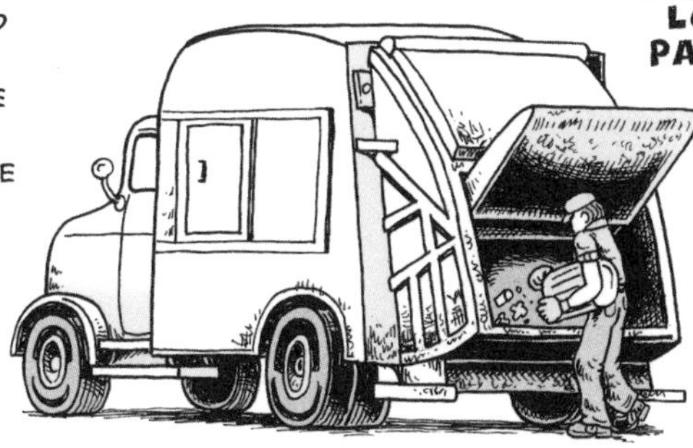

THE REAR-LOADING PACKER WAS DEVELOPED BY GARWOOD. IT REVOLUTIONIZED TRASH PICKUP BY COMBINING THE PACKER WITH A LOW HOPPER THAT MADE IT EASIER FOR GARBAGEMEN TO EMPTY CANS. WHEN THE HOPPER WAS FULL, A POWERFUL TAILGATE BLADE SWEPT THE TRASH UP INTO THE INTERNAL HOLD, WHERE IT WAS PACKED. IT WAS SUCH AN EFFECTIVE AND EFFICIENT DESIGN, IT IS STILL THE STANDARD MORE THAN 75 YEARS LATER.

THE GARWOOD AND ITS MANY IMITATORS BECAME AN ALMOST UNIVERSAL CHOICE FOR MUNICIPAL GARBAGE COLLECTION BY 1960. MORE THAN A MILLION OF THESE TYPES OF TRUCKS WERE ON THE STREETS IN AMERICAN CITIES AND TOWNS.

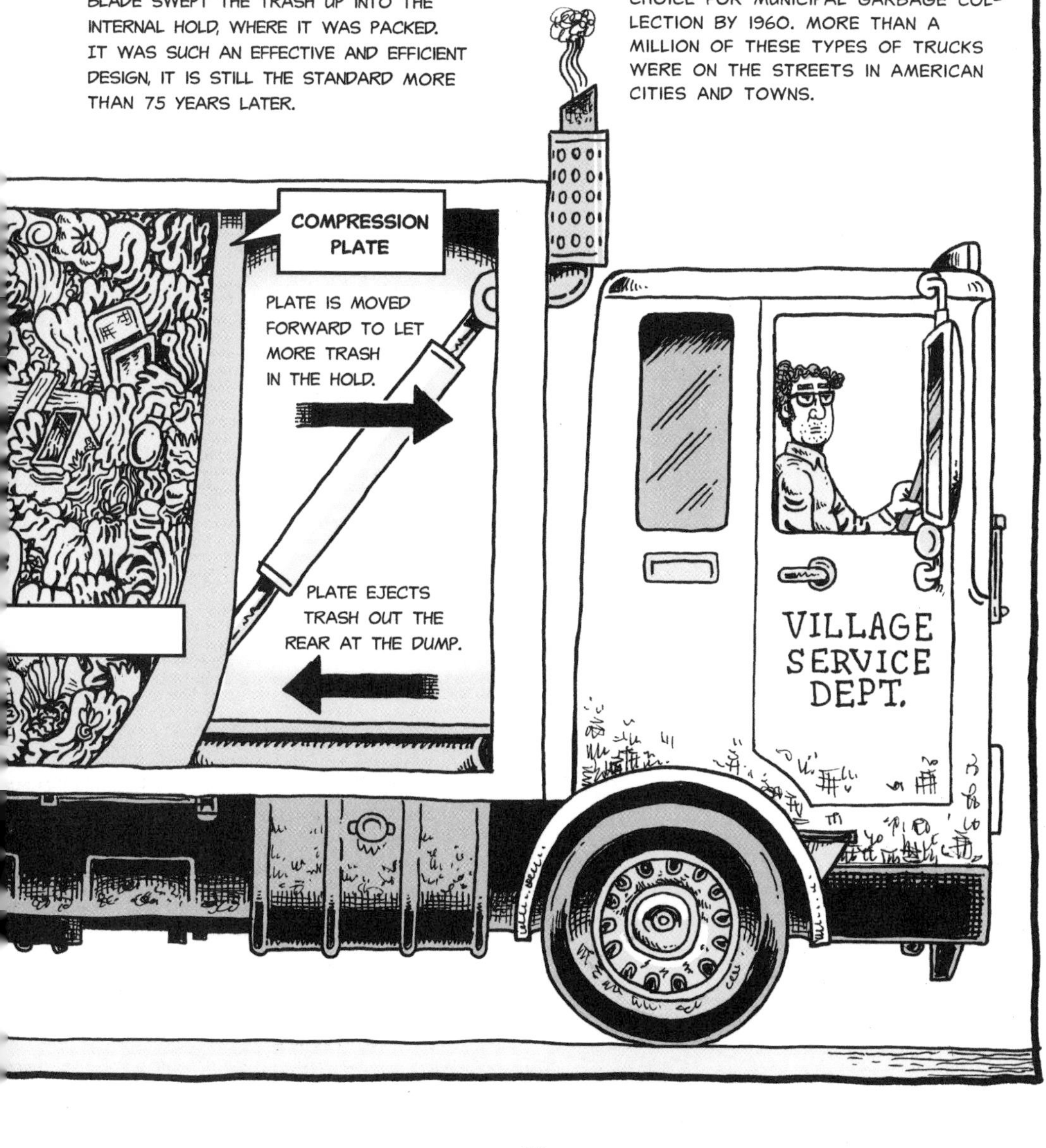
COMPRESSION PLATE
PLATE IS MOVED FORWARD TO LET MORE TRASH IN THE HOLD.
PLATE EJECTS TRASH OUT THE REAR AT THE DUMP.
VILLAGE SERVICE DEPT.

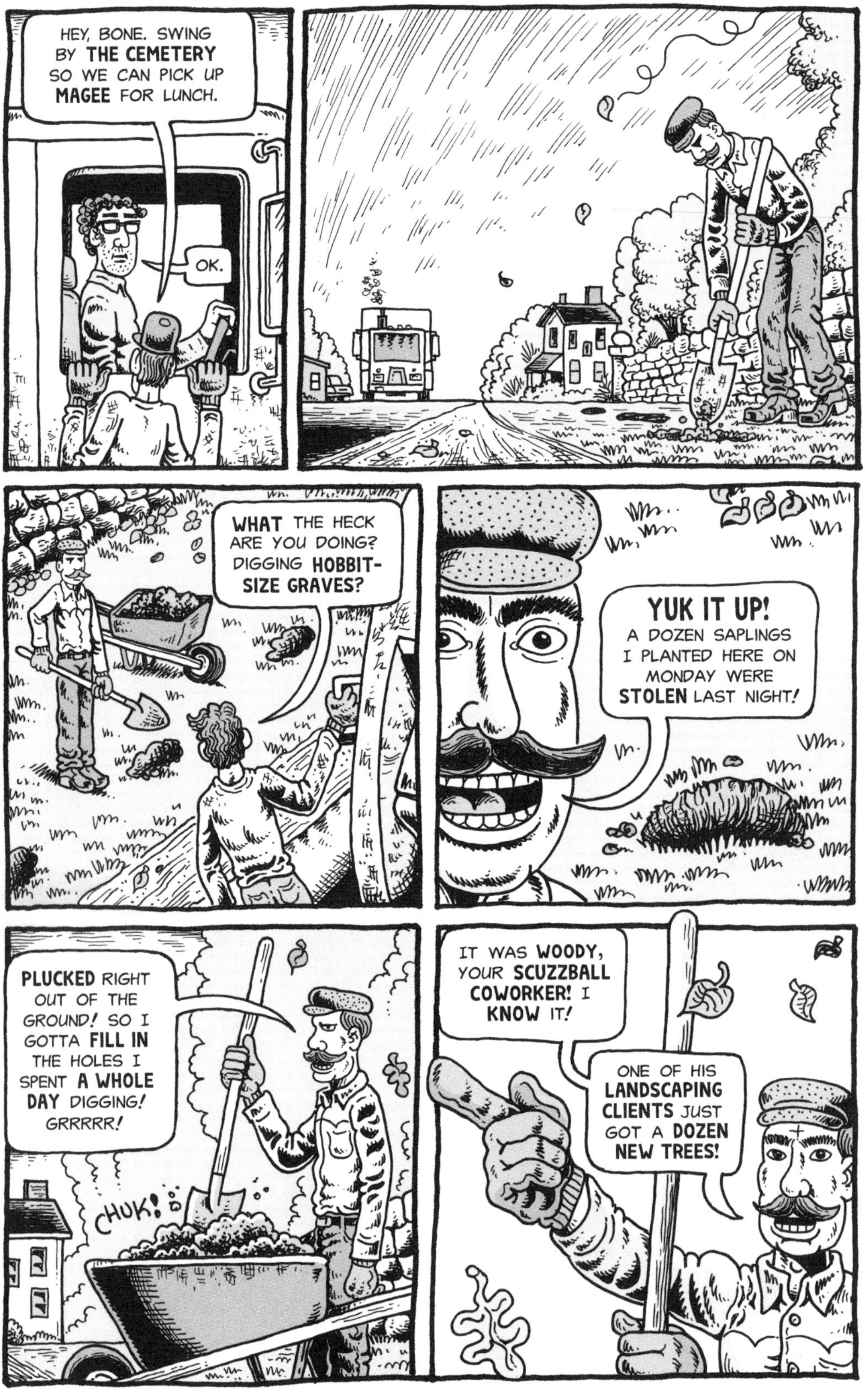
HEY, BONE. SWING BY THE CEMETERY SO WE CAN PICK UP MAGEE FOR LUNCH.
OK.
WHAT THE HECK ARE YOU DOING? DIGGING HOBBIT-SIZE GRAVES?
YUK IT UP! A DOZEN SAPLINGS I PLANTED HERE ON MONDAY WERE STOLEN LAST NIGHT!
PLUCKED RIGHT OUT OF THE GROUND! SO I GOTTA FILL IN THE HOLES I SPENT A WHOLE DAY DIGGING! GRRRRR!
CHUK!
IT WAS WOODY, YOUR SCUZZBALL COWORKER! I KNOW IT!
ONE OF HIS LANDSCAPING CLIENTS JUST GOT A DOZEN NEW TREES!

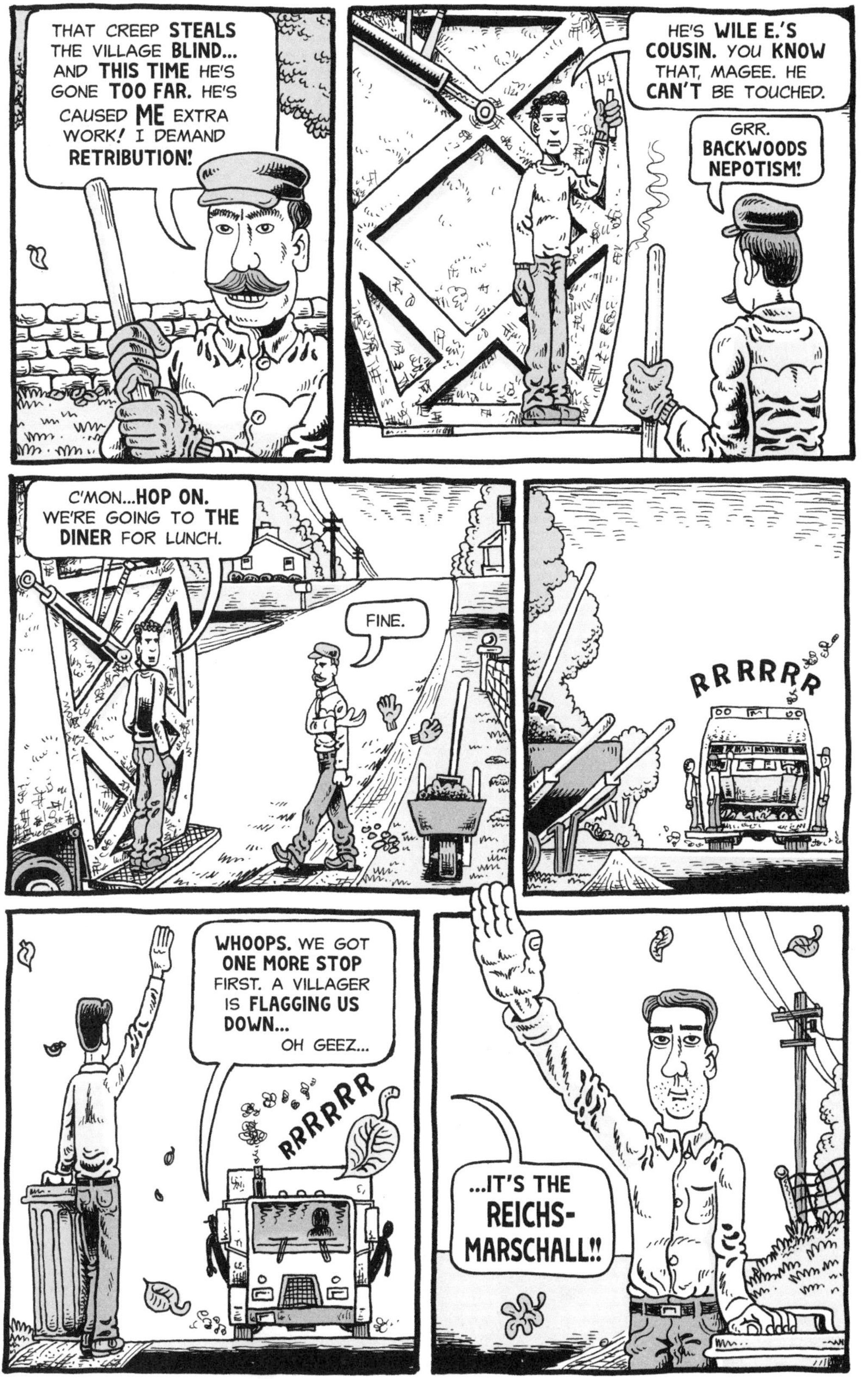
THAT CREEP STEALS THE VILLAGE BLIND... AND THIS TIME HE'S GONE TOO FAR. HE'S CAUSED ME EXTRA WORK! I DEMAND RETRIBUTION!
HE'S WILE E.'S COUSIN. YOU KNOW THAT, MAGEE. HE CAN'T BE TOUCHED.
GRR. BACKWOODS NEPOTISM!
C'MON...HOP ON. WE'RE GOING TO THE DINER FOR LUNCH.
FINE.
RRRRRR
WHOOPS. WE GOT ONE MORE STOP FIRST. A VILLAGER IS FLAGGING US DOWN...
OH GEEZ...
RRRRRR
...IT'S THE REICHS-MARSCHALL!!

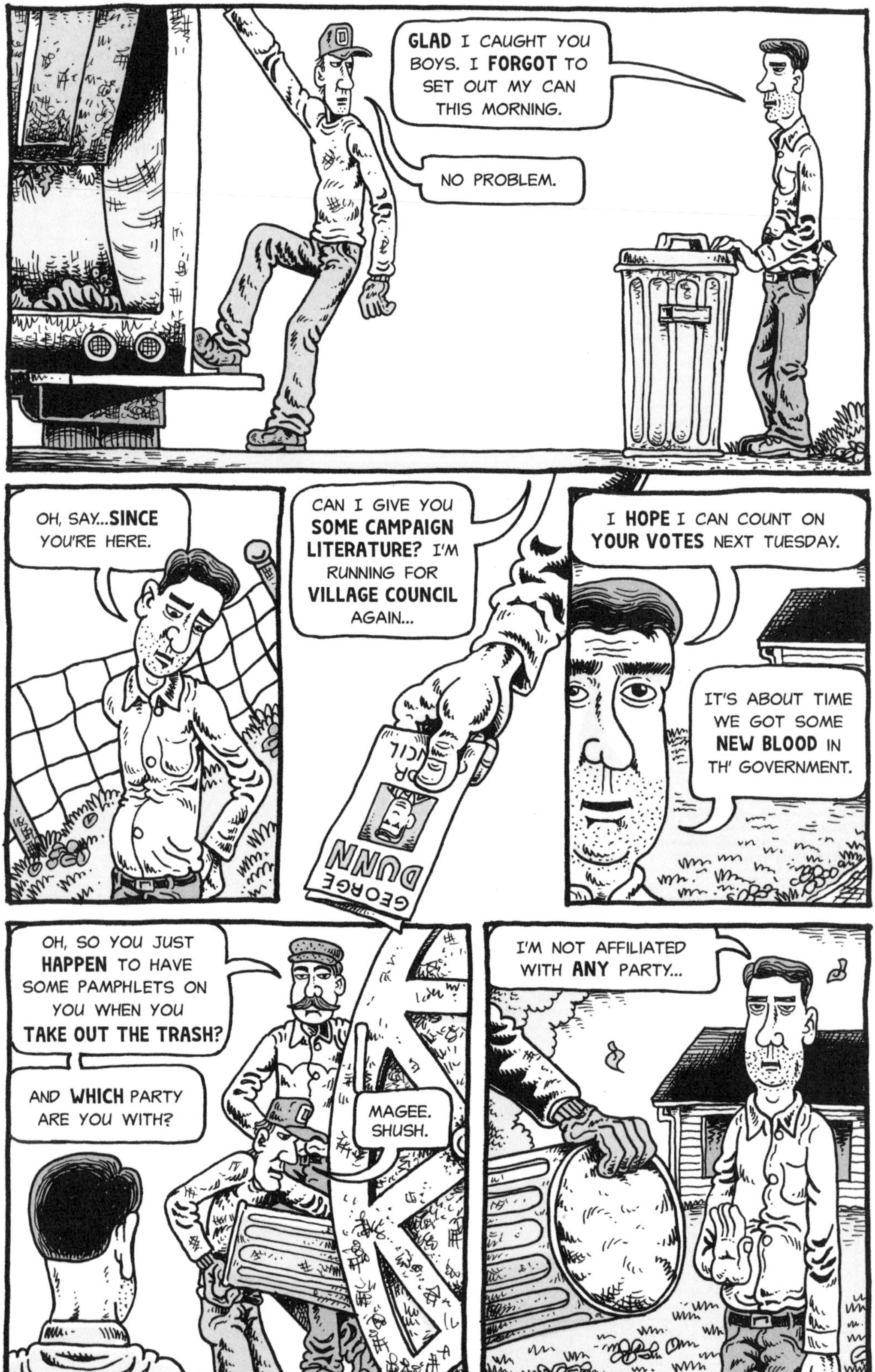
GLAD I CAUGHT YOU BOYS. I FORGOT TO SET OUT MY CAN THIS MORNING.
NO PROBLEM.
OH, SAY...SINCE YOU'RE HERE.
CAN I GIVE YOU SOME CAMPAIGN LITERATURE? I'M RUNNING FOR VILLAGE COUNCIL AGAIN...
GEORGE DUNN
I HOPE I CAN COUNT ON YOUR VOTES NEXT TUESDAY.
IT'S ABOUT TIME WE GOT SOME NEW BLOOD IN TH' GOVERNMENT.
OH, SO YOU JUST HAPPEN TO HAVE SOME PAMPHLETS ON YOU WHEN YOU TAKE OUT THE TRASH?
AND WHICH PARTY ARE YOU WITH?
MAGEE. SHUSH.
I'M NOT AFFILIATED WITH ANY PARTY...

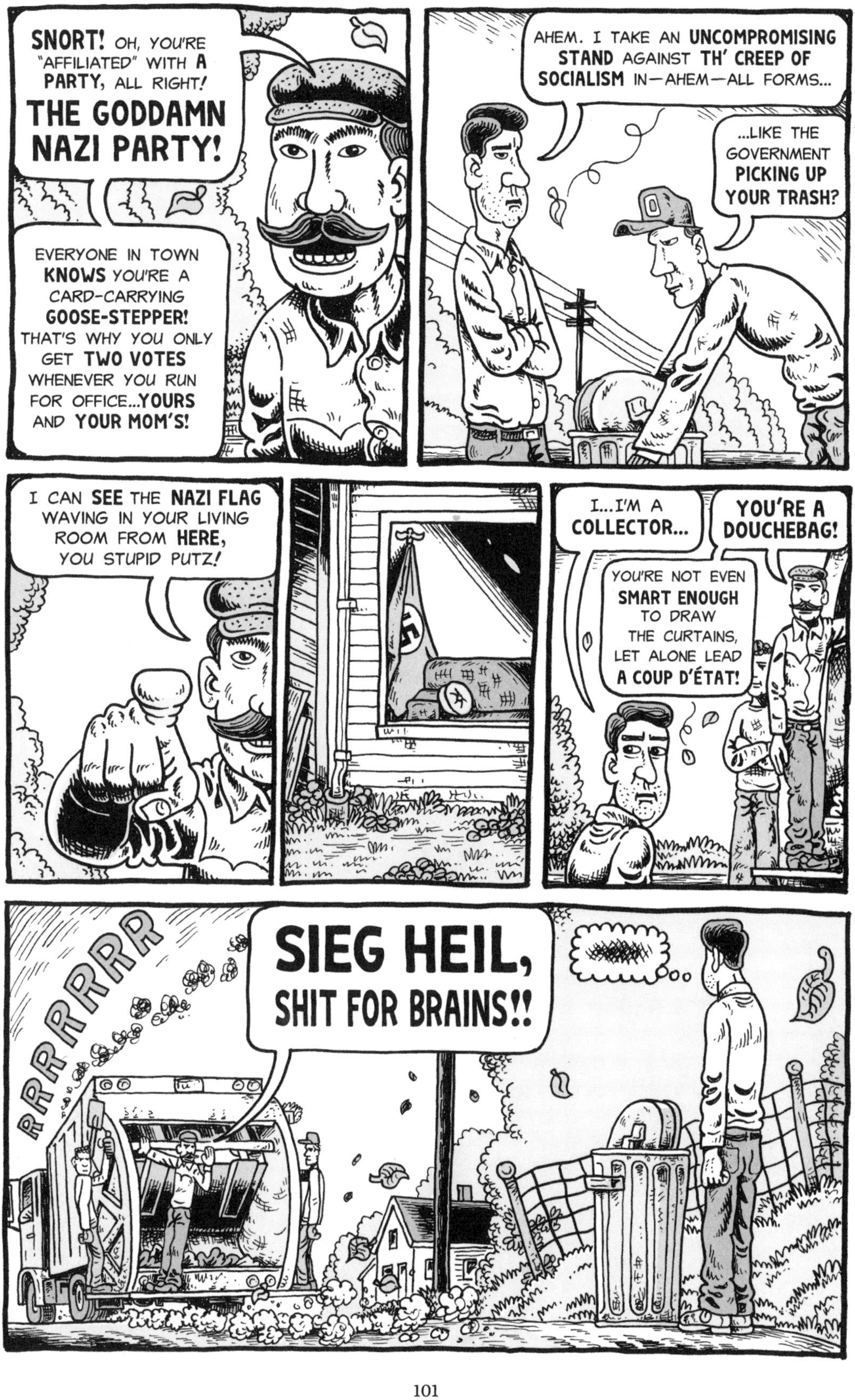
SNORT! OH, YOU'RE "AFFILIATED" WITH A PARTY, ALL RIGHT! THE GODDAMN NAZI PARTY!
EVERYONE IN TOWN KNOWS YOU'RE A CARD-CARRYING GOOSE-STEPPER! THAT'S WHY YOU ONLY GET TWO VOTES WHENEVER YOU RUN FOR OFFICE...YOURS AND YOUR MOM'S!
AHEM. I TAKE AN UNCOMPROMISING STAND AGAINST TH' CREEP OF SOCIALISM IN—AHEM—ALL FORMS...
...LIKE THE GOVERNMENT PICKING UP YOUR TRASH?
I CAN SEE THE NAZI FLAG WAVING IN YOUR LIVING ROOM FROM HERE, YOU STUPID PUTZ!
I...I'M A COLLECTOR...
YOU'RE A DOUCHEBAG!
YOU'RE NOT EVEN SMART ENOUGH TO DRAW THE CURTAINS, LET ALONE LEAD A COUP D'ÉTAT!
RRRRRRRR
SIEG HEIL, SHIT FOR BRAINS!!

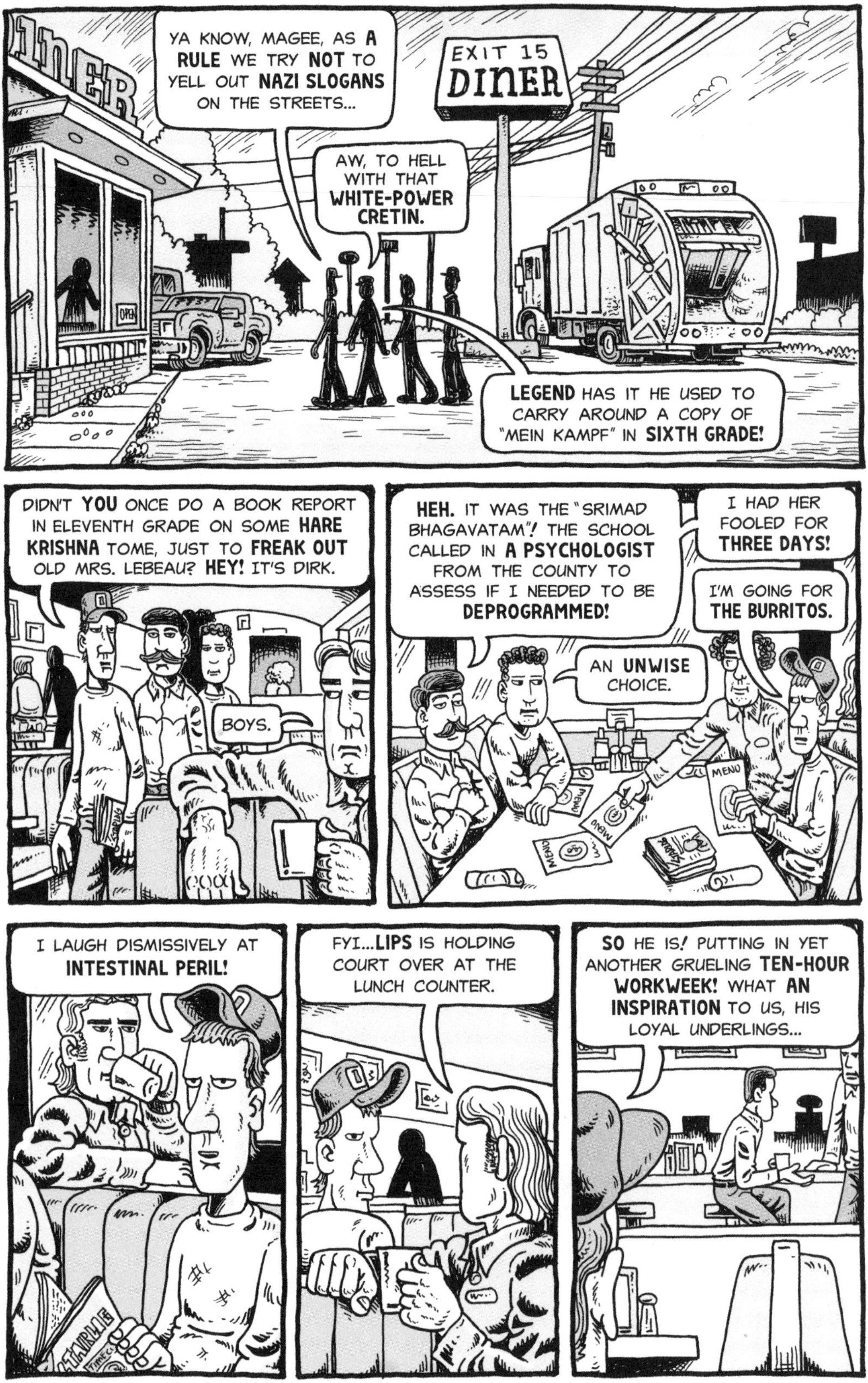
DINER
YA KNOW, MAGEE, AS A RULE WE TRY NOT TO YELL OUT NAZI SLOGANS ON THE STREETS...
EXIT 15 DINER
AW, TO HELL WITH THAT WHITE-POWER CRETIN.
OPEN
LEGEND HAS IT HE USED TO CARRY AROUND A COPY OF "MEIN KAMPF" IN SIXTH GRADE!
DIDN'T YOU ONCE DO A BOOK REPORT IN ELEVENTH GRADE ON SOME HARE KRISHNA TOME, JUST TO FREAK OUT OLD MRS. LEBEAU? HEY! IT'S DIRK.
BOYS.
HEH. IT WAS THE "SRIMAD BHAGAVATAM"! THE SCHOOL CALLED IN A PSYCHOLOGIST FROM THE COUNTY TO ASSESS IF I NEEDED TO BE DEPROGRAMMED!
I HAD HER FOOLED FOR THREE DAYS!
I'M GOING FOR THE BURRITOS.
AN UNWISE CHOICE.
MENU
I LAUGH DISMISSIVELY AT INTESTINAL PERIL!
FYI...LIPS IS HOLDING COURT OVER AT THE LUNCH COUNTER.
SO HE IS! PUTTING IN YET ANOTHER GRUELING TEN-HOUR WORKWEEK! WHAT AN INSPIRATION TO US, HIS LOYAL UNDERLINGS...

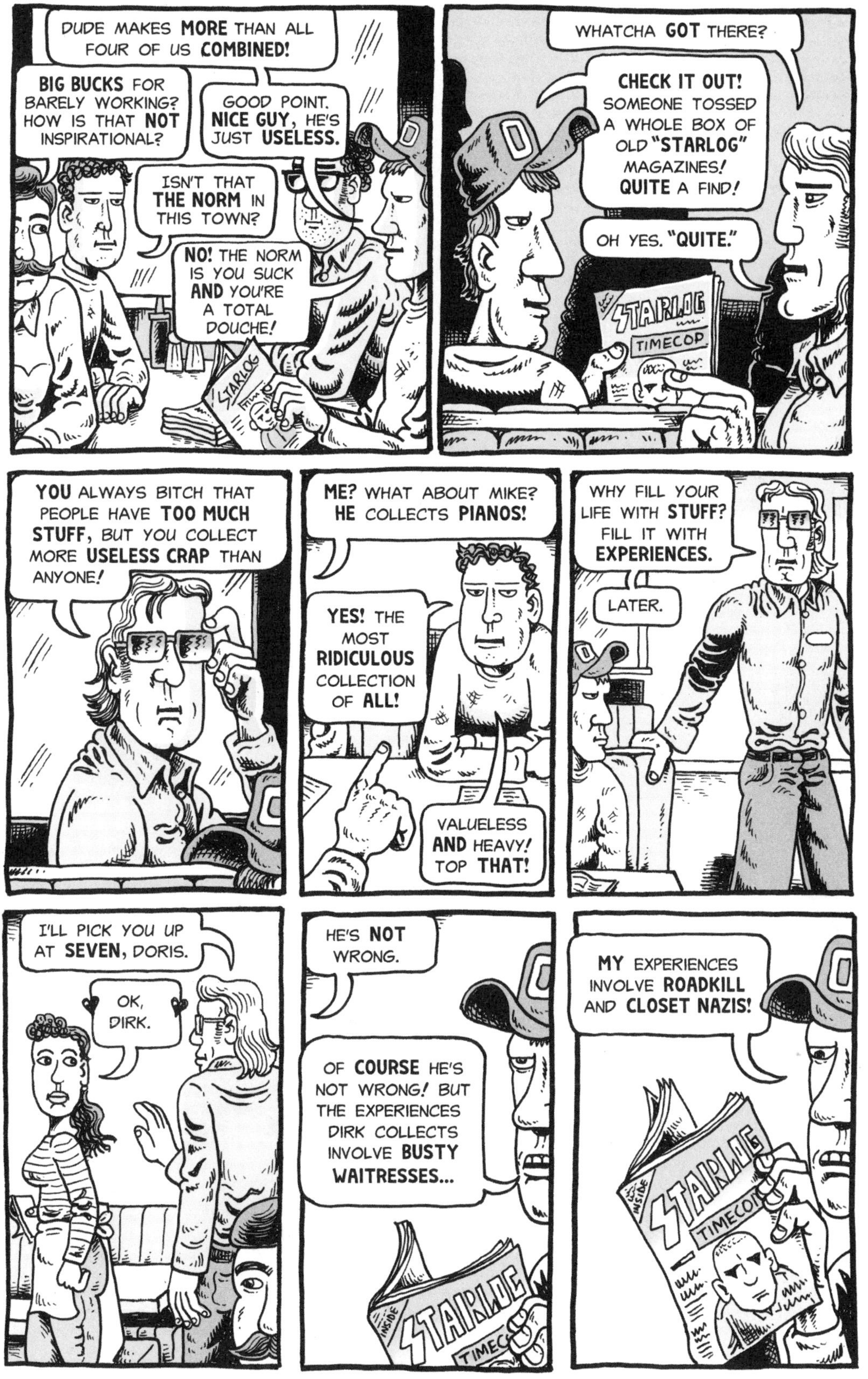
DUDE MAKES MORE THAN ALL FOUR OF US COMBINED!
BIG BUCKS FOR BARELY WORKING? HOW IS THAT NOT INSPIRATIONAL?
GOOD POINT. NICE GUY, HE'S JUST USELESS.
ISN'T THAT THE NORM IN THIS TOWN?
NO! THE NORM IS YOU SUCK AND YOU'RE A TOTAL DOUCHE!
STARLOG
WHATCHA GOT THERE?
CHECK IT OUT! SOMEONE TOSSED A WHOLE BOX OF OLD "STARLOG" MAGAZINES! QUITE A FIND!
OH YES. "QUITE."
STARLOG
TIMECOP
YOU ALWAYS BITCH THAT PEOPLE HAVE TOO MUCH STUFF, BUT YOU COLLECT MORE USELESS CRAP THAN ANYONE!
ME? WHAT ABOUT MIKE? HE COLLECTS PIANOS!
YES! THE MOST RIDICULOUS COLLECTION OF ALL!
VALUELESS AND HEAVY! TOP THAT!
WHY FILL YOUR LIFE WITH STUFF? FILL IT WITH EXPERIENCES.
LATER.
I'LL PICK YOU UP AT SEVEN, DORIS.
OK, DIRK.
HE'S NOT WRONG.
OF COURSE HE'S NOT WRONG! BUT THE EXPERIENCES DIRK COLLECTS INVOLVE BUSTY WAITRESSES...
INSIDE
STARLOG
MY EXPERIENCES INVOLVE ROADKILL AND CLOSET NAZIS!
INSIDE
STARLOG
TIMECO

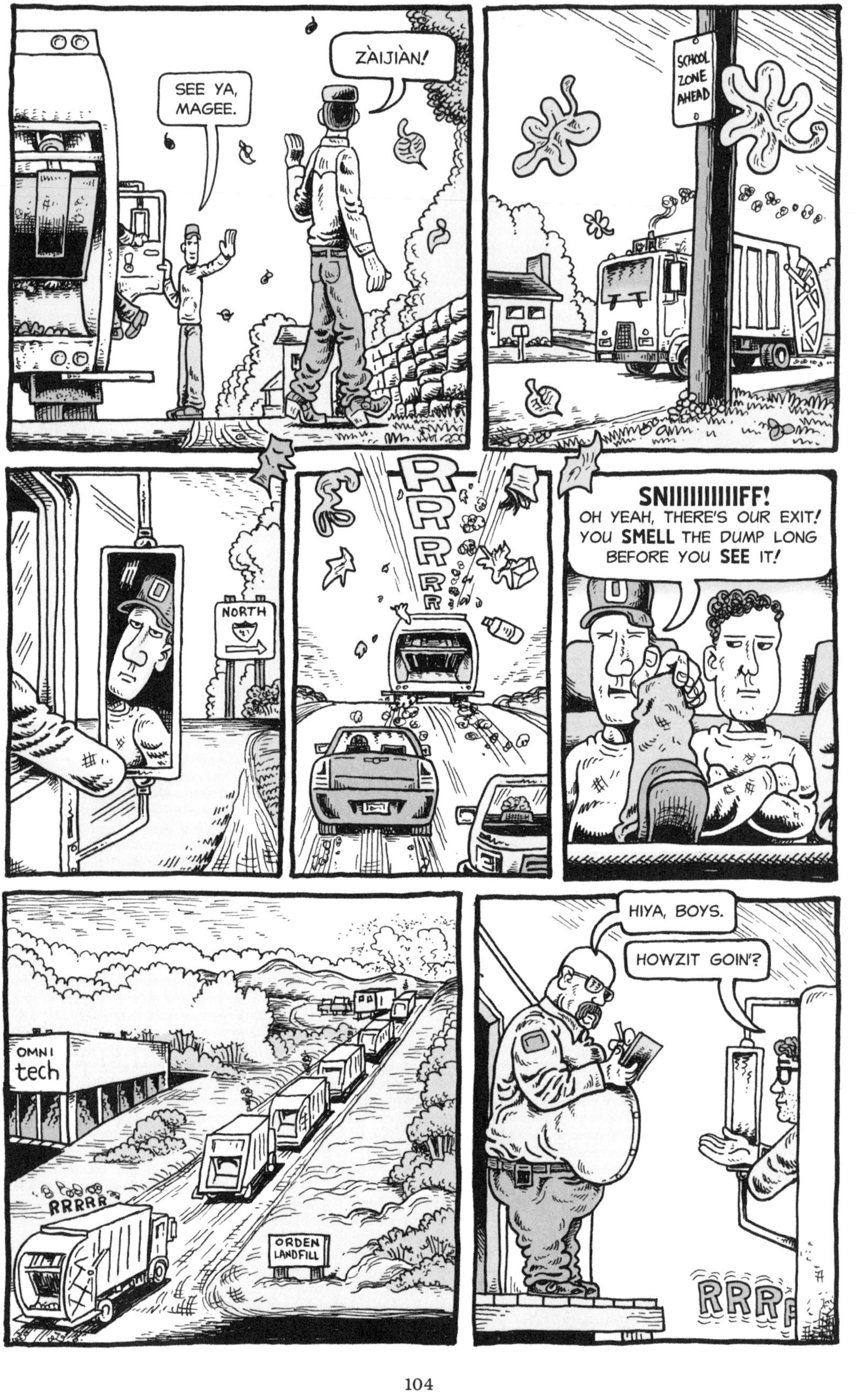
SEE YA, MAGEE.
ZÀIJIÀN!
SCHOOL ZONE AHEAD
NORTH
RRRRR
SNIIIIIIIIIIFF!
OH YEAH, THERE'S OUR EXIT! YOU SMELL THE DUMP LONG BEFORE YOU SEE IT!
OMNI tech
RRRRR
ORDEN LANDFILL
HIYA, BOYS.
HOWZIT GOIN'?
RRRR

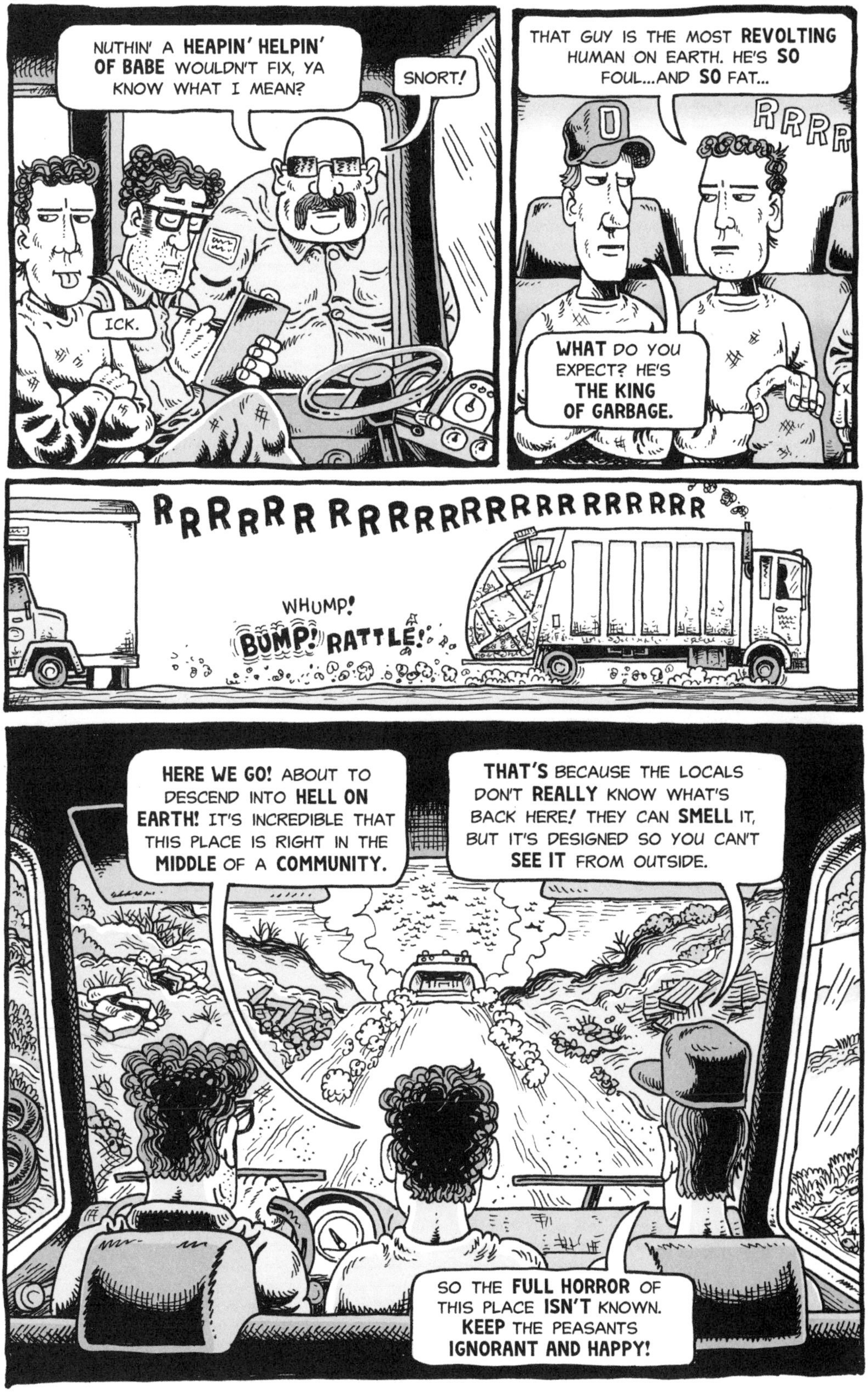
NUTHIN' A HEAPIN' HELPIN' OF BABE WOULDN'T FIX, YA KNOW WHAT I MEAN?
SNORT!
ICK.
THAT GUY IS THE MOST REVOLTING HUMAN ON EARTH. HE'S SO FOUL...AND SO FAT...
RRRRR
WHAT DO YOU EXPECT? HE'S THE KING OF GARBAGE.
RRRRRR RRRRRRRRRRRRRRRRR
WHUMP!
BUMP! RATTLE!
HERE WE GO! ABOUT TO DESCEND INTO HELL ON EARTH! IT'S INCREDIBLE THAT THIS PLACE IS RIGHT IN THE MIDDLE OF A COMMUNITY.
THAT'S BECAUSE THE LOCALS DON'T REALLY KNOW WHAT'S BACK HERE! THEY CAN SMELL IT, BUT IT'S DESIGNED SO YOU CAN'T SEE IT FROM OUTSIDE.
SO THE FULL HORROR OF THIS PLACE ISN'T KNOWN. KEEP THE PEASANTS IGNORANT AND HAPPY!

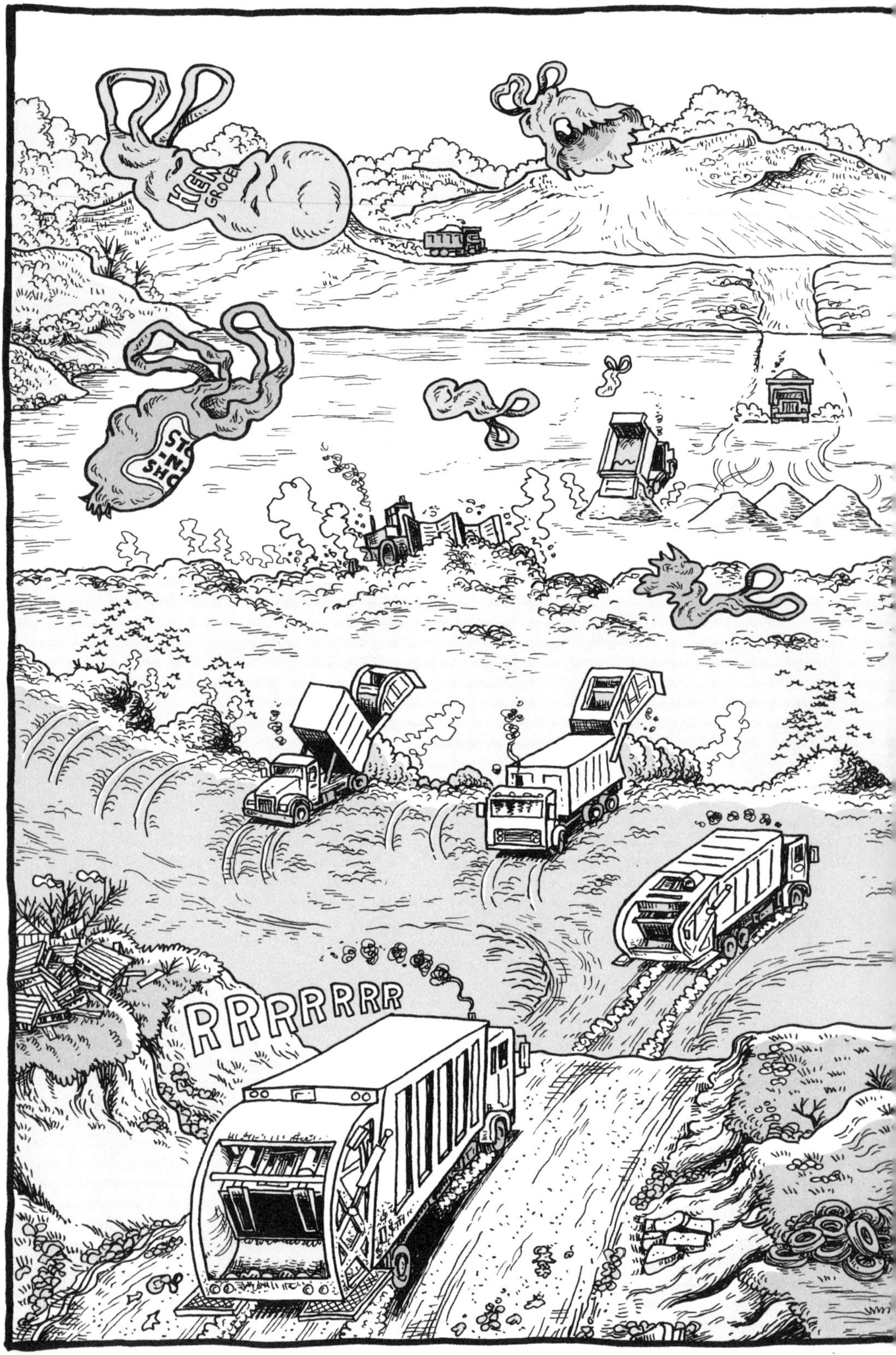
RRRRRRR

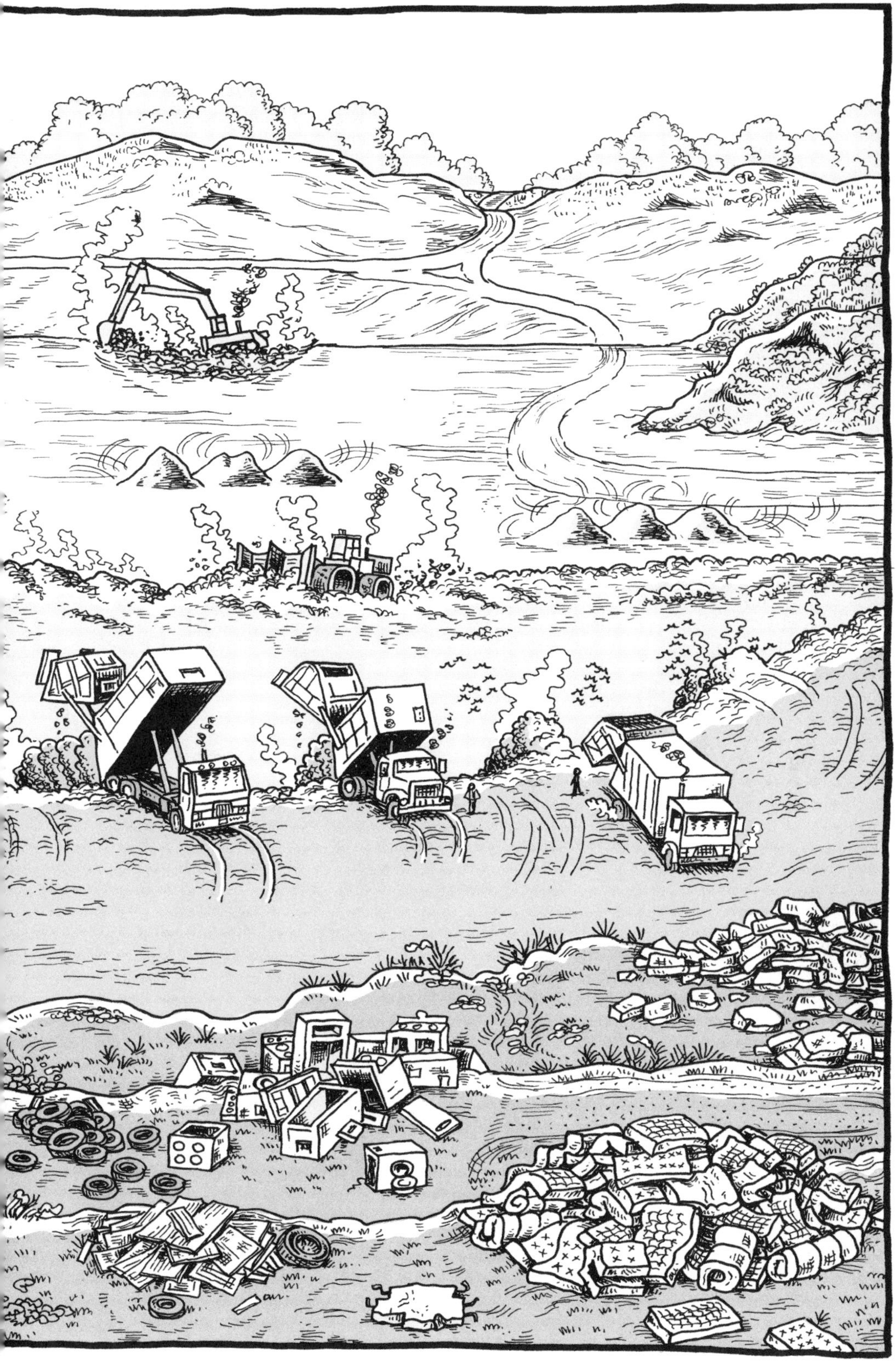

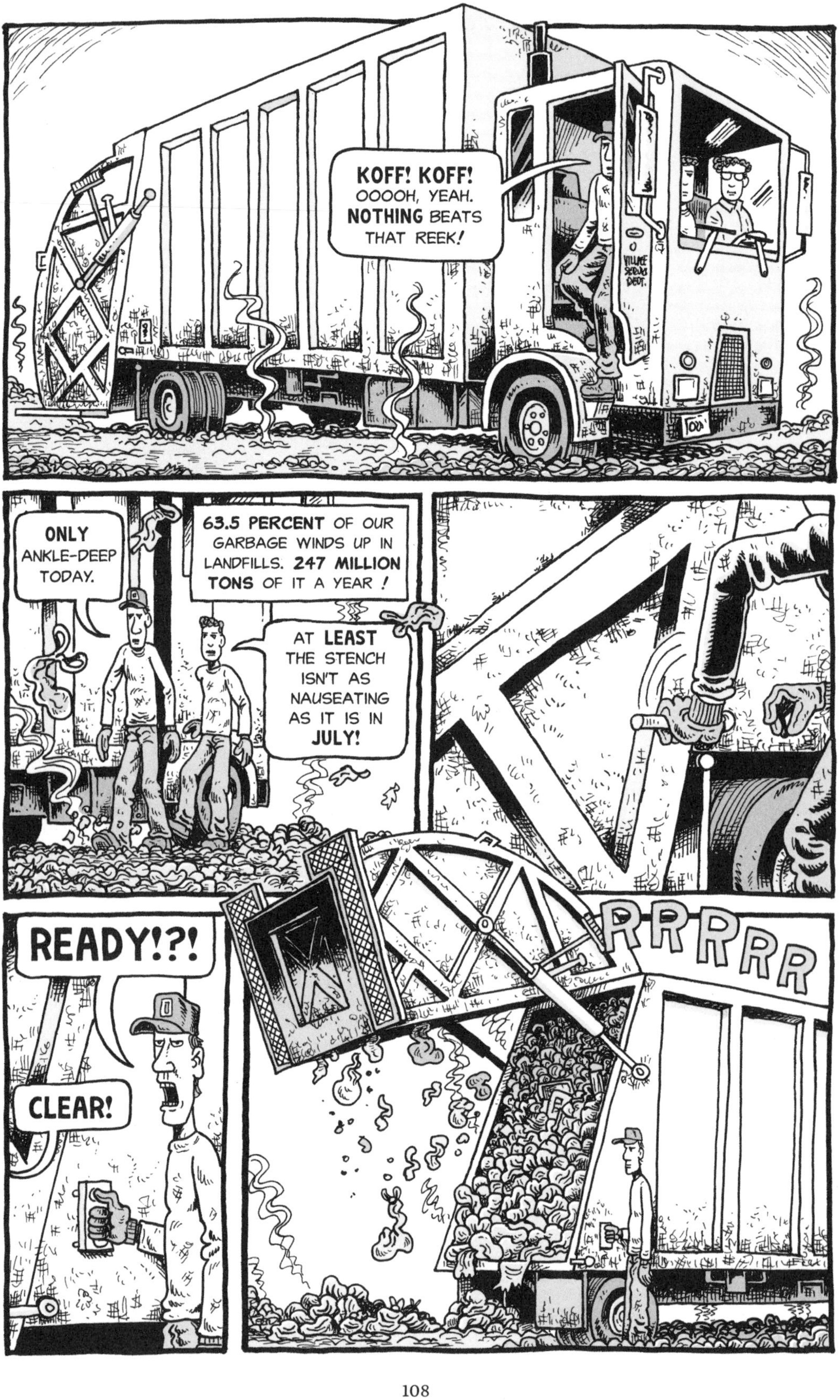
KOFF! KOFF! OOOOH, YEAH. NOTHING BEATS THAT REEK!
VILLAGE SERVICES DEPT.
ONLY ANKLE-DEEP TODAY.
63.5 PERCENT OF OUR GARBAGE WINDS UP IN LANDFILLS. 247 MILLION TONS OF IT A YEAR!
AT LEAST THE STENCH ISN'T AS NAUSEATING AS IT IS IN JULY!
READY!?!
CLEAR!
RRRRR

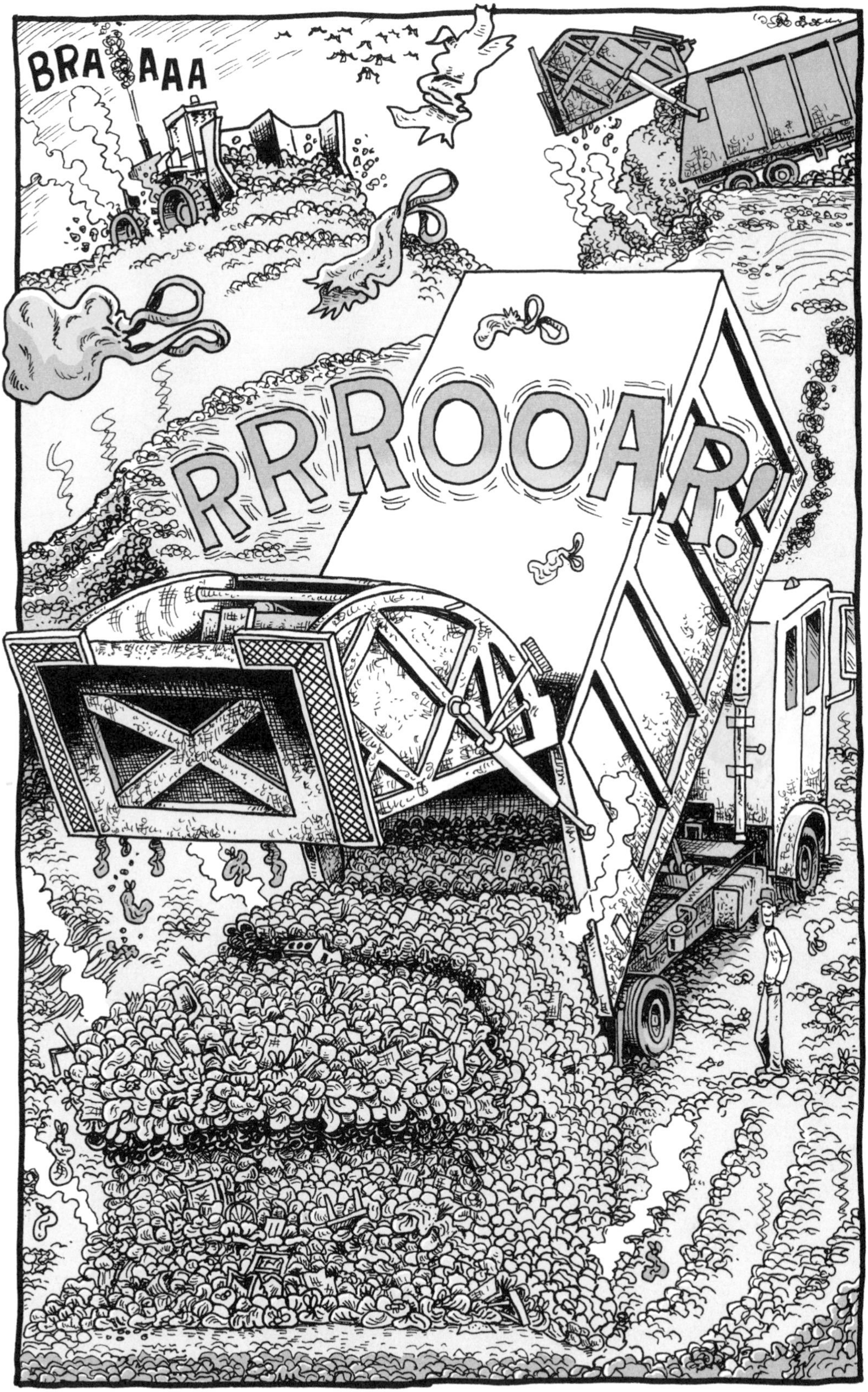
BRAAAA
RRROOAR!

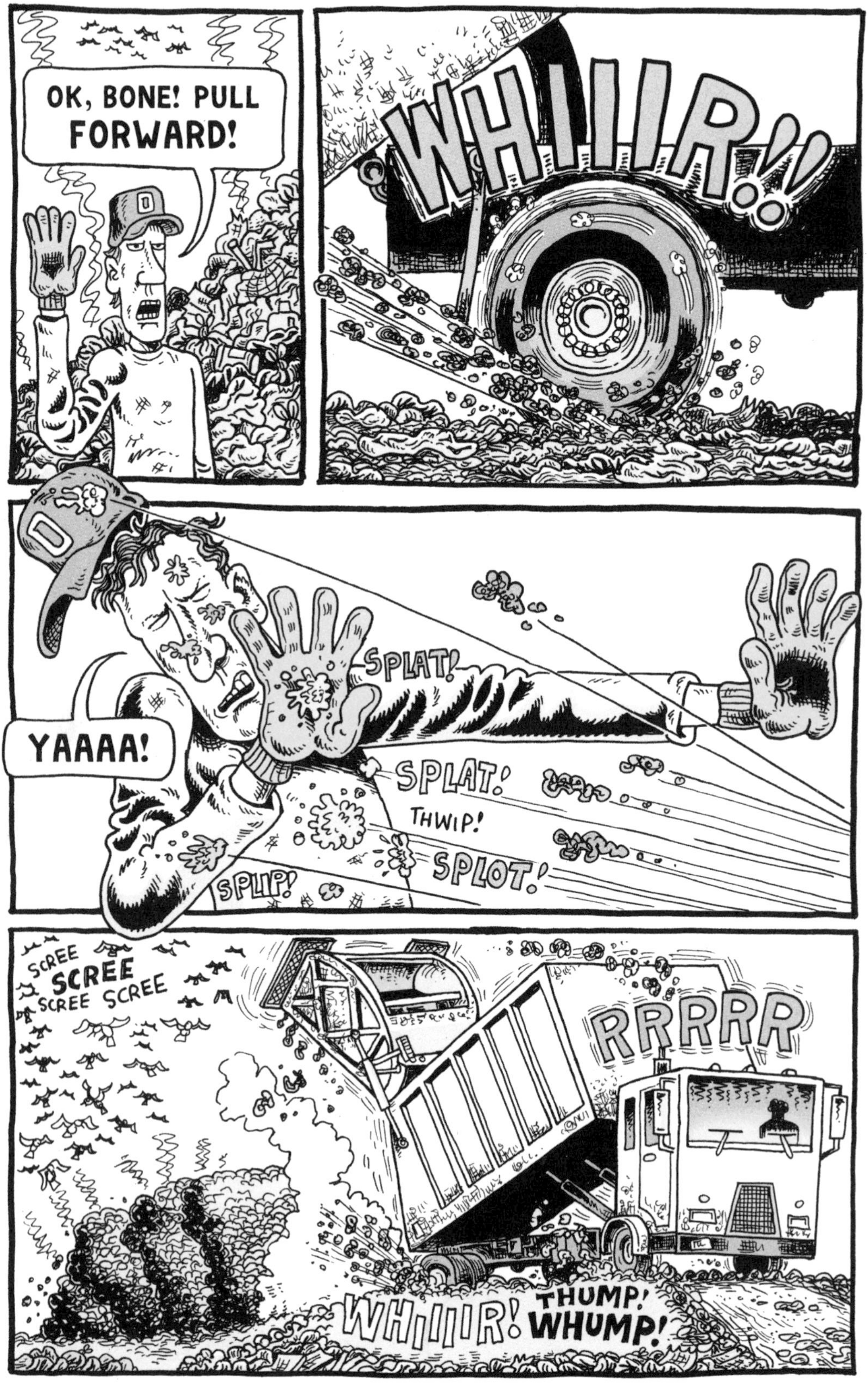
OK, BONE! PULL FORWARD!
WHIIIR!!
YAAAA!
SPLAT!
SPLAT!
THWIP!
SPLOT!
SPLIP!
SCREE
SCREE
SCREE SCREE
RRRRR
WHIIIR!
THUMP!
WHUMP!

THE NUMBER OF MUNICIPAL LANDFILLS HAS **DECREASED** CONSIDERABLY OVER THE PAST FEW DECADES. THAT'S **MISLEADING**, THOUGH, AS **THE SIZE** OF THE REMAINING LANDFILLS HAS **GREATLY INCREASED**.

BLECH!

HA! YOU GOT IT GOOD, EH?

HOW **BIG** ARE OUR LANDFILLS NOW? THE E.P.A. HASN'T COMPILED THAT INFORMATION, FOR OBVIOUS POLITICAL REASONS. BUT **HERE'S** A CLUE. IN 2008, THE **SALTON CITY LANDFILL** IN SOUTHERN CALIFORNIA WAS **EIGHT ACRES.** BY 2012, IT HAD EXPANDED...**TO 287 ACRES!** AND THE DEPTH OF THE FILL WAS INCREASED FROM **45 FEET** TO **250 FEET!**

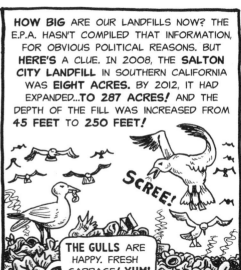

SCREE!

THE GULLS ARE HAPPY. FRESH GARBAGE! YUM!

AW, MAN! WHERE DID THOSE **CANS OF PAINT** COME FROM?

SOME **MORON** MUST HAVE **HIDDEN THEM** IN GARBAGE BAGS!

SCREEE!

BANG!

THE DICK **COULDN'T** WAIT FOR **TOXIC TRASH DAY?**

RRRRR RR

SORRY, GUYS. LUNCH BUFFET IS **OVER.**

LANDFILLS

HERE'S HOW THESE THINGS ARE BUILT

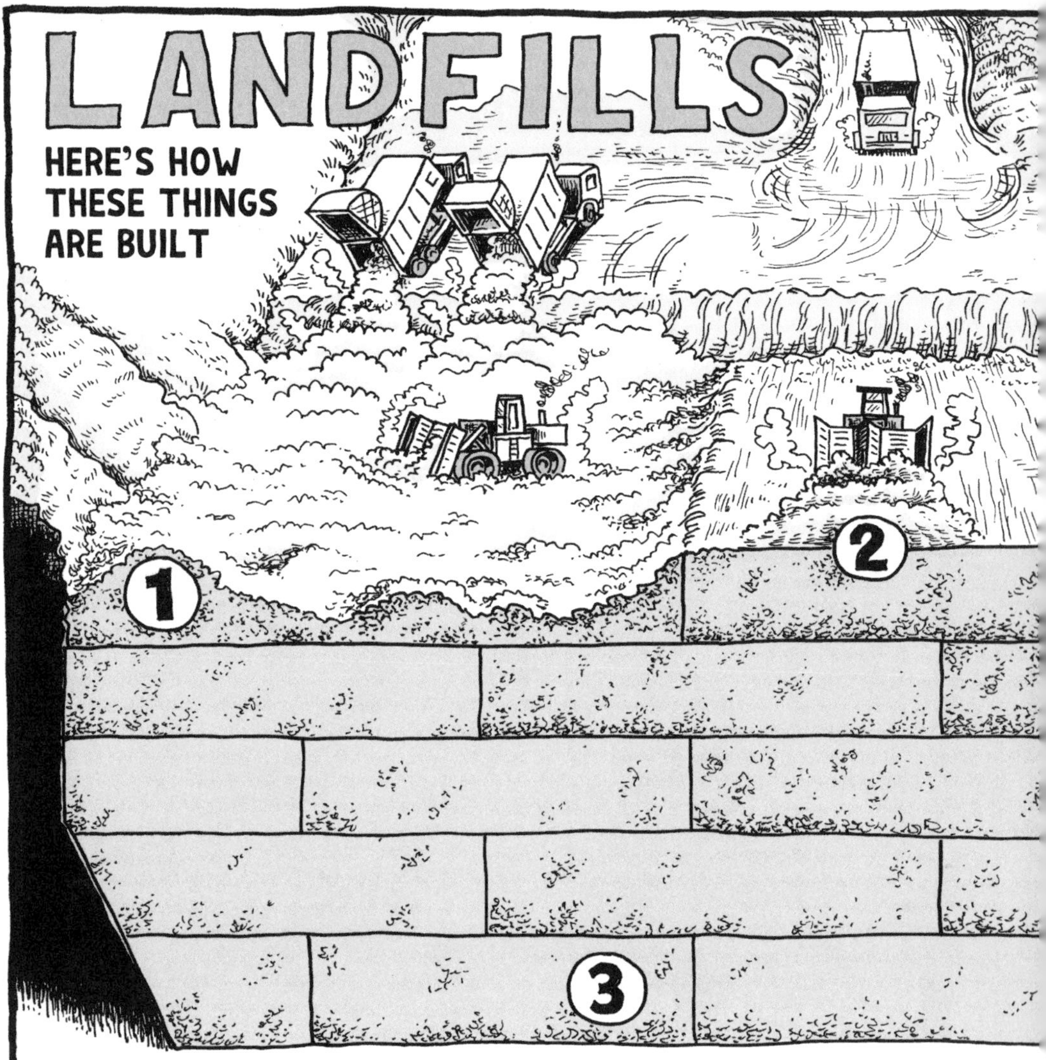
1
2
3

1 A MODERN LANDFILL IS CONSTRUCTED OF **"CELLS,"** 50 FEET SQUARE BY 15 FEET DEEP. TRUCKS DEPOSIT THEIR FRESH LOADS ALL INTO ONE CELL, WHERE LARGE BULLDOZERS COMPRESS THE TRASH. IT'S LIKE A STACK OF GARBAGE LEGOS.

2 WHEN **A CELL IS FULL** AT THE END OF THE DAY, IT'S COVERED WITH **SIX INCHES OF DIRT** TO LIMIT THE GHASTLY SMELL AND THE BLOWING GARBAGE AND TO CONFOUND VERMIN AND BIRDS. YEAH, DOESN'T REALLY WORK AS WELL AS THE LANDFILL OPERATORS CLAIM, BUT IT IS A HUGE IMPROVEMENT OVER THE OLD METHOD OF JUST DUMPING GARBAGE IN A VAST OPEN HOLE.

BUT EVEN LANDFILLS CONSTRUCTED WITH CELLS HAVE AREAS OF EXPOSED TRASH AND PILES OF DEBRIS. IT'S **NOT** A TIDY OPERATION.

3 SO WHAT OCCURS **INSIDE** A LANDFILL? THINK ALL THAT GARBAGE DECOMPOSES INTO **HARMLESS COMPOST?** WRONG! NO LIGHT, NO CIRCULATING AIR, LITTLE DECOMPOSITION. A **DISPOSABLE DIAPER** WON'T BREAK DOWN FOR HUNDREDS OF YEARS. WE DON'T HAVE AN EXACT TIME, BECAUSE DISPOSABLE DIAPERS HAVE ONLY BEEN AROUND SINCE **1947** AND HAVEN'T **EVEN BEGUN** TO BREAK DOWN INSIDE LANDFILLS!

THE BIGGEST LANDFILL HAZARDS ARE **METHANE GAS** AND **LEAKS.** BURIED TRASH PRODUCES LARGE POCKETS OF **EXPLOSIVE METHANE GAS.** IT'S ALSO ONE OF THE WORST GREENHOUSE GASES. MOST LANDFILLS **SIPHON OUT** AND **COLLECT** THE METHANE FOR USE IN POWER GENERATION.

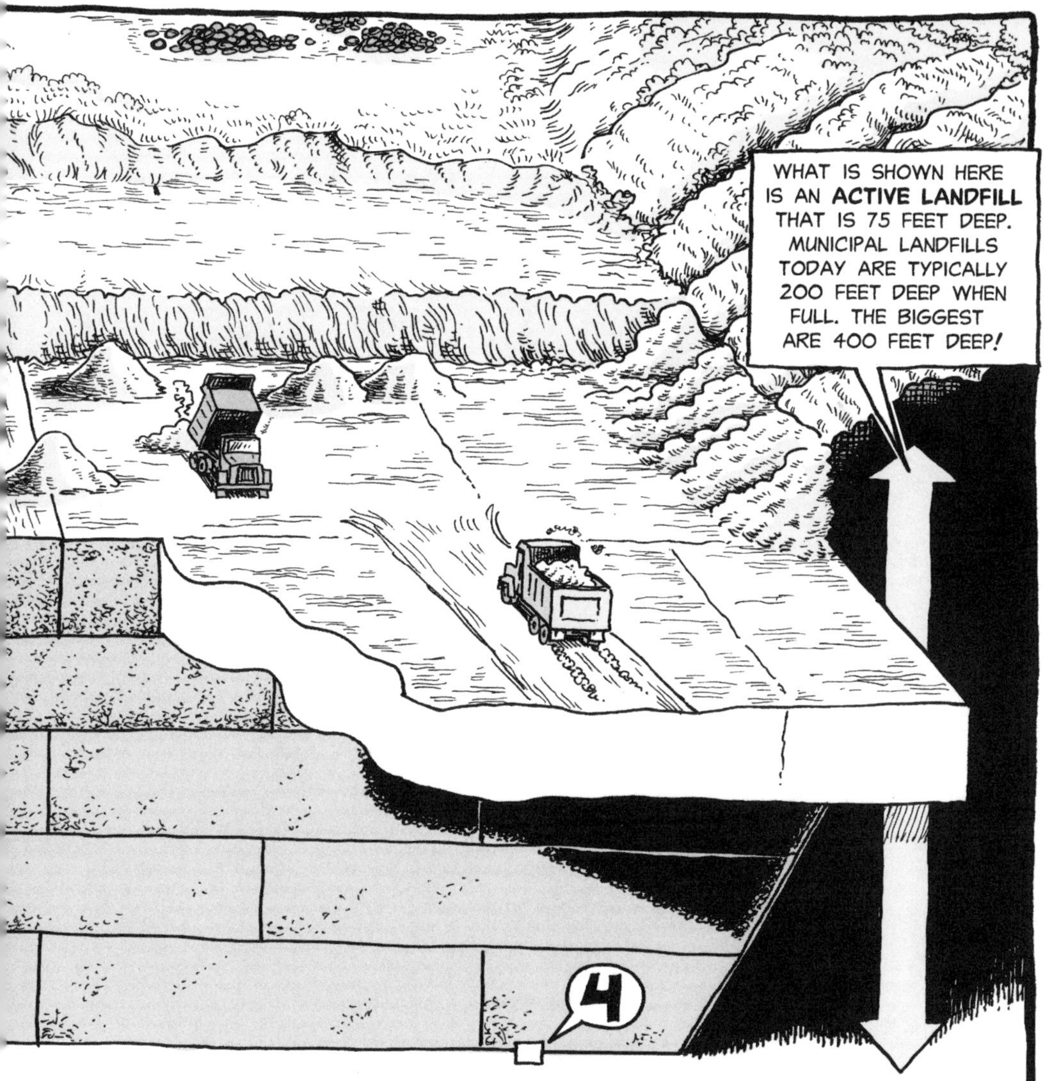
WHAT IS SHOWN HERE IS AN **ACTIVE LANDFILL** THAT IS 75 FEET DEEP. MUNICIPAL LANDFILLS TODAY ARE TYPICALLY 200 FEET DEEP WHEN FULL. THE BIGGEST ARE 400 FEET DEEP!
4

IT'S QUITE PROFITABLE. LANDFILLS PRODUCE METHANE FOR **30 TO 50 YEARS** AFTER THEY ARE FULL AND STOP ACCEPTING NEW TRASH.

4 **LEAKS** ARE THE BIGGER ISSUE. THE TOXIC LIQUID THAT IS SQUEEZED OUT OF BURIED TRASH IS CALLED **LEACHATE.** IT'S A MIX OF SOLVENTS, HOUSEHOLD CHEMICALS, OLD PAINT, AND FAR NASTIER (AND OFTEN ILLEGAL) STUFF THAT WINDS UP IN LANDFILLS. THIS ALL SEEPS DOWNWARD, HEADING STRAIGHT FOR **THE WATER TABLE.** MODERN LANDFILLS HAVE A TOUTED **"LINER,"** DESIGNED TO KEEP THIS CARCINOGENIC SOUP CONTAINED. THEY'RE LAUGHABLY INEFFECTIVE. THE E.P.A. DISCOVERED IN 1989 THAT **ALL** LANDFILL LINERS LEAK. **EVERY ONE!** A TEN-ACRE LANDFILL LEAKS UP TO 3,000 GALLONS OF TOXIC LEACHATE A YEAR, EASILY ENOUGH TO POISON A GROUNDWATER DRINKING SUPPLY.

THE LANDFILL LINER

A. A GEOTEXTILE STOPS TRASH FROM CLOGGING THE LEACHATE PIPE SYSTEM.

B. TWO FEET OF GRAVEL CONTAINS THE NETWORK OF PERFORATED PVC PIPES TO COLLECT LEACHATE, WHICH IS PUMPED OUT.

C. PLASTIC MEMBRANE 1/32" TO 1/8" THICK.

D. TWO FEET OF COMPACTED CLAY.

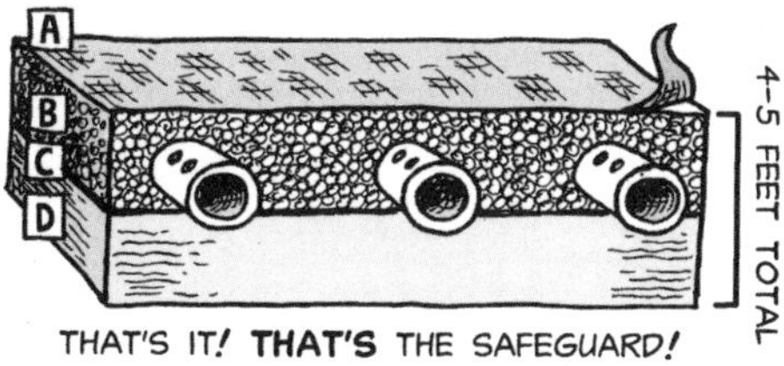
A
B
C
D
4–5 FEET TOTAL

THAT'S IT! **THAT'S** THE SAFEGUARD!

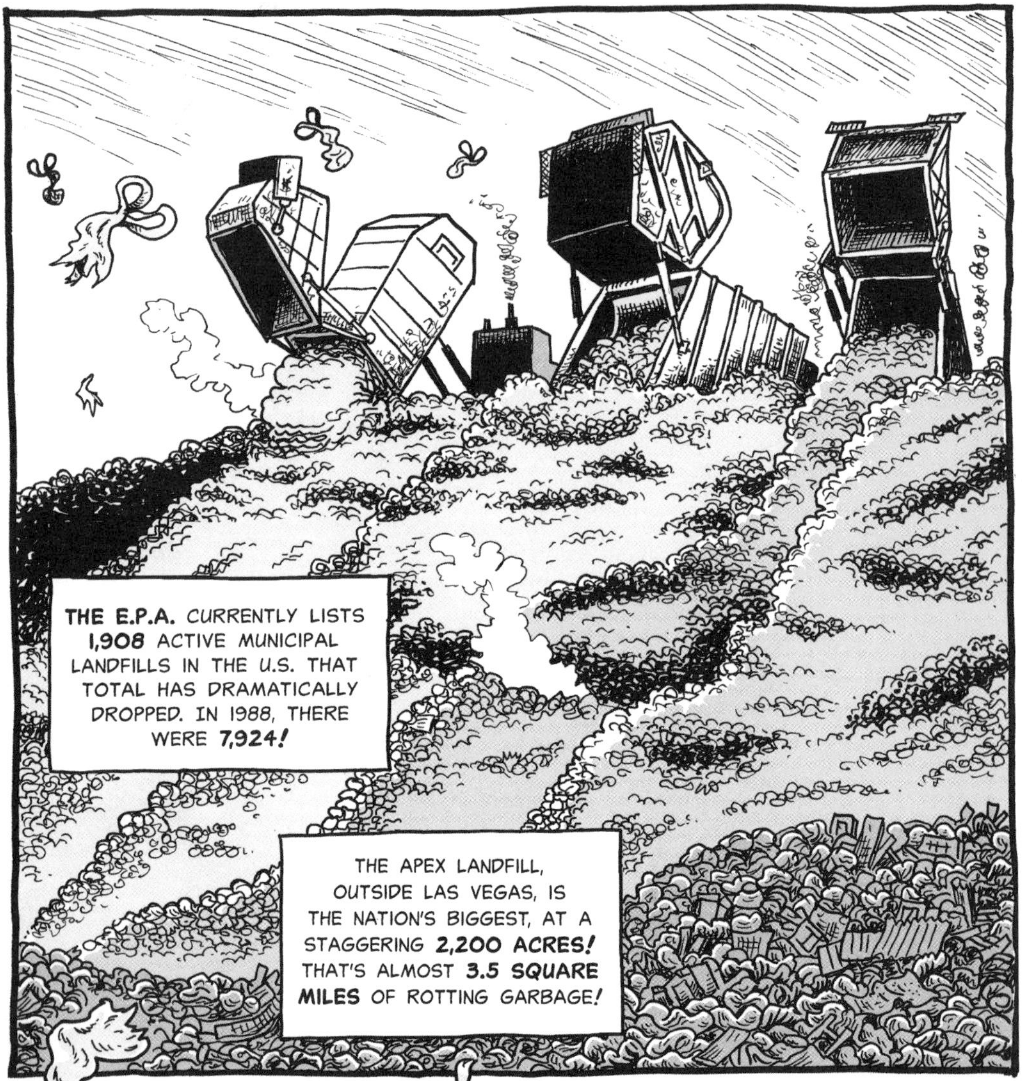
THE E.P.A. CURRENTLY LISTS 1,908 ACTIVE MUNICIPAL LANDFILLS IN THE U.S. THAT TOTAL HAS DRAMATICALLY DROPPED. IN 1988, THERE WERE 7,924!
THE APEX LANDFILL, OUTSIDE LAS VEGAS, IS THE NATION'S BIGGEST, AT A STAGGERING 2,200 ACRES! THAT'S ALMOST 3.5 SQUARE MILES OF ROTTING GARBAGE!

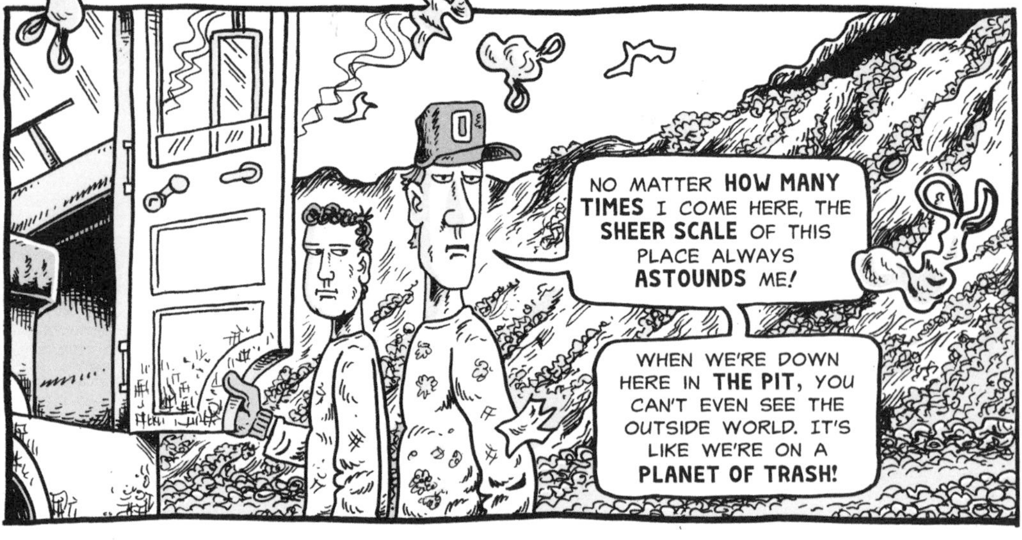
NO MATTER HOW MANY TIMES I COME HERE, THE SHEER SCALE OF THIS PLACE ALWAYS ASTOUNDS ME!
WHEN WE'RE DOWN HERE IN THE PIT, YOU CAN'T EVEN SEE THE OUTSIDE WORLD. IT'S LIKE WE'RE ON A PLANET OF TRASH!

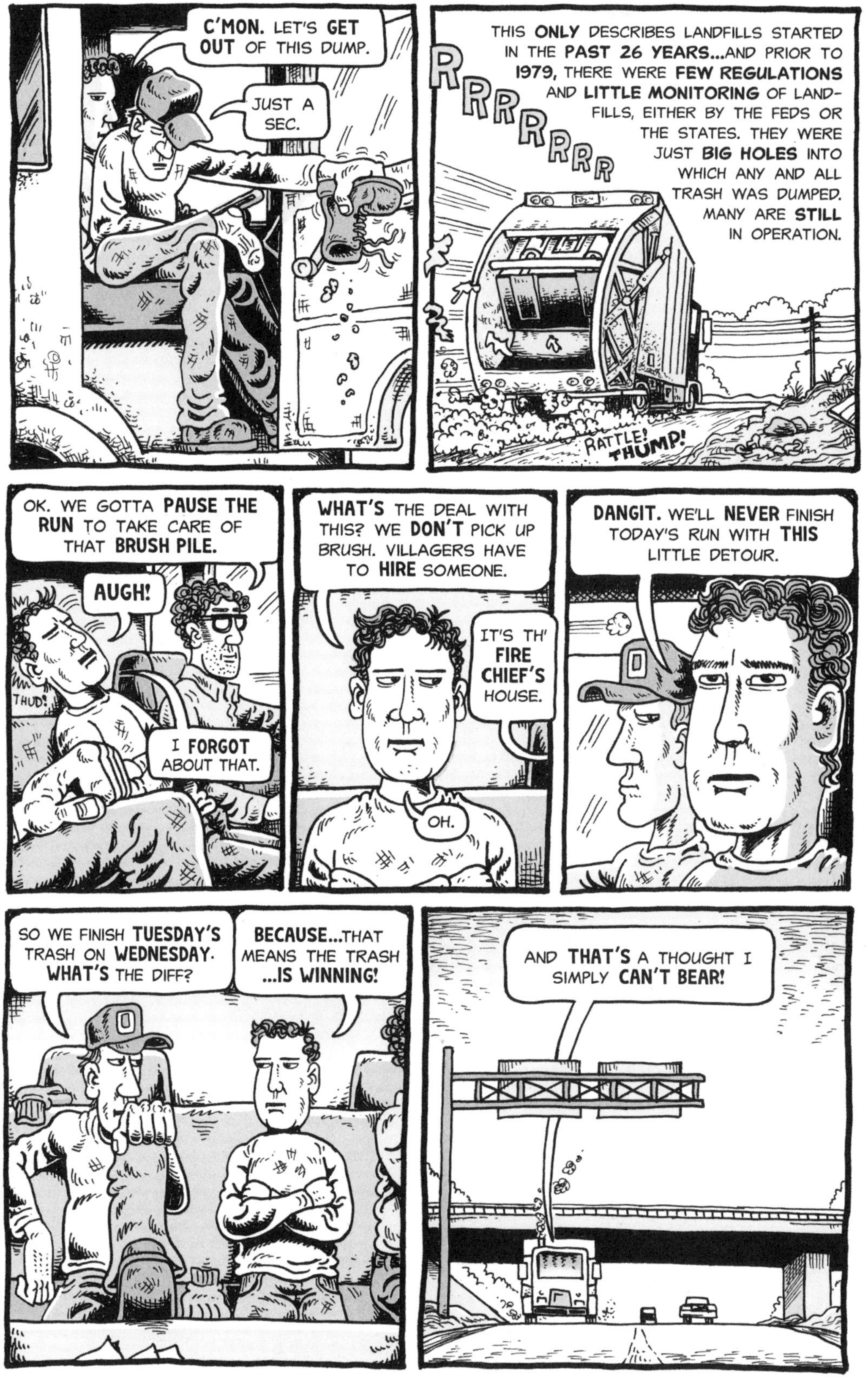
C'MON. LET'S GET OUT OF THIS DUMP.
JUST A SEC.
RRRRRRRR
THIS ONLY DESCRIBES LANDFILLS STARTED IN THE PAST 26 YEARS...AND PRIOR TO 1979, THERE WERE FEW REGULATIONS AND LITTLE MONITORING OF LANDFILLS, EITHER BY THE FEDS OR THE STATES. THEY WERE JUST BIG HOLES INTO WHICH ANY AND ALL TRASH WAS DUMPED. MANY ARE STILL IN OPERATION.
RATTLE! THUMP!
OK. WE GOTTA PAUSE THE RUN TO TAKE CARE OF THAT BRUSH PILE.
AUGH!
THUD!
I FORGOT ABOUT THAT.
WHAT'S THE DEAL WITH THIS? WE DON'T PICK UP BRUSH. VILLAGERS HAVE TO HIRE SOMEONE.
IT'S TH' FIRE CHIEF'S HOUSE.
OH.
DANGIT. WE'LL NEVER FINISH TODAY'S RUN WITH THIS LITTLE DETOUR.
SO WE FINISH TUESDAY'S TRASH ON WEDNESDAY. WHAT'S THE DIFF?
BECAUSE...THAT MEANS THE TRASH ...IS WINNING!
AND THAT'S A THOUGHT I SIMPLY CAN'T BEAR!

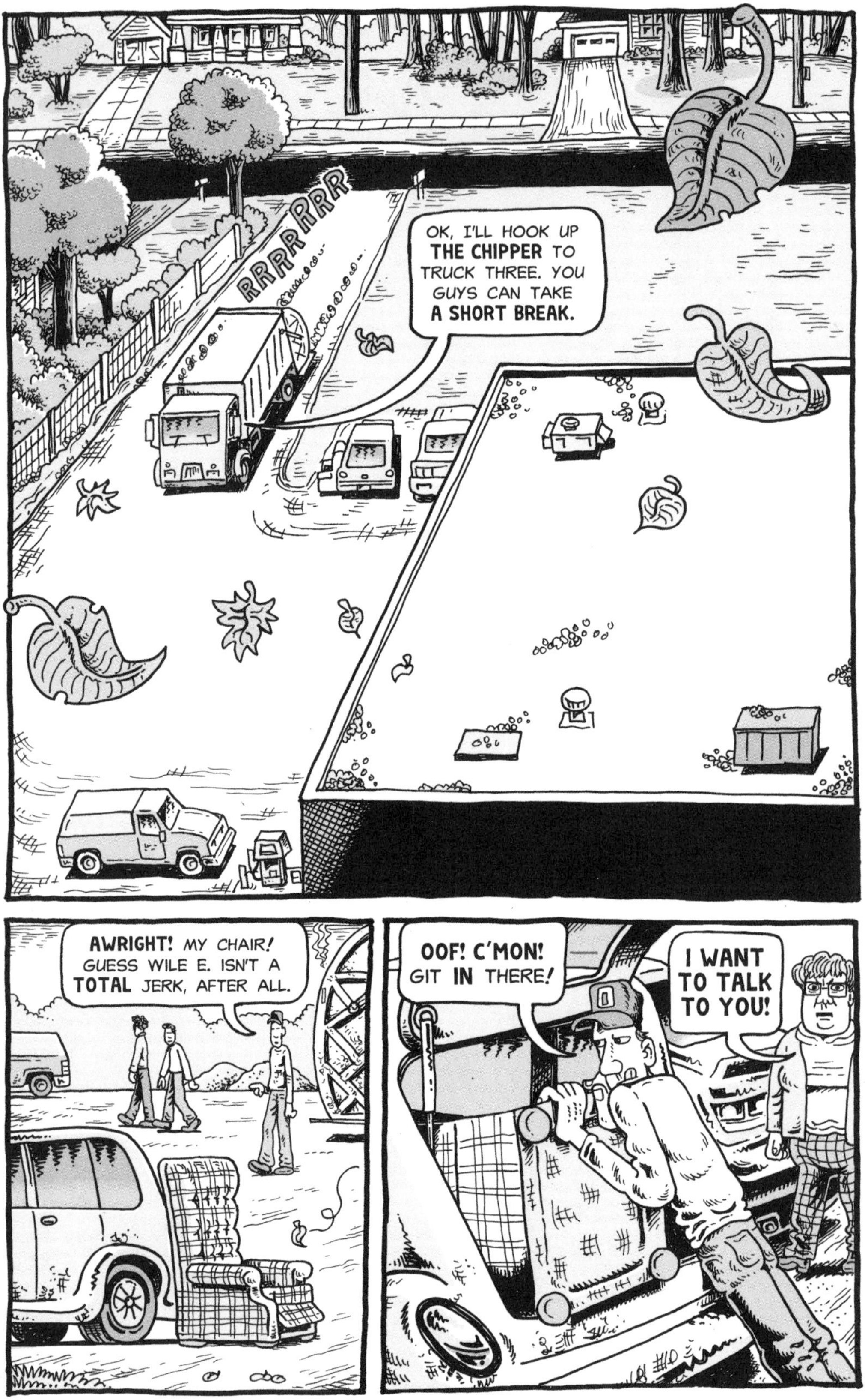
RRRRRRR
OK, I'LL HOOK UP THE CHIPPER TO TRUCK THREE. YOU GUYS CAN TAKE A SHORT BREAK.
AWRIGHT! MY CHAIR! GUESS WILE E. ISN'T A TOTAL JERK, AFTER ALL.
OOF! C'MON! GIT IN THERE!
I WANT TO TALK TO YOU!

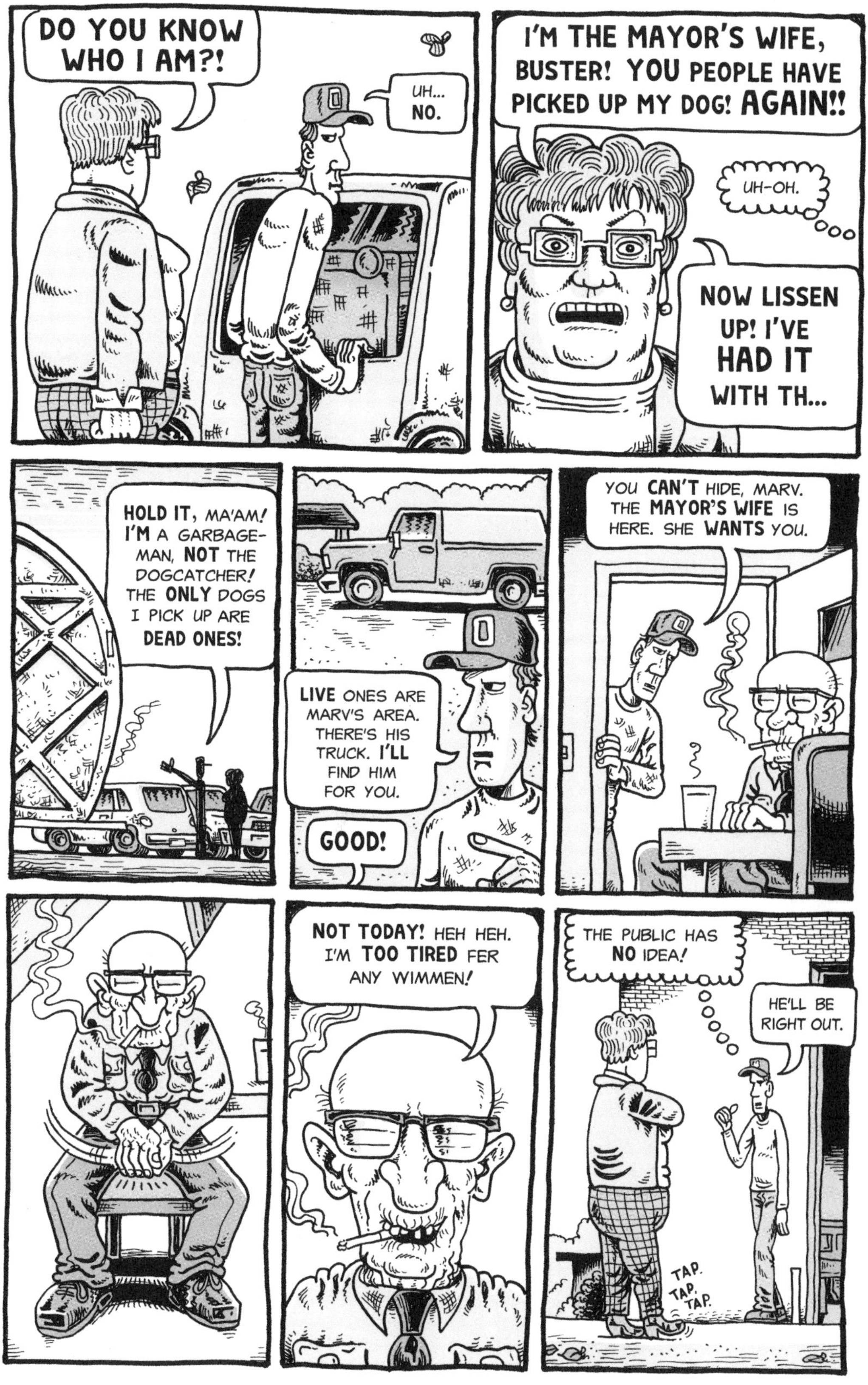
DO YOU KNOW WHO I AM?!
UH... NO.
I'M THE MAYOR'S WIFE, BUSTER! YOU PEOPLE HAVE PICKED UP MY DOG! AGAIN!!
UH-OH.
NOW LISSEN UP! I'VE HAD IT WITH TH...
HOLD IT, MA'AM! I'M A GARBAGEMAN, NOT THE DOGCATCHER! THE ONLY DOGS I PICK UP ARE DEAD ONES!
LIVE ONES ARE MARV'S AREA. THERE'S HIS TRUCK. I'LL FIND HIM FOR YOU.
GOOD!
YOU CAN'T HIDE, MARV. THE MAYOR'S WIFE IS HERE. SHE WANTS YOU.
NOT TODAY! HEH HEH. I'M TOO TIRED FER ANY WIMMEN!
THE PUBLIC HAS NO IDEA!
HE'LL BE RIGHT OUT.
TAP. TAP. TAP.

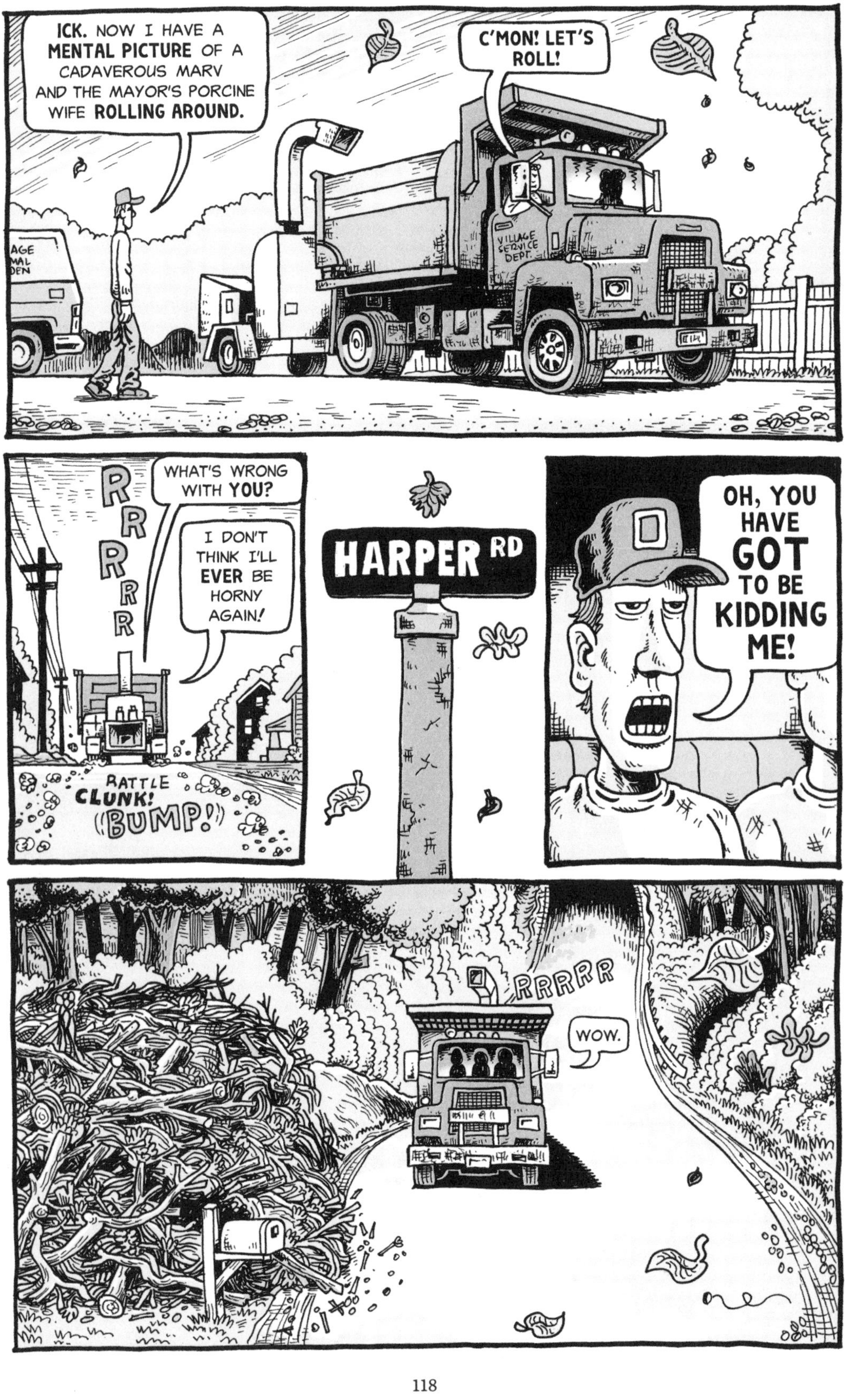
ICK. NOW I HAVE A MENTAL PICTURE OF A CADAVEROUS MARV AND THE MAYOR'S PORCINE WIFE ROLLING AROUND.
C'MON! LET'S ROLL!
VILLAGE SERVICE DEPT.
RRRRR
WHAT'S WRONG WITH YOU?
I DON'T THINK I'LL EVER BE HORNY AGAIN!
RATTLE
CLUNK!
BUMP!
HARPER RD
OH, YOU HAVE GOT TO BE KIDDING ME!
RRRRR
WOW.

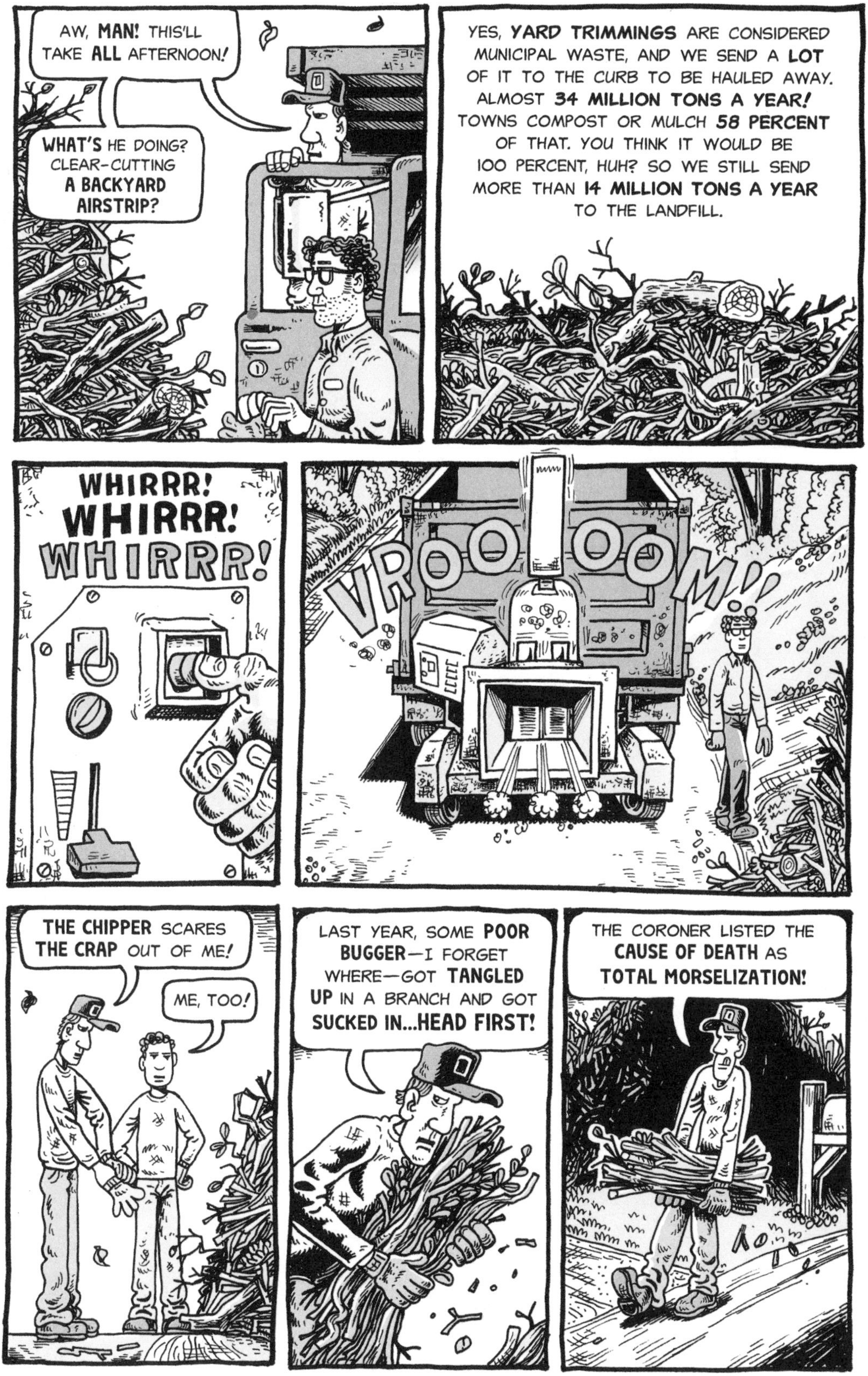
AW, MAN! THIS'LL TAKE ALL AFTERNOON!
WHAT'S HE DOING? CLEAR-CUTTING A BACKYARD AIRSTRIP?
YES, YARD TRIMMINGS ARE CONSIDERED MUNICIPAL WASTE, AND WE SEND A LOT OF IT TO THE CURB TO BE HAULED AWAY. ALMOST 34 MILLION TONS A YEAR! TOWNS COMPOST OR MULCH 58 PERCENT OF THAT. YOU THINK IT WOULD BE 100 PERCENT, HUH? SO WE STILL SEND MORE THAN 14 MILLION TONS A YEAR TO THE LANDFILL.
WHIRRR! WHIRRR! WHIRRR!
VROOOOOM!!
THE CHIPPER SCARES THE CRAP OUT OF ME!
ME, TOO!
LAST YEAR, SOME POOR BUGGER—I FORGET WHERE—GOT TANGLED UP IN A BRANCH AND GOT SUCKED IN...HEAD FIRST!
THE CORONER LISTED THE CAUSE OF DEATH AS TOTAL MORSELIZATION!

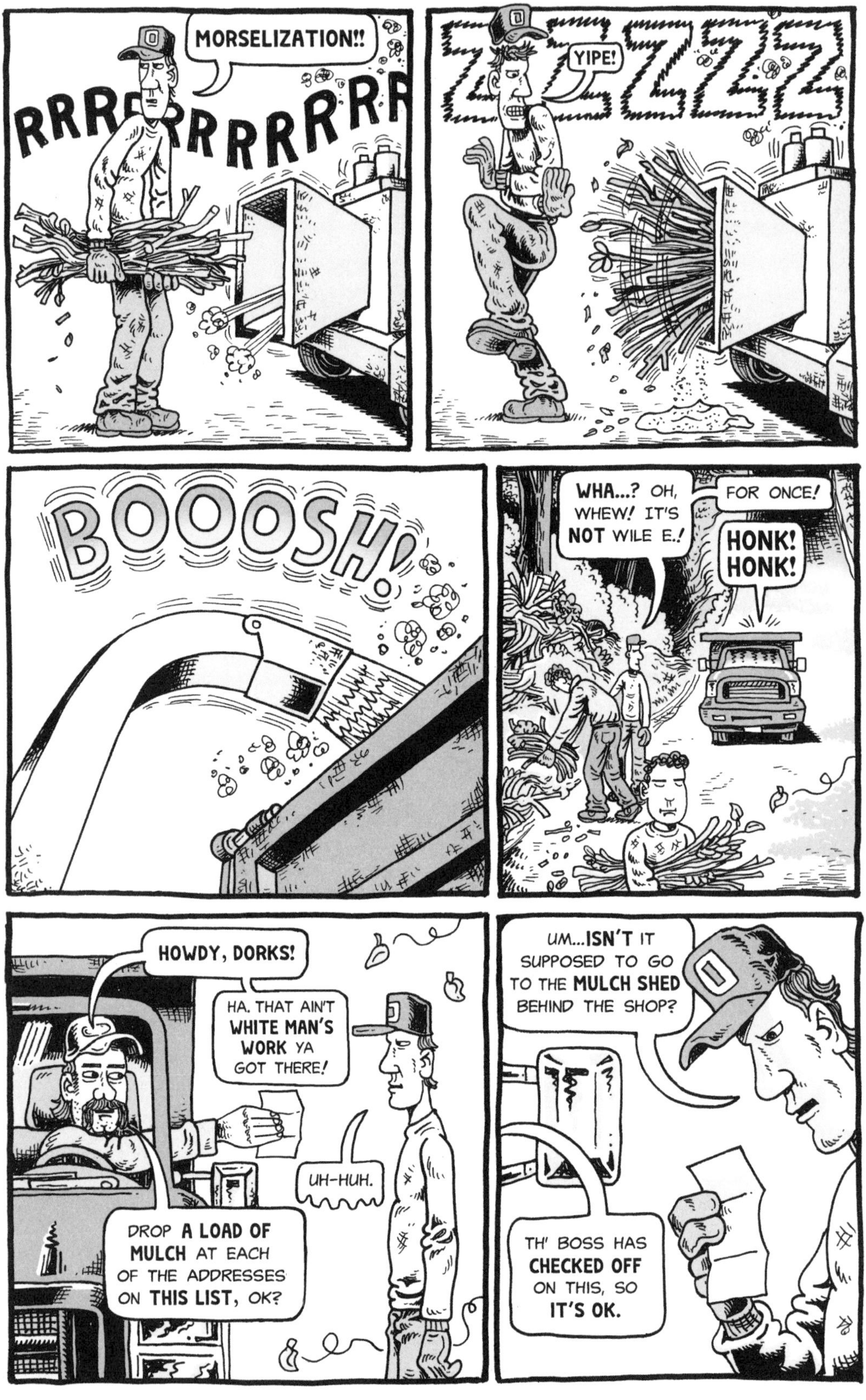
MORSELIZATION!!
RRRRRRRRRRRR
YIPE!
ZZZZZ
BOOOSH!
WHA...? OH, WHEW! IT'S NOT WILE E.!
FOR ONCE!
HONK! HONK!
HOWDY, DORKS!
HA. THAT AIN'T WHITE MAN'S WORK YA GOT THERE!
UH-HUH.
DROP A LOAD OF MULCH AT EACH OF THE ADDRESSES ON THIS LIST, OK?
UM...ISN'T IT SUPPOSED TO GO TO THE MULCH SHED BEHIND THE SHOP?
TH' BOSS HAS CHECKED OFF ON THIS, SO IT'S OK.

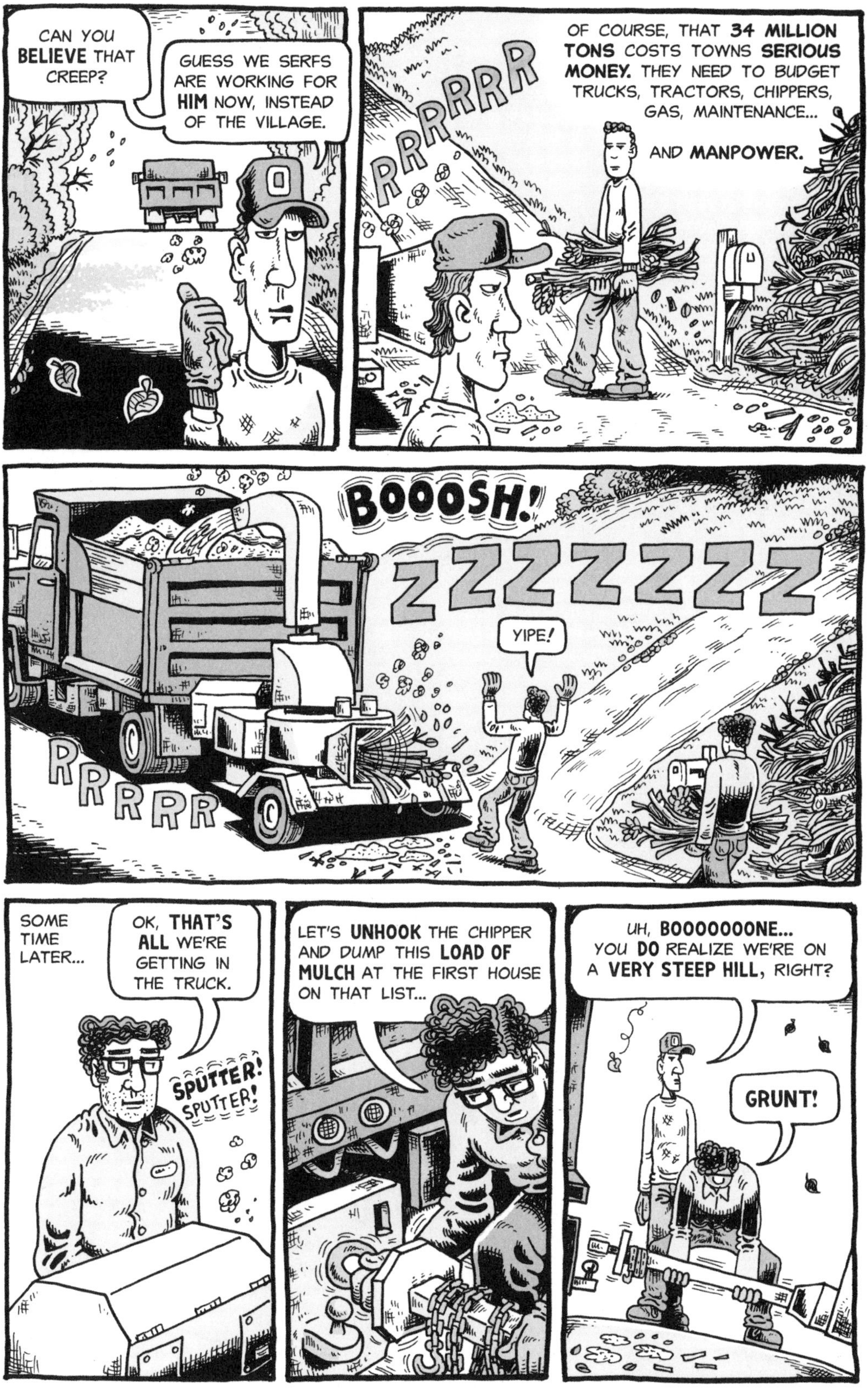
CAN YOU BELIEVE THAT CREEP?
GUESS WE SERFS ARE WORKING FOR HIM NOW, INSTEAD OF THE VILLAGE.
RRRRRRRR
OF COURSE, THAT 34 MILLION TONS COSTS TOWNS SERIOUS MONEY. THEY NEED TO BUDGET TRUCKS, TRACTORS, CHIPPERS, GAS, MAINTENANCE...
AND MANPOWER.
BOOOSH!
ZZZZZZZ
YIPE!
RRRRR
SOME TIME LATER...
OK, THAT'S ALL WE'RE GETTING IN THE TRUCK.
SPUTTER! SPUTTER!
LET'S UNHOOK THE CHIPPER AND DUMP THIS LOAD OF MULCH AT THE FIRST HOUSE ON THAT LIST...
UH, BOOOOOOONE... YOU DO REALIZE WE'RE ON A VERY STEEP HILL, RIGHT?
GRUNT!

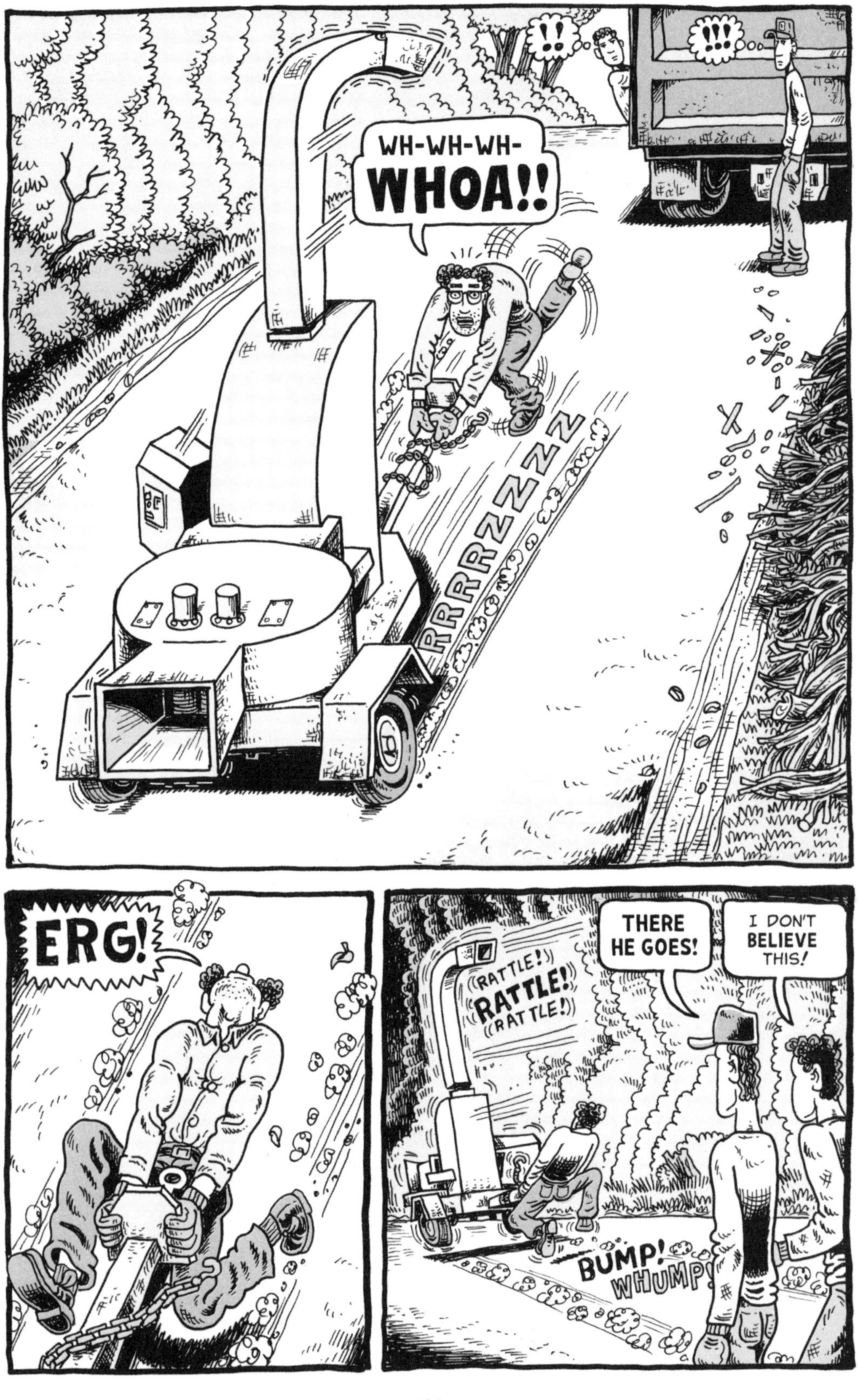
!!
!!!
WH-WH-WH-
WHOA!!
RRRRRZZZZZZ
ERG!
RATTLE!
RATTLE!
RATTLE!
THERE HE GOES!
I DON'T BELIEVE THIS!
BUMP!
WHUMP!

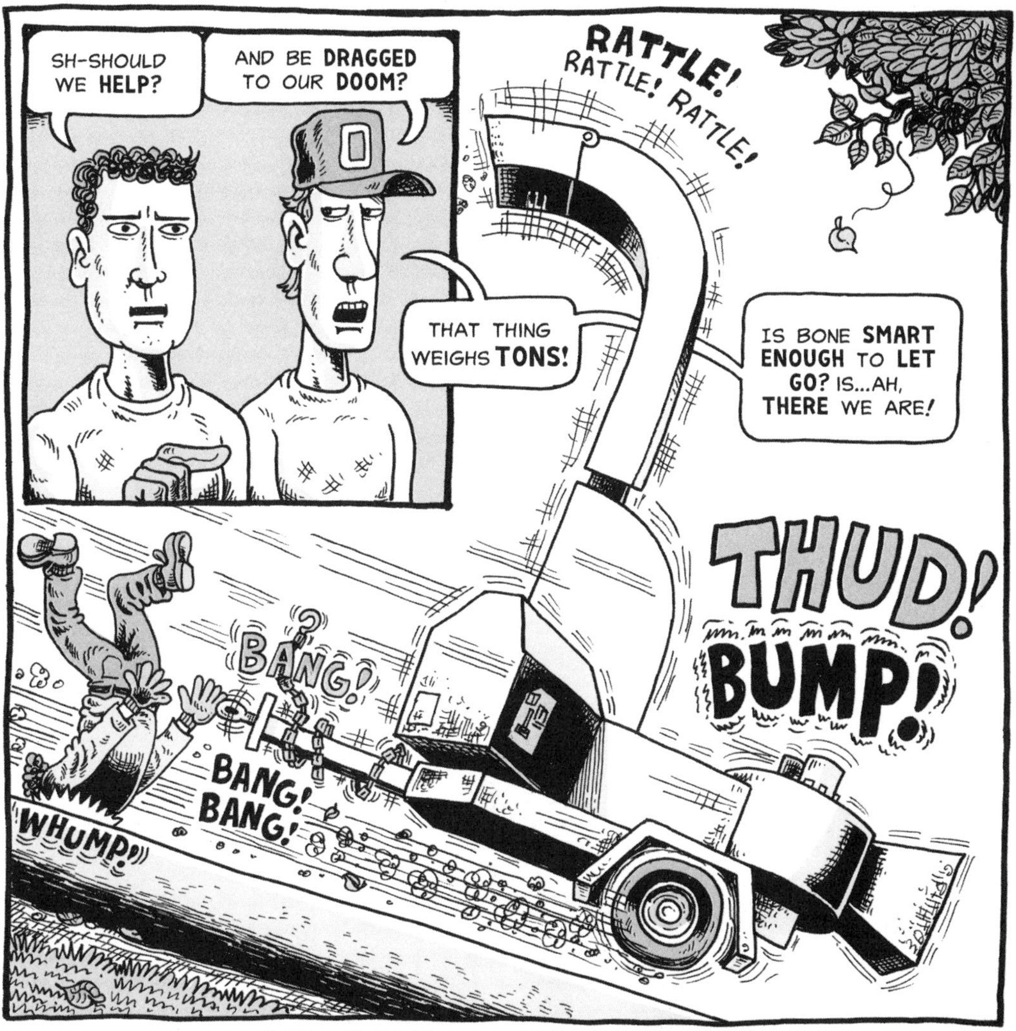
SH-SHOULD WE HELP?
AND BE DRAGGED TO OUR DOOM?
THAT THING WEIGHS TONS!
IS BONE SMART ENOUGH TO LET GO? IS...AH, THERE WE ARE!
RATTLE! RATTLE! RATTLE!
THUD!
BUMP!
BANG!
BANG! BANG!
WHUMP!

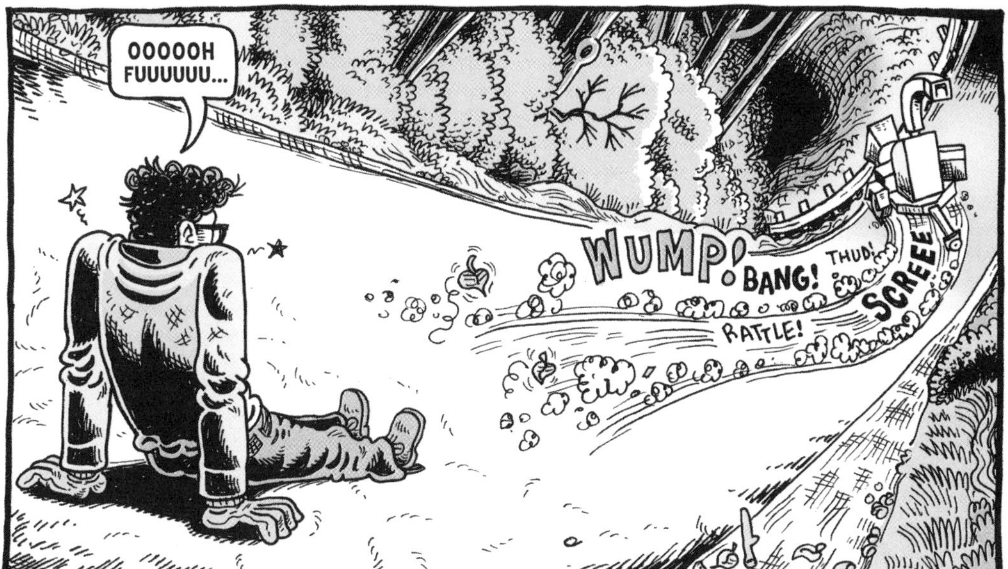
OOOOOH FUUUUUU...
WUMP!
BANG!
THUD!
SCREEE
RATTLE!

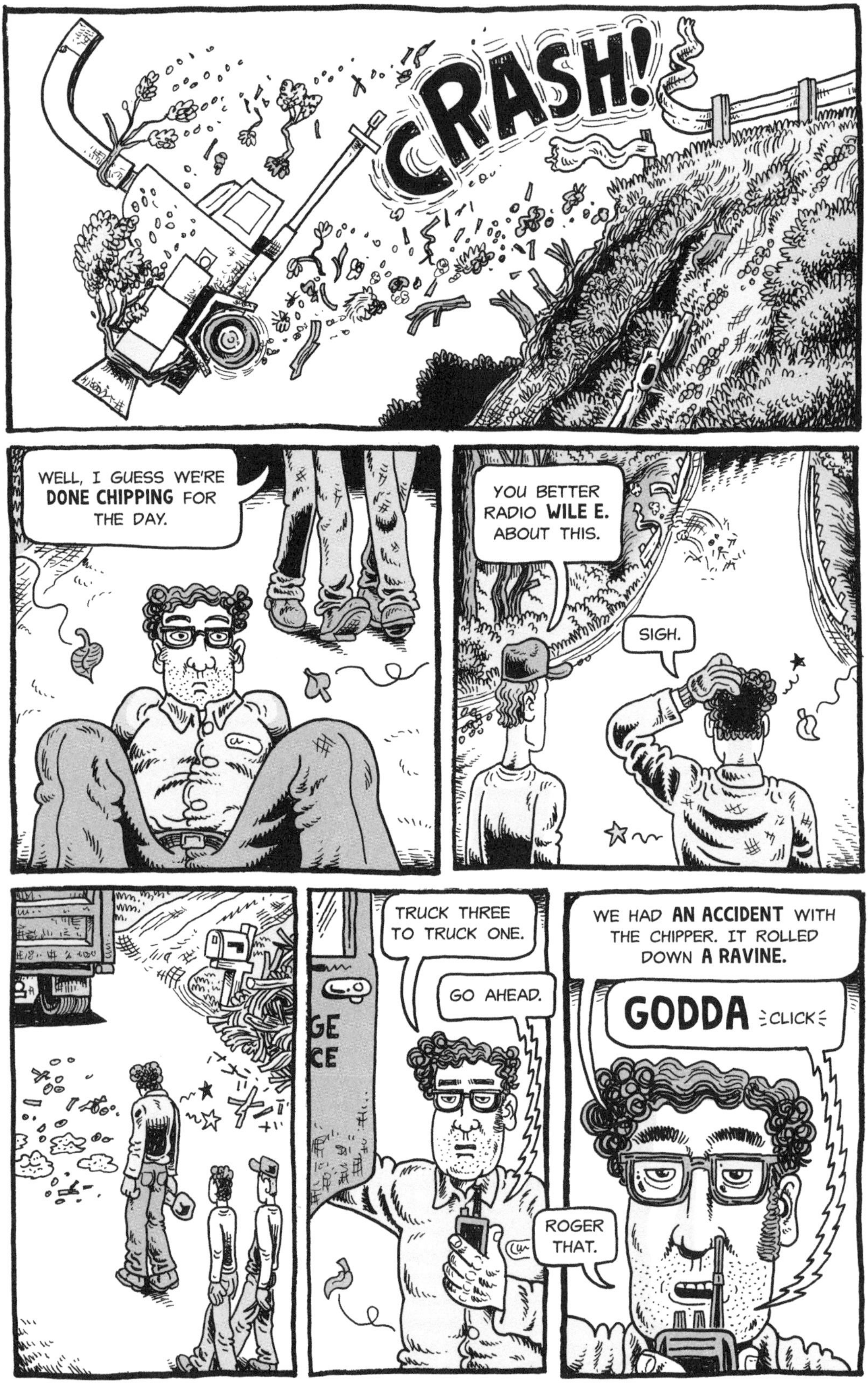
CRASH!
WELL, I GUESS WE'RE DONE CHIPPING FOR THE DAY.
YOU BETTER RADIO WILE E. ABOUT THIS.
SIGH.
TRUCK THREE TO TRUCK ONE.
GO AHEAD.
ROGER THAT.
WE HAD AN ACCIDENT WITH THE CHIPPER. IT ROLLED DOWN A RAVINE.
GODDA
CLICK

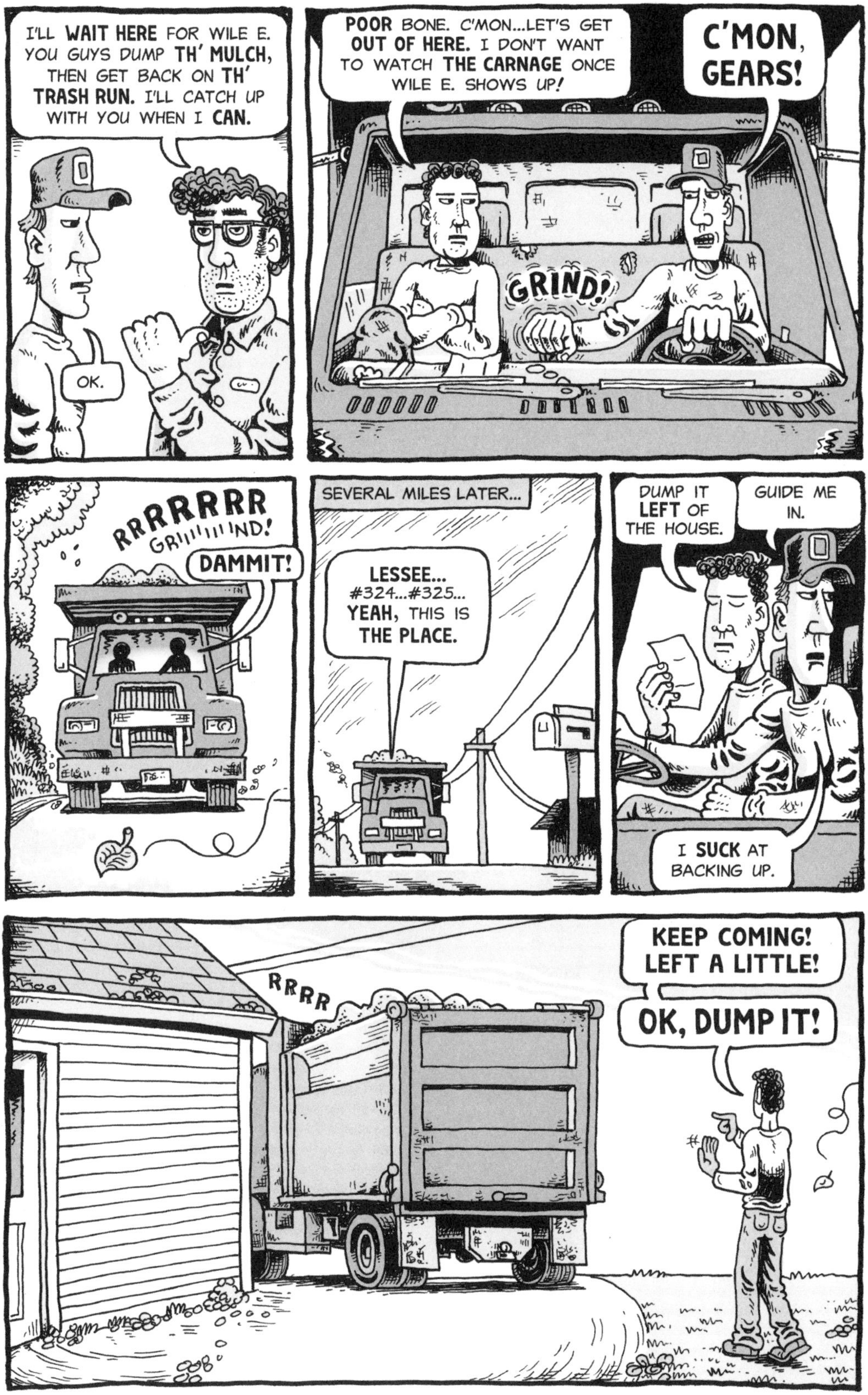
I'LL WAIT HERE FOR WILE E. YOU GUYS DUMP TH' MULCH, THEN GET BACK ON TH' TRASH RUN. I'LL CATCH UP WITH YOU WHEN I CAN.
OK.
POOR BONE. C'MON...LET'S GET OUT OF HERE. I DON'T WANT TO WATCH THE CARNAGE ONCE WILE E. SHOWS UP!
C'MON, GEARS!
GRIND!
RRRRRRR GRIIIIIIND!
DAMMIT!
SEVERAL MILES LATER...
LESSEE... #324...#325... YEAH, THIS IS THE PLACE.
DUMP IT LEFT OF THE HOUSE.
GUIDE ME IN.
I SUCK AT BACKING UP.
RRRR
KEEP COMING! LEFT A LITTLE!
OK, DUMP IT!

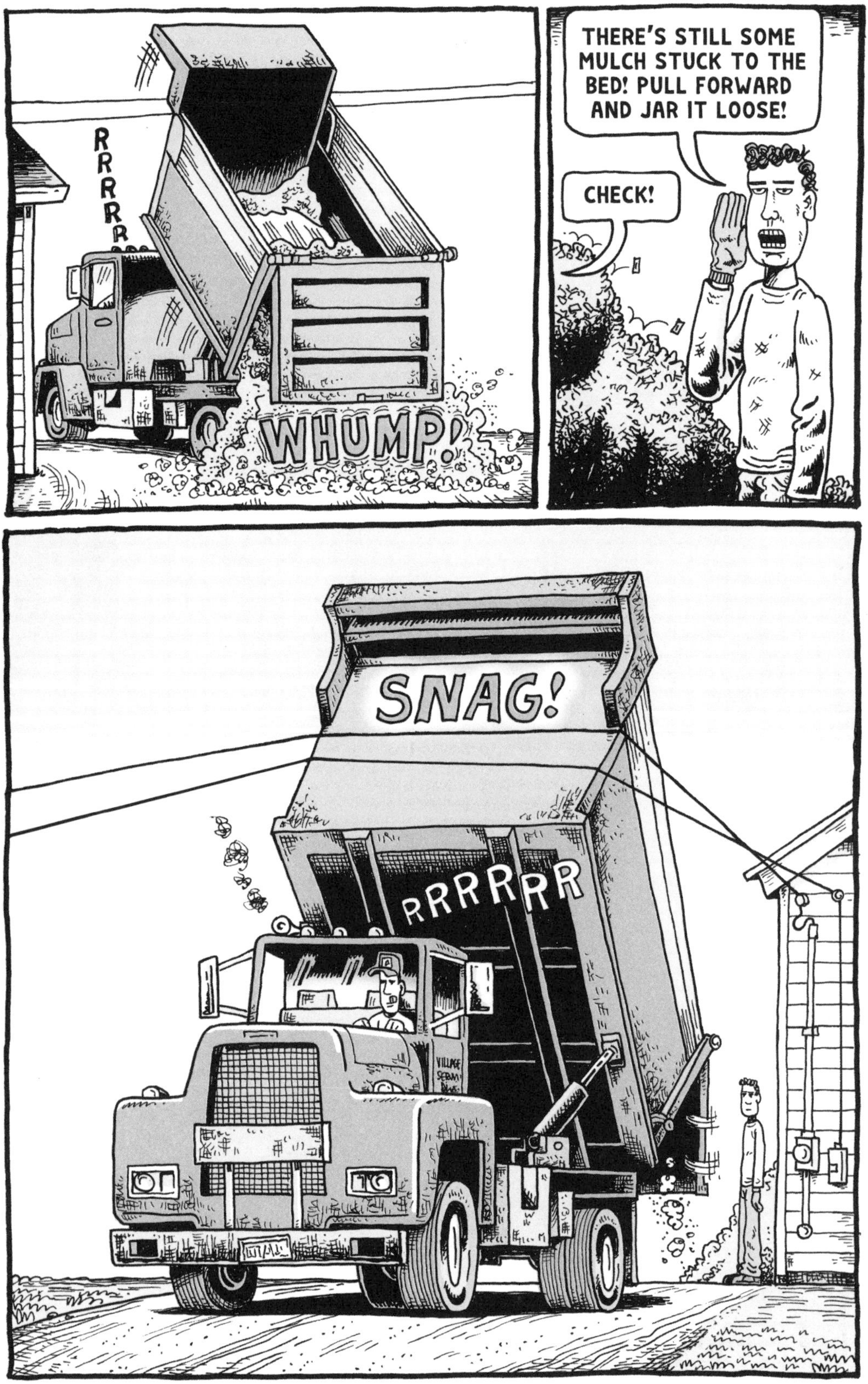
RRRRR
WHUMP!
THERE'S STILL SOME MULCH STUCK TO THE BED! PULL FORWARD AND JAR IT LOOSE!
CHECK!
SNAG!
RRRRRR

OH CRAP! THE LINES! STOP!
SNAP!
RIIIIIP!
POP!
YAAAAA!
THWAP!
BZZAT!
ZZZZZZ
GREAT.
VILLAGE SERVICE DEPT.

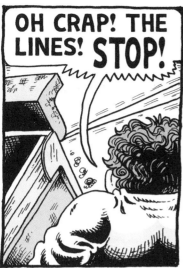

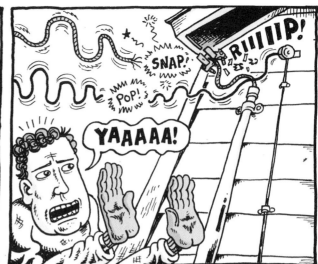

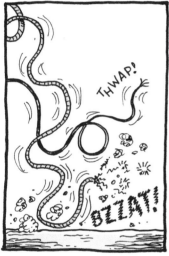

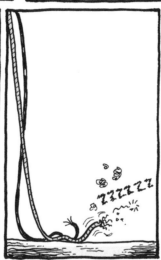

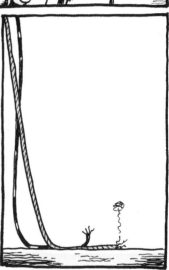

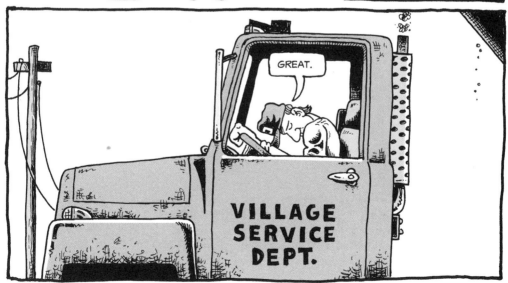

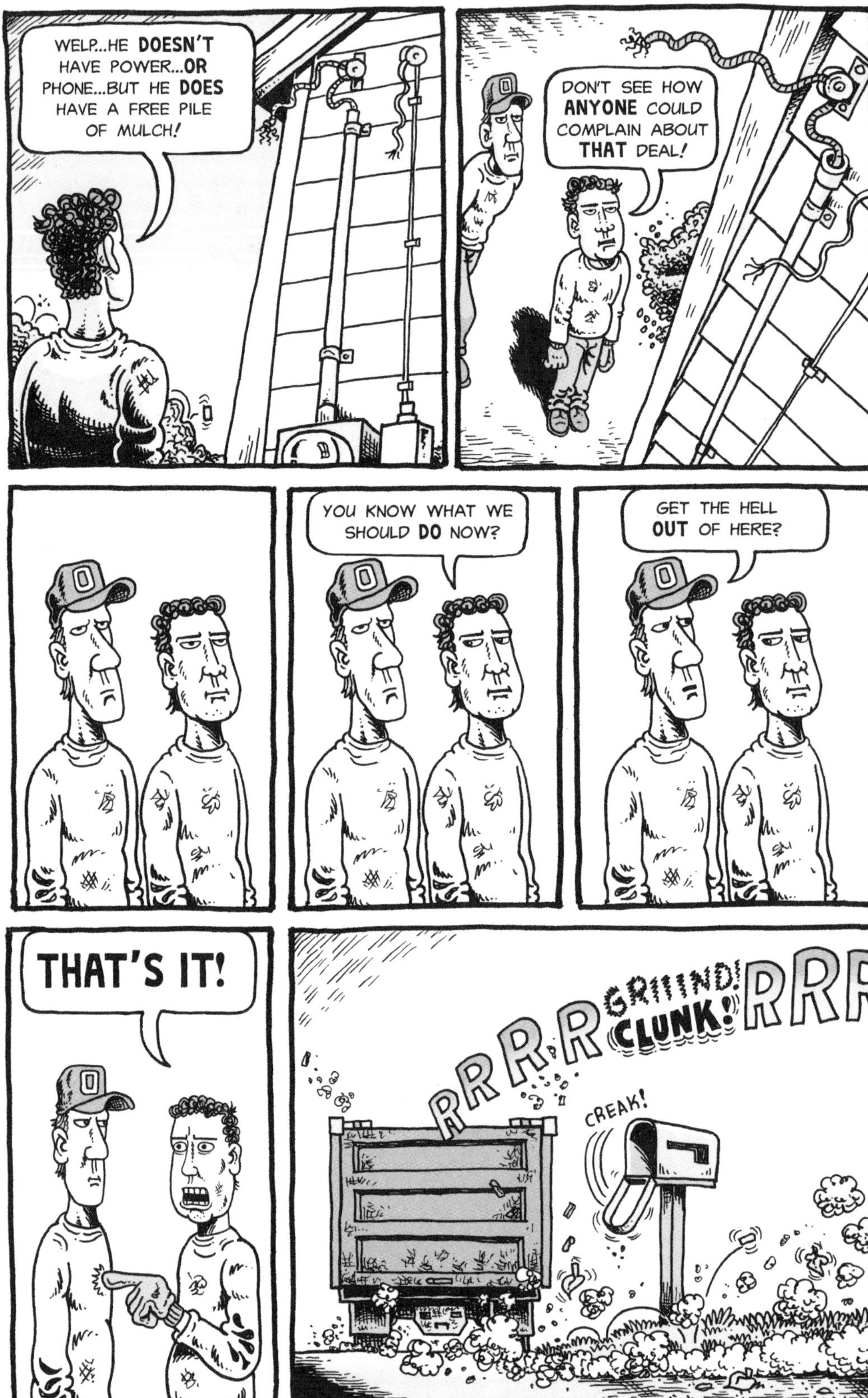
WELP...HE DOESN'T HAVE POWER...OR PHONE...BUT HE DOES HAVE A FREE PILE OF MULCH!
DON'T SEE HOW ANYONE COULD COMPLAIN ABOUT THAT DEAL!
YOU KNOW WHAT WE SHOULD DO NOW?
GET THE HELL OUT OF HERE?
THAT'S IT!
RRRR
GRIIIND!
CLUNK!
RRR
CREAK!

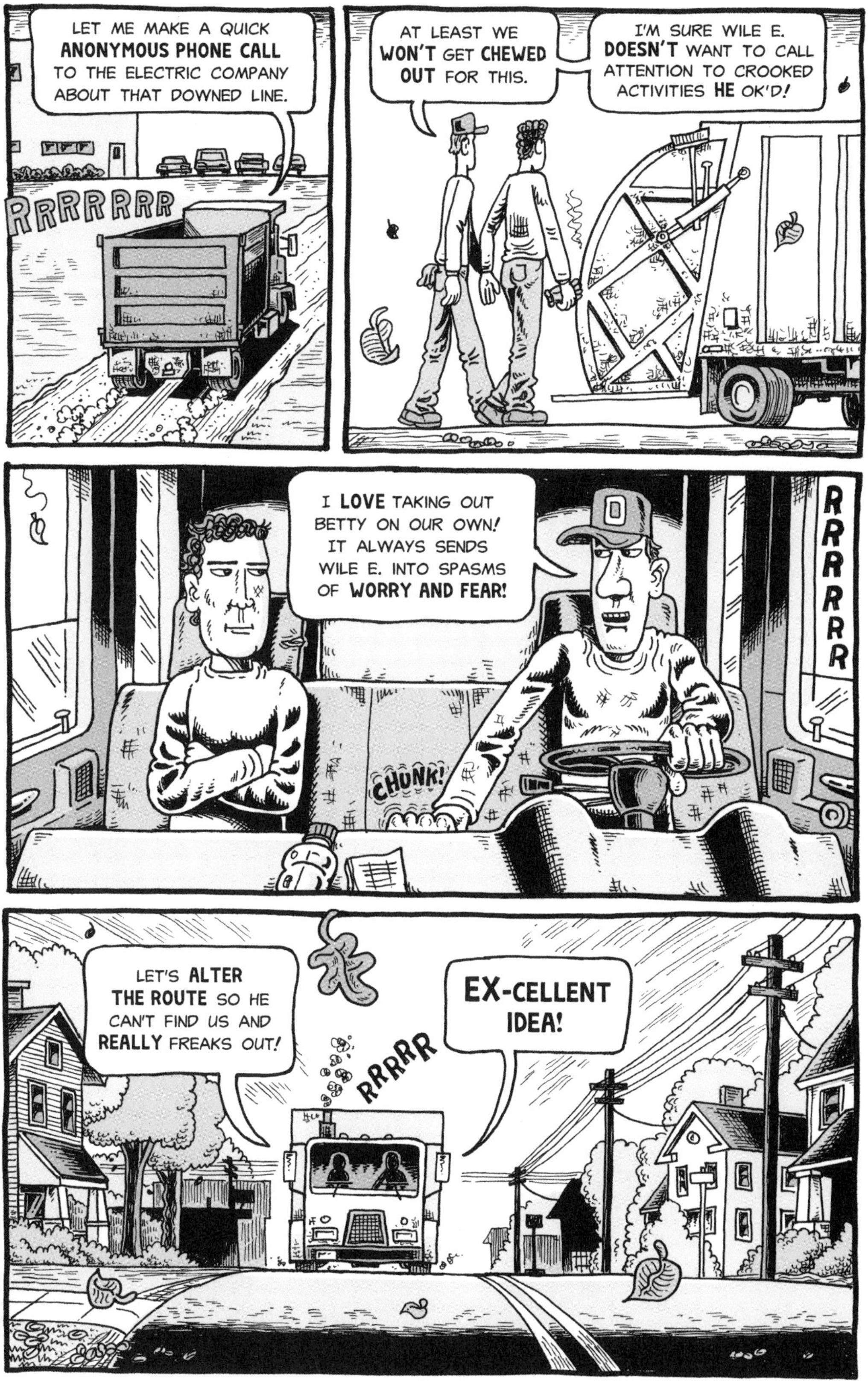
LET ME MAKE A QUICK ANONYMOUS PHONE CALL TO THE ELECTRIC COMPANY ABOUT THAT DOWNED LINE.
RRRRRRR
AT LEAST WE WON'T GET CHEWED OUT FOR THIS.
I'M SURE WILE E. DOESN'T WANT TO CALL ATTENTION TO CROOKED ACTIVITIES HE OK'D!
I LOVE TAKING OUT BETTY ON OUR OWN! IT ALWAYS SENDS WILE E. INTO SPASMS OF WORRY AND FEAR!
RRRRRR
CHUNK!
LET'S ALTER THE ROUTE SO HE CAN'T FIND US AND REALLY FREAKS OUT!
RRRRR
EX-CELLENT IDEA!

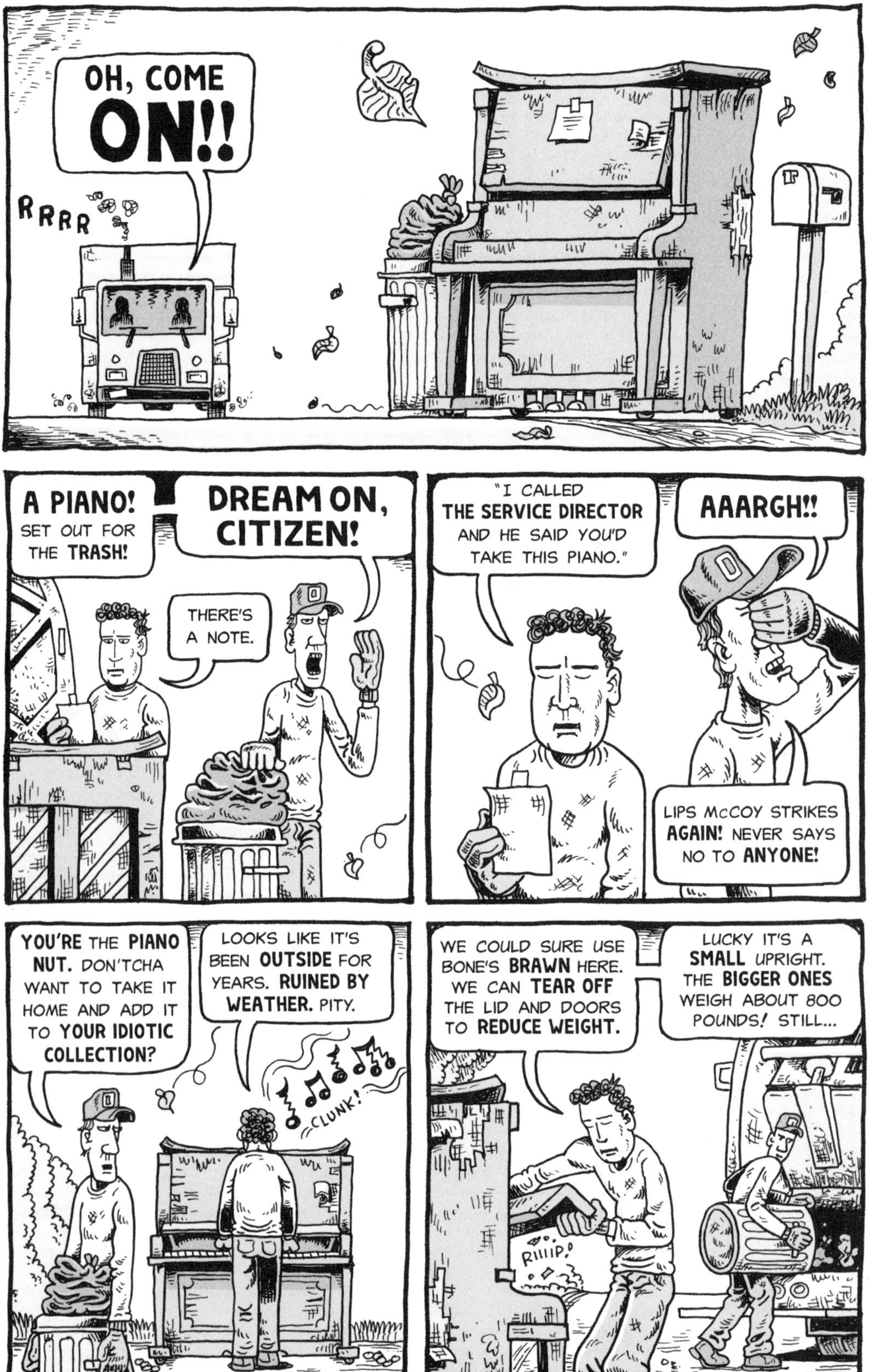

RRRR
OH, COME ON!!
A PIANO! SET OUT FOR THE TRASH!
DREAM ON, CITIZEN!
THERE'S A NOTE.
"I CALLED THE SERVICE DIRECTOR AND HE SAID YOU'D TAKE THIS PIANO."
AAARGH!!
LIPS McCOY STRIKES AGAIN! NEVER SAYS NO TO ANYONE!
YOU'RE THE PIANO NUT. DON'TCHA WANT TO TAKE IT HOME AND ADD IT TO YOUR IDIOTIC COLLECTION?
LOOKS LIKE IT'S BEEN OUTSIDE FOR YEARS. RUINED BY WEATHER. PITY.
CLUNK!
WE COULD SURE USE BONE'S BRAWN HERE. WE CAN TEAR OFF THE LID AND DOORS TO REDUCE WEIGHT.
LUCKY IT'S A SMALL UPRIGHT. THE BIGGER ONES WEIGH ABOUT 800 POUNDS! STILL...
RIIIP!

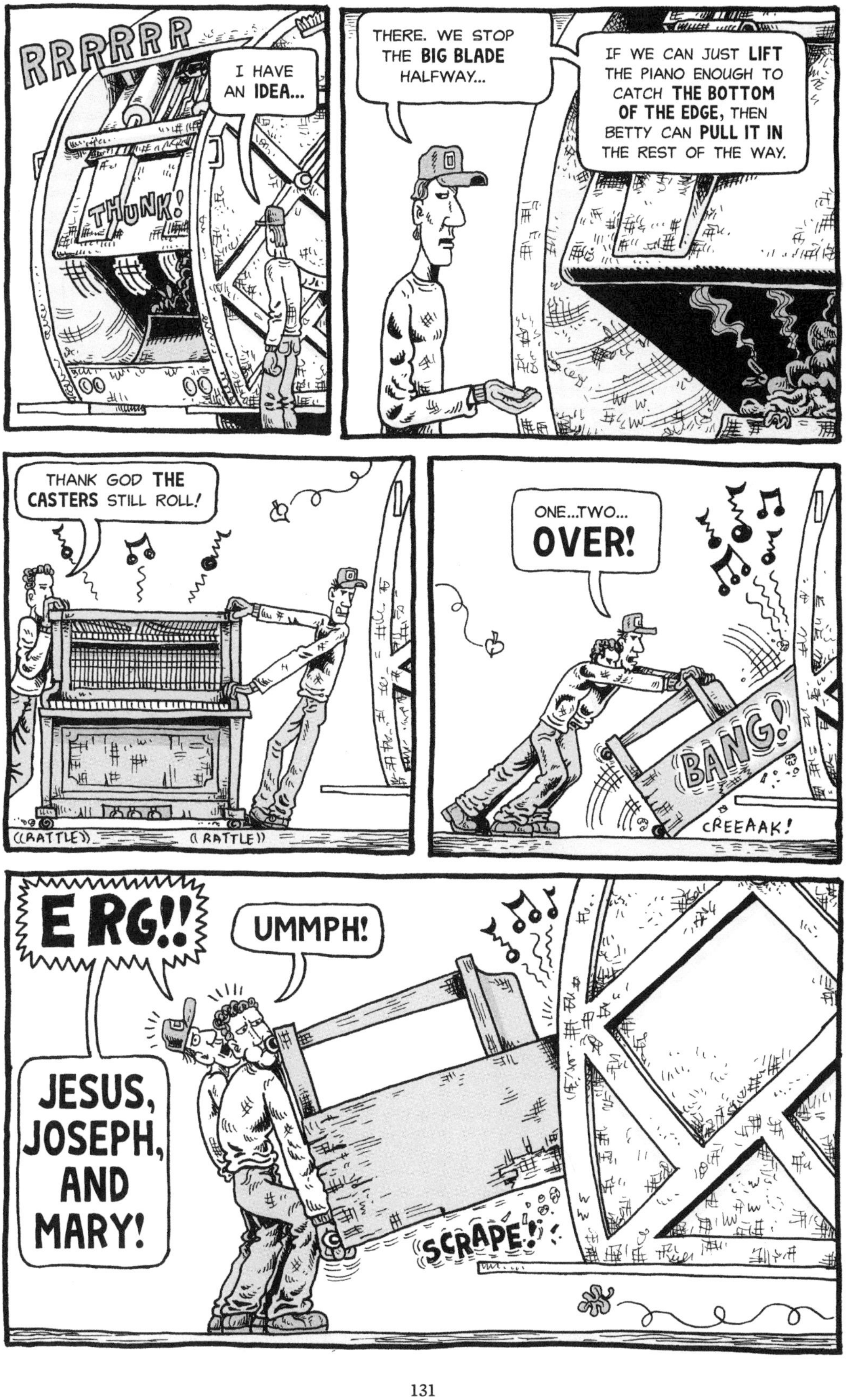
RRRRRR
I HAVE AN IDEA...
THUNK!
THERE. WE STOP THE BIG BLADE HALFWAY...
IF WE CAN JUST LIFT THE PIANO ENOUGH TO CATCH THE BOTTOM OF THE EDGE, THEN BETTY CAN PULL IT IN THE REST OF THE WAY.
THANK GOD THE CASTERS STILL ROLL!
((RATTLE))
((RATTLE))
ONE...TWO... OVER!
BANG!
CREEAAK!
ERG!!
UMMPH!
JESUS, JOSEPH, AND MARY!
SCRAPE!

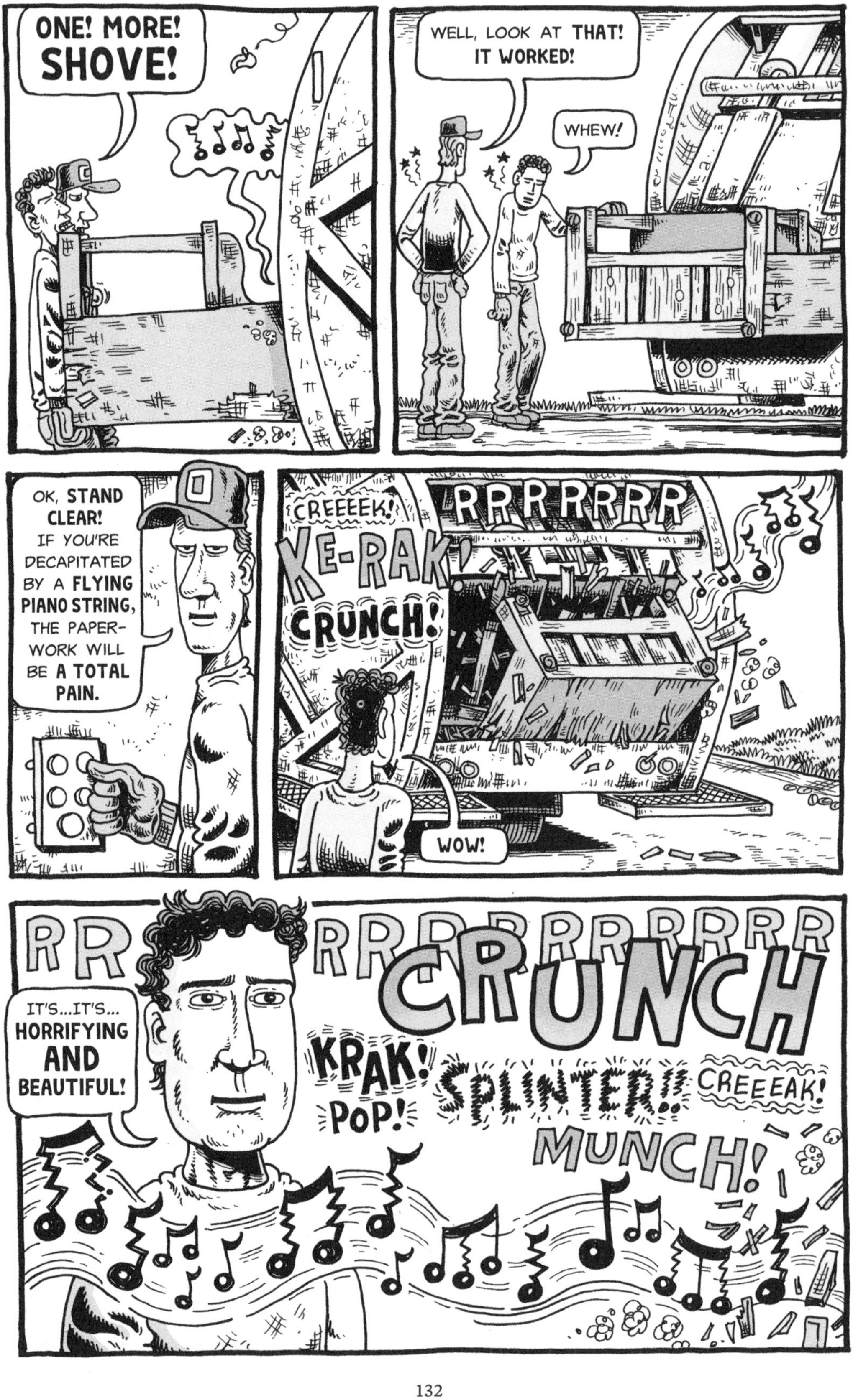
ONE! MORE! SHOVE!
WELL, LOOK AT THAT! IT WORKED!
WHEW!
OK, STAND CLEAR! IF YOU'RE DECAPITATED BY A FLYING PIANO STRING, THE PAPER-WORK WILL BE A TOTAL PAIN.
CREEEEK!
RRRRRRRR
KE-RAK!
CRUNCH!
WOW!
RR RRRRRRRRRRR
CRUNCH
IT'S...IT'S... HORRIFYING AND BEAUTIFUL!
KRAK!
POP!
SPLINTER!!
CREEEAK!
MUNCH!

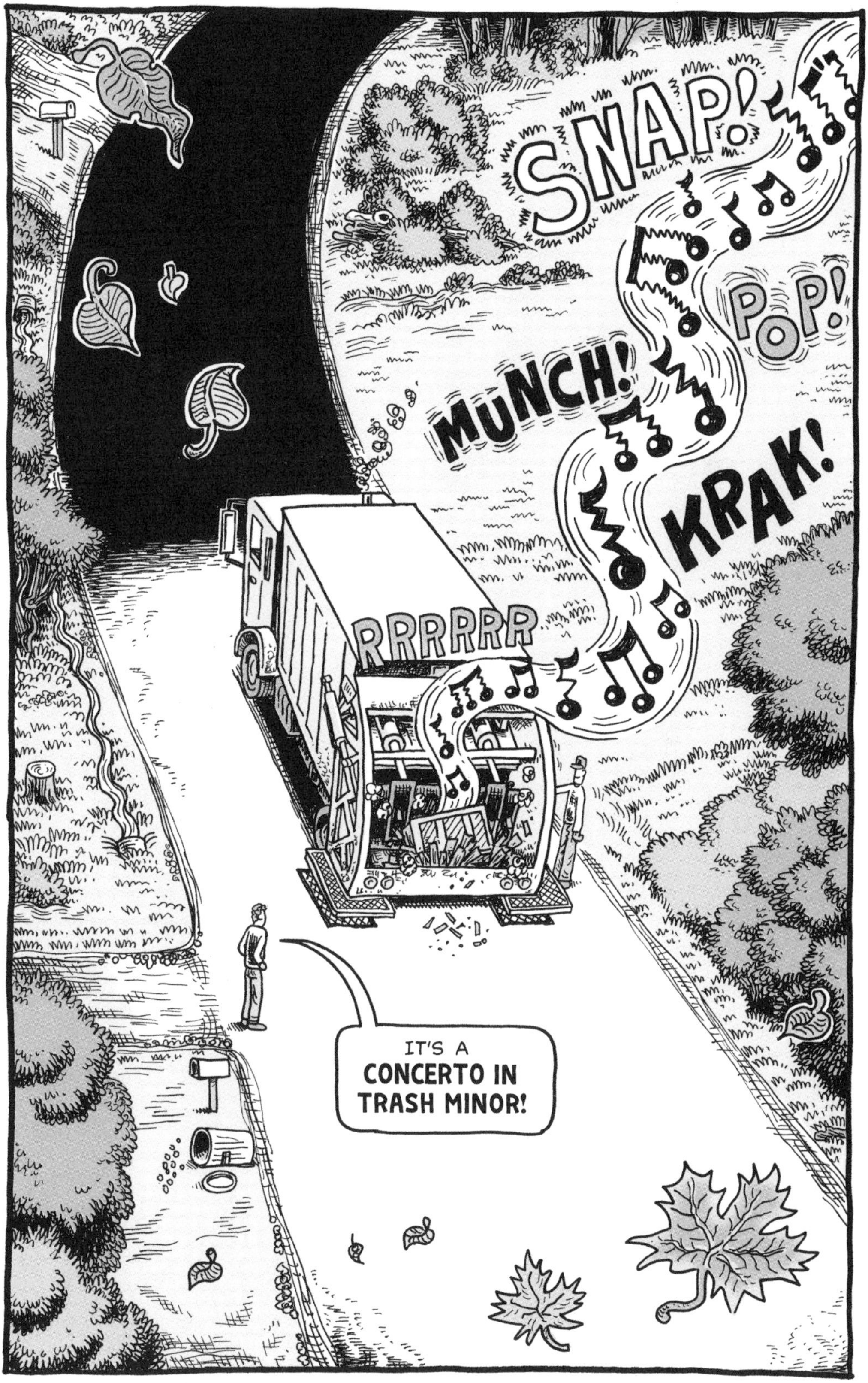
SNAP!
POP!
MUNCH!
KRAK!
RRRRRR
IT'S A CONCERTO IN TRASH MINOR!

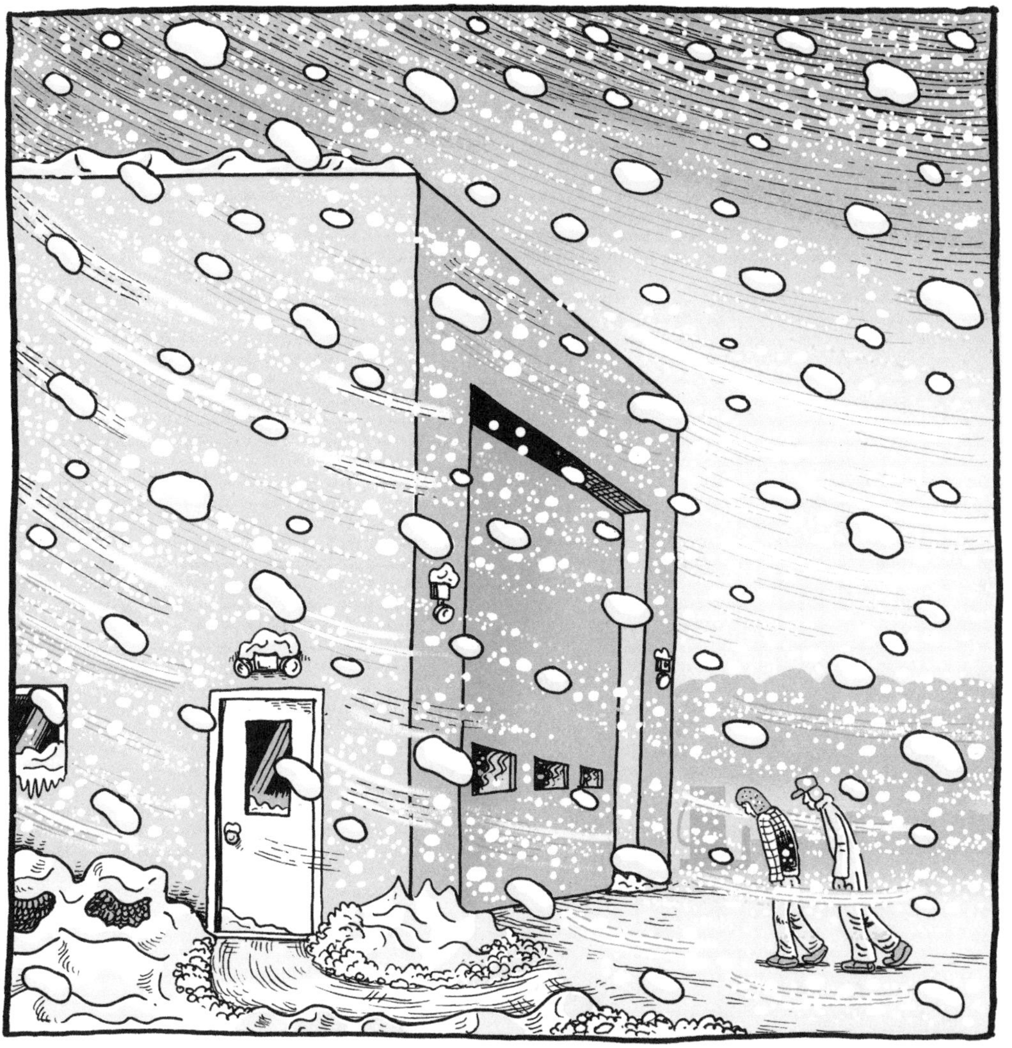

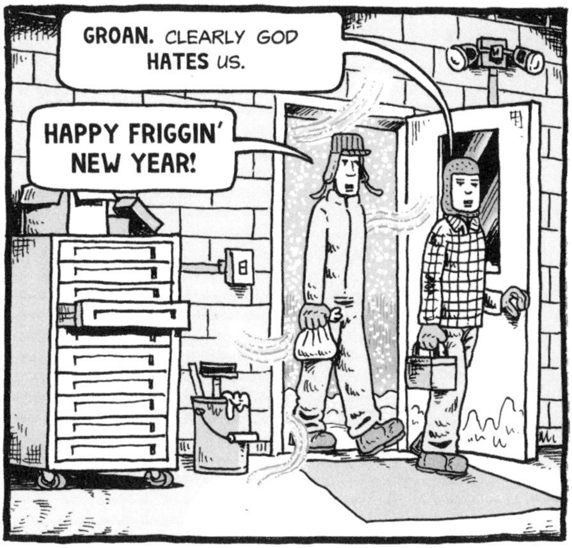
HAPPY FRIGGIN' NEW YEAR!
GROAN. CLEARLY GOD HATES US.

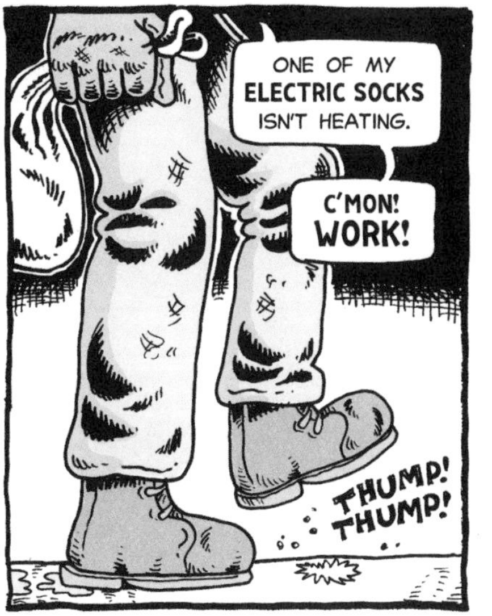
ONE OF MY ELECTRIC SOCKS ISN'T HEATING.
C'MON! WORK!
THUMP! THUMP!

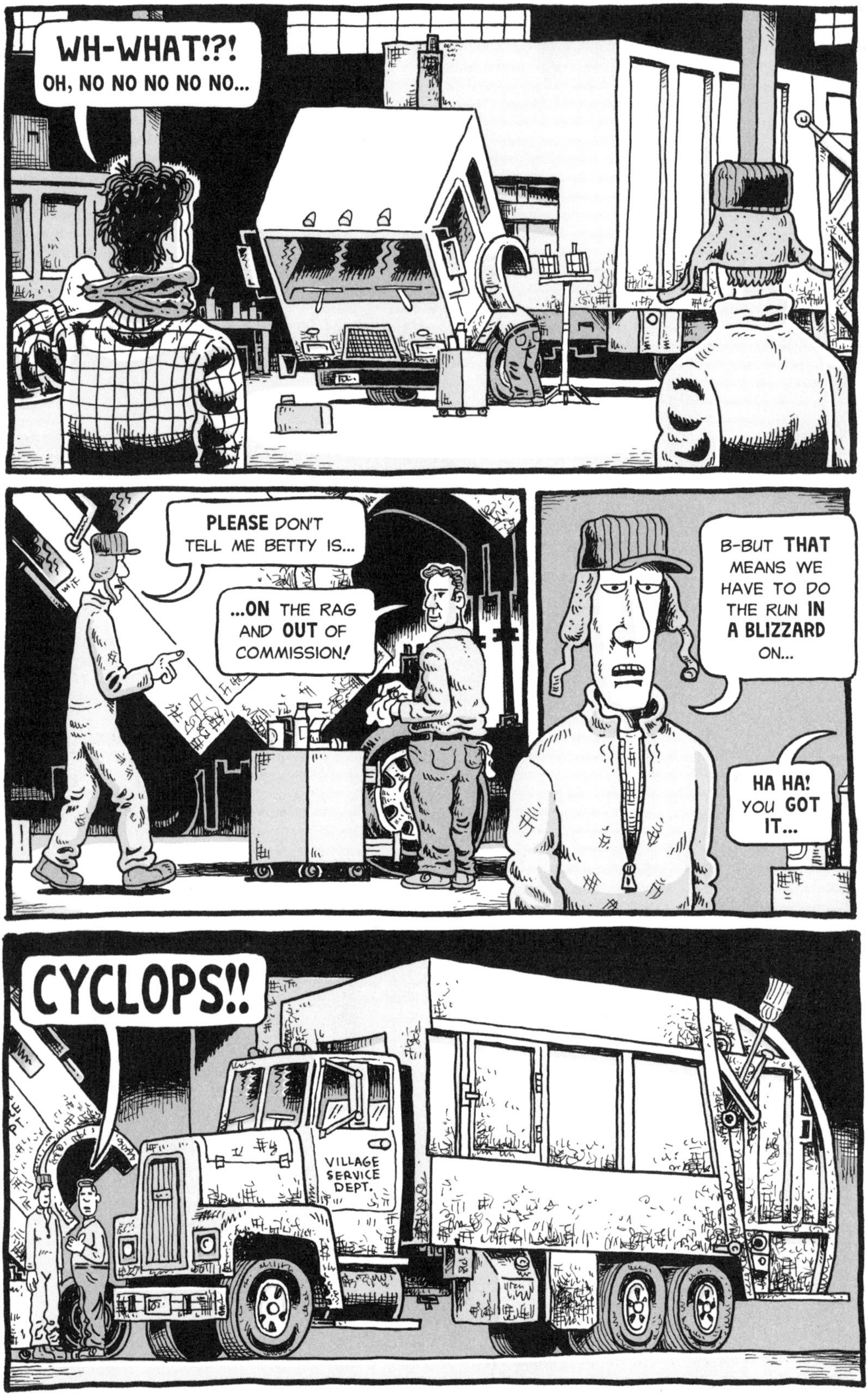
WH-WHAT!?! OH, NO NO NO NO NO...
PLEASE DON'T TELL ME BETTY IS...
...ON THE RAG AND OUT OF COMMISSION!
B-BUT THAT MEANS WE HAVE TO DO THE RUN IN A BLIZZARD ON...
HA HA! YOU GOT IT...
CYCLOPS!!
VILLAGE SERVICE DEPT.

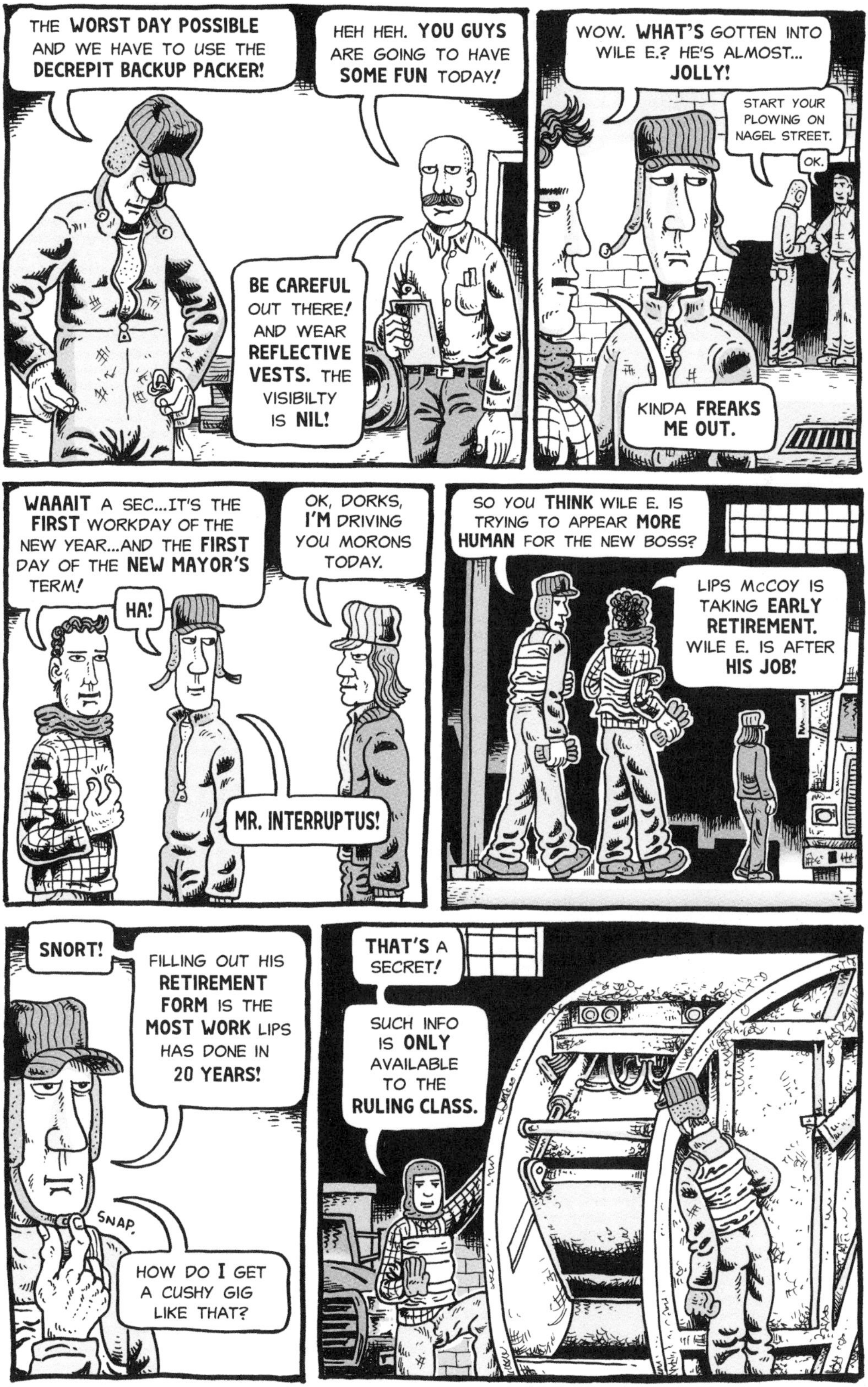
THE WORST DAY POSSIBLE AND WE HAVE TO USE THE DECREPIT BACKUP PACKER!
HEH HEH. YOU GUYS ARE GOING TO HAVE SOME FUN TODAY!
BE CAREFUL OUT THERE! AND WEAR REFLECTIVE VESTS. THE VISIBILTY IS NIL!
WOW. WHAT'S GOTTEN INTO WILE E.? HE'S ALMOST... JOLLY!
START YOUR PLOWING ON NAGEL STREET.
OK.
KINDA FREAKS ME OUT.
WAAAIT A SEC...IT'S THE FIRST WORKDAY OF THE NEW YEAR...AND THE FIRST DAY OF THE NEW MAYOR'S TERM!
HA!
OK, DORKS, I'M DRIVING YOU MORONS TODAY.
MR. INTERRUPTUS!
SO YOU THINK WILE E. IS TRYING TO APPEAR MORE HUMAN FOR THE NEW BOSS?
LIPS McCOY IS TAKING EARLY RETIREMENT. WILE E. IS AFTER HIS JOB!
SNORT!
FILLING OUT HIS RETIREMENT FORM IS THE MOST WORK LIPS HAS DONE IN 20 YEARS!
SNAP.
HOW DO I GET A CUSHY GIG LIKE THAT?
THAT'S A SECRET!
SUCH INFO IS ONLY AVAILABLE TO THE RULING CLASS.

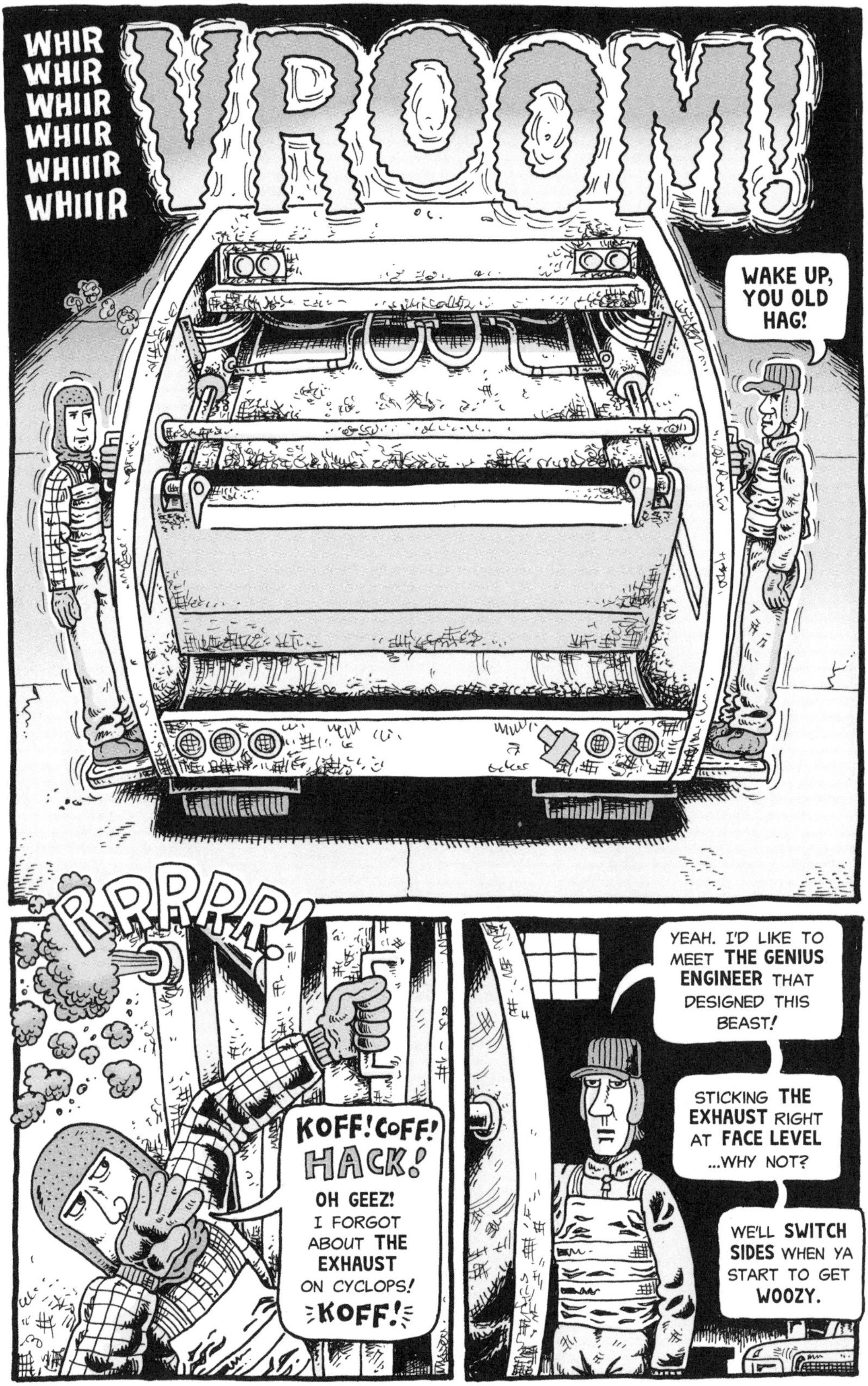
WHIR
WHIR
WHIIR
WHIIR
WHIIIR
WHIIIR
VROOM!
WAKE UP, YOU OLD HAG!
RRRRR!
KOFF! COFF! HACK!
OH GEEZ! I FORGOT ABOUT THE EXHAUST ON CYCLOPS! KOFF!
YEAH. I'D LIKE TO MEET THE GENIUS ENGINEER THAT DESIGNED THIS BEAST!
STICKING THE EXHAUST RIGHT AT FACE LEVEL ...WHY NOT?
WE'LL SWITCH SIDES WHEN YA START TO GET WOOZY.

RRRR **GRIIIIND CLUNK!** RRRRRRR

KOFF! HACK!

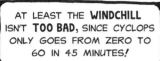
AT LEAST THE **WINDCHILL** ISN'T **TOO BAD,** SINCE CYCLOPS ONLY GOES FROM ZERO TO 60 IN 45 MINUTES!

RRRRR
CLANK!
RRRRR

THE TRASH IS ALL **TOTALLY** BURIED.

RRRRR

OOF! EERG! C'MON!

GET! OUTTA! **THERE!**

BAM! BAM! BAM!

YEP, IT'S GONNA BE A **LOOOONG** DAY.

SCRAAPE!

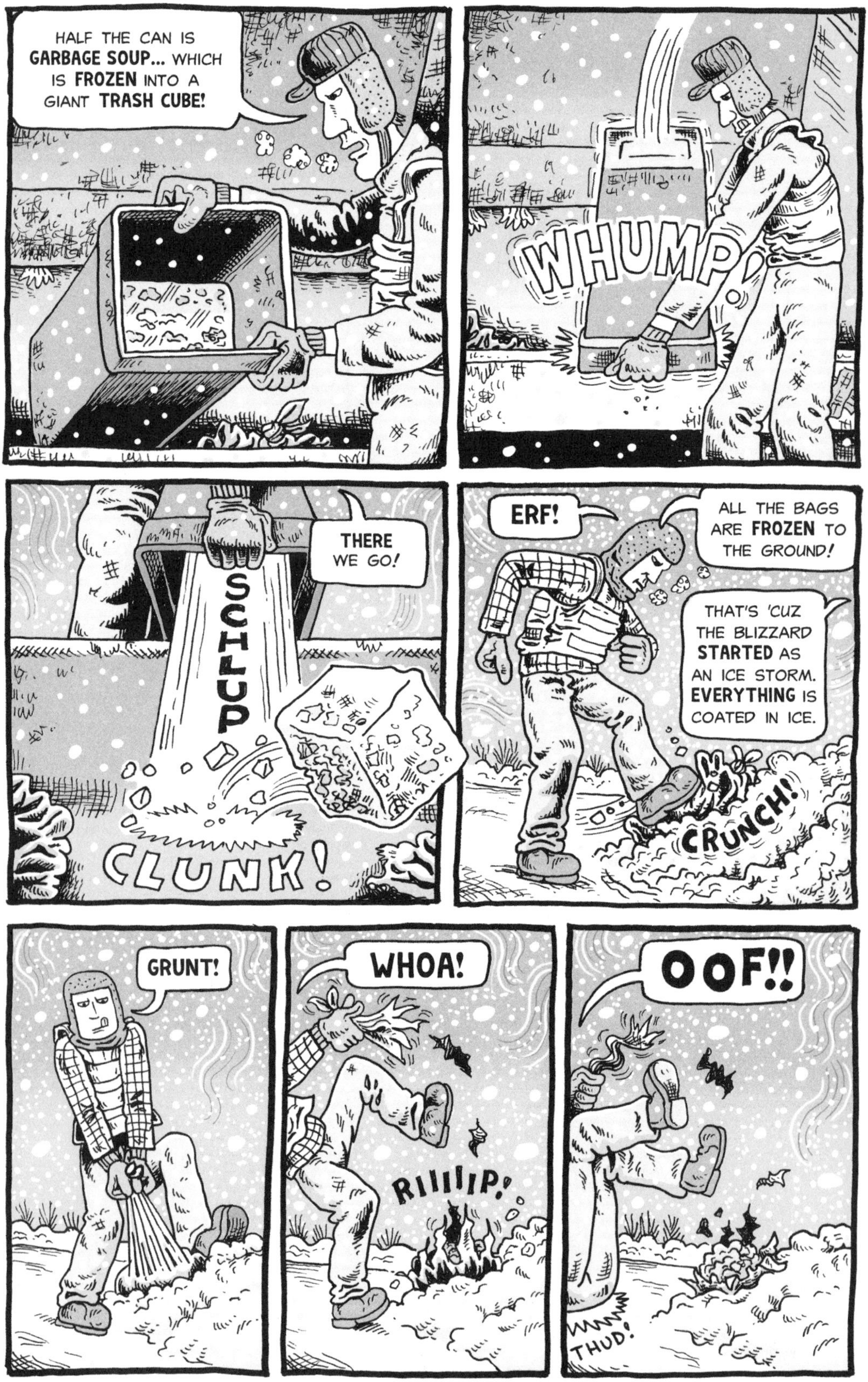
HALF THE CAN IS GARBAGE SOUP... WHICH IS FROZEN INTO A GIANT TRASH CUBE!
WHUMP!
THERE WE GO!
SCHLUP
CLUNK!
ERF!
ALL THE BAGS ARE FROZEN TO THE GROUND!
THAT'S 'CUZ THE BLIZZARD STARTED AS AN ICE STORM. EVERYTHING IS COATED IN ICE.
CRUNCH!
GRUNT!
WHOA!
RIIIIIP!
OOF!!
THUD!

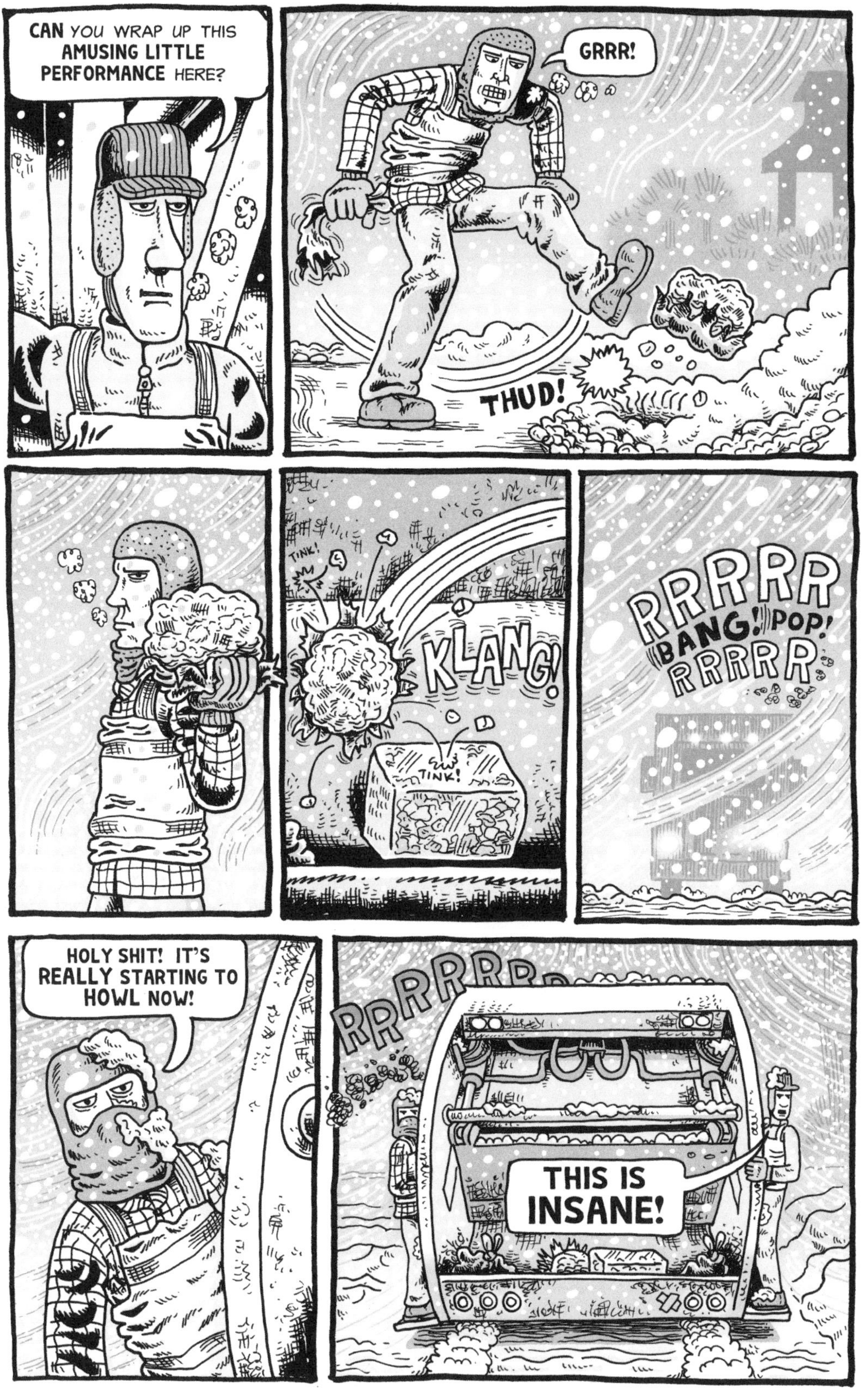
CAN YOU WRAP UP THIS AMUSING LITTLE PERFORMANCE HERE?
GRRR!
THUD!
TINK!
KLANG!
TINK!
RRRRR
BANG! POP!
RRRRR
HOLY SHIT! IT'S REALLY STARTING TO HOWL NOW!
RRRRRR
THIS IS INSANE!

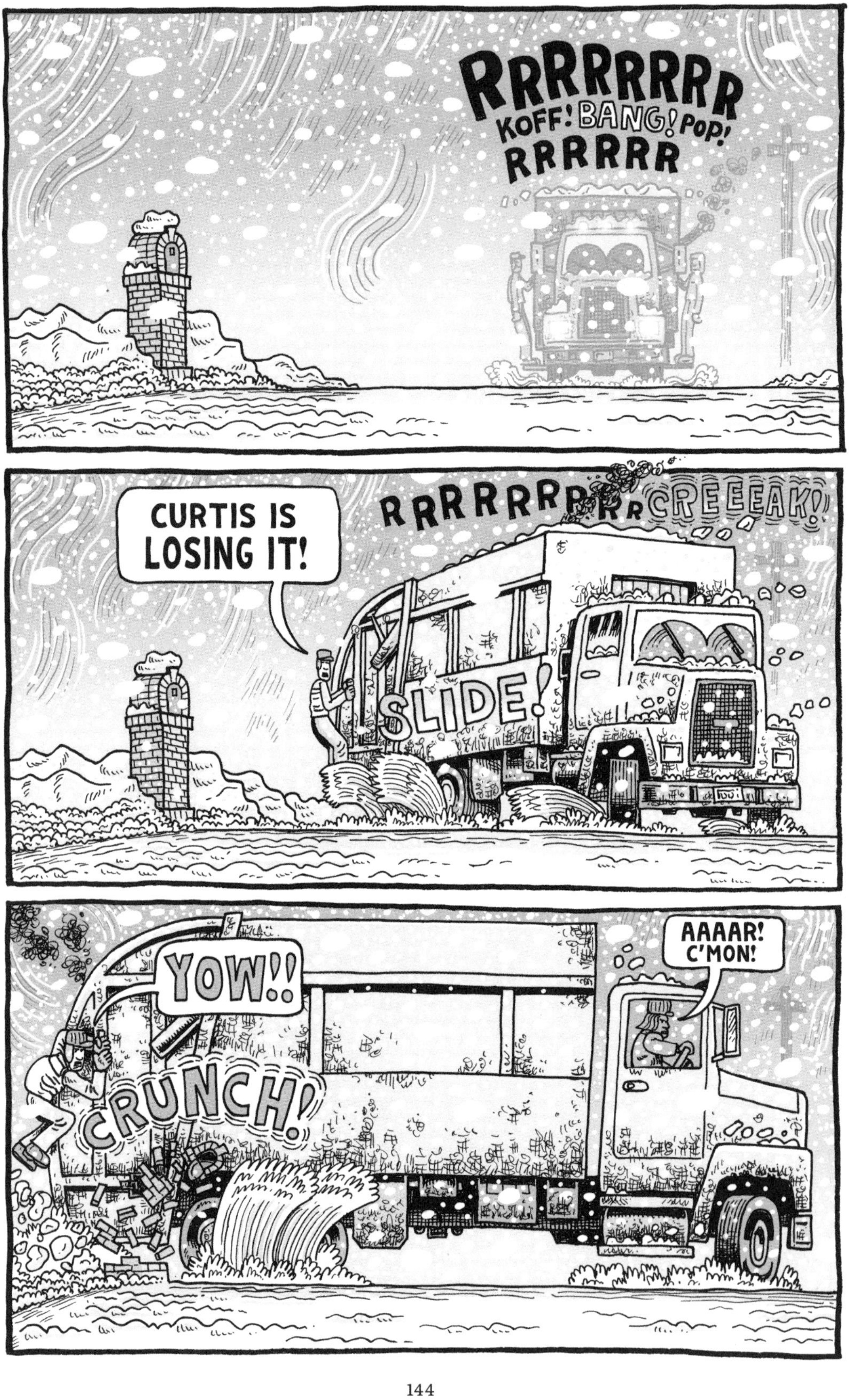
RRRRRRRR
KOFF! BANG! POP!
RRRRRR
RRRRRRRRR
CREEEAK!
CURTIS IS LOSING IT!
SLIDE!
YOW!!
AAAAR! C'MON!
CRUNCH!

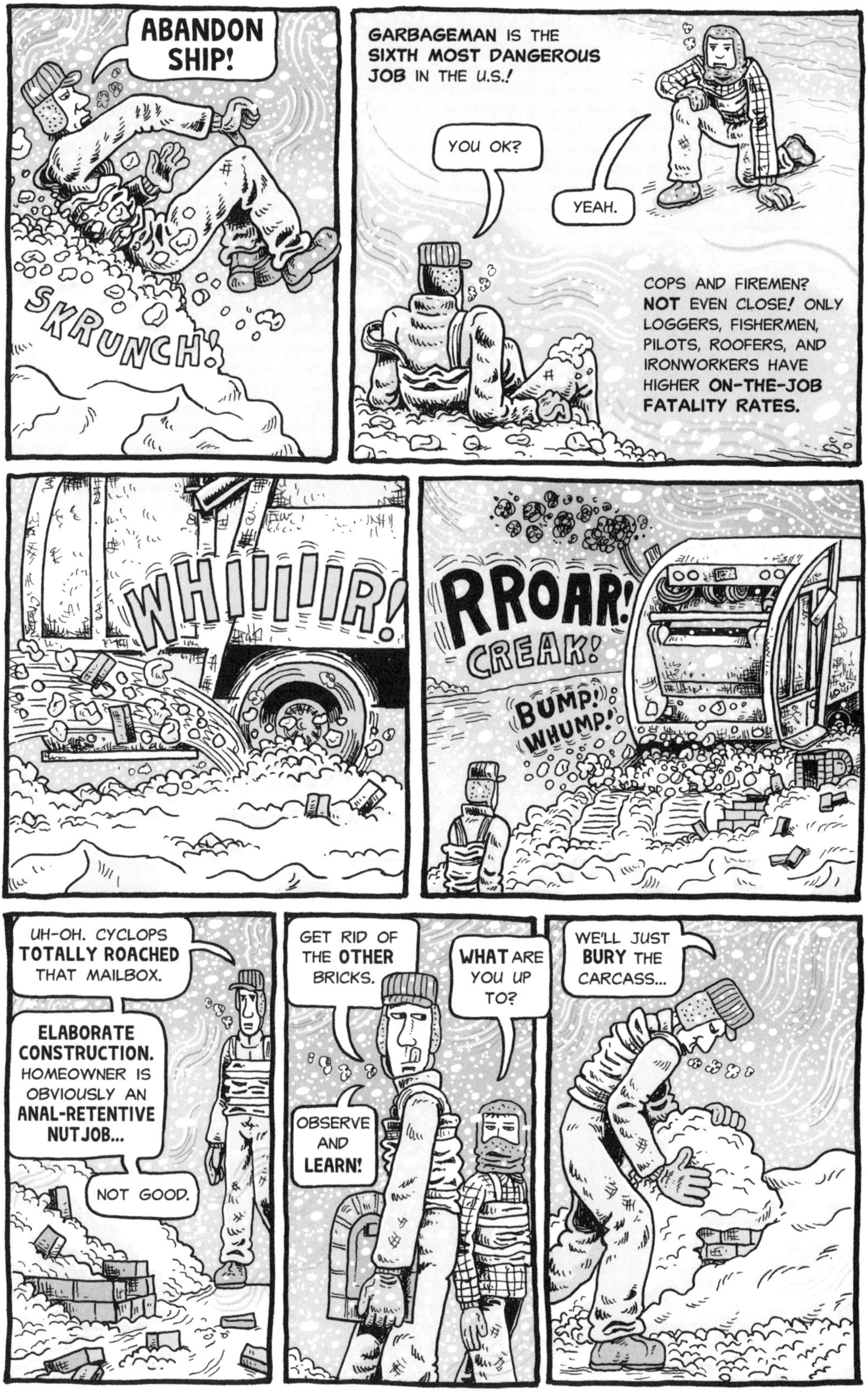

ABANDON SHIP!
SKRUNCH!
GARBAGEMAN IS THE SIXTH MOST DANGEROUS JOB IN THE U.S.!
YOU OK?
YEAH.
COPS AND FIREMEN? NOT EVEN CLOSE! ONLY LOGGERS, FISHERMEN, PILOTS, ROOFERS, AND IRONWORKERS HAVE HIGHER ON-THE-JOB FATALITY RATES.
WHIIIIIR!
RROAR!
CREAK!
BUMP! WHUMP!
UH-OH. CYCLOPS TOTALLY ROACHED THAT MAILBOX.
ELABORATE CONSTRUCTION. HOMEOWNER IS OBVIOUSLY AN ANAL-RETENTIVE NUTJOB...
NOT GOOD.
GET RID OF THE OTHER BRICKS.
WHAT ARE YOU UP TO?
OBSERVE AND LEARN!
WE'LL JUST BURY THE CARCASS...

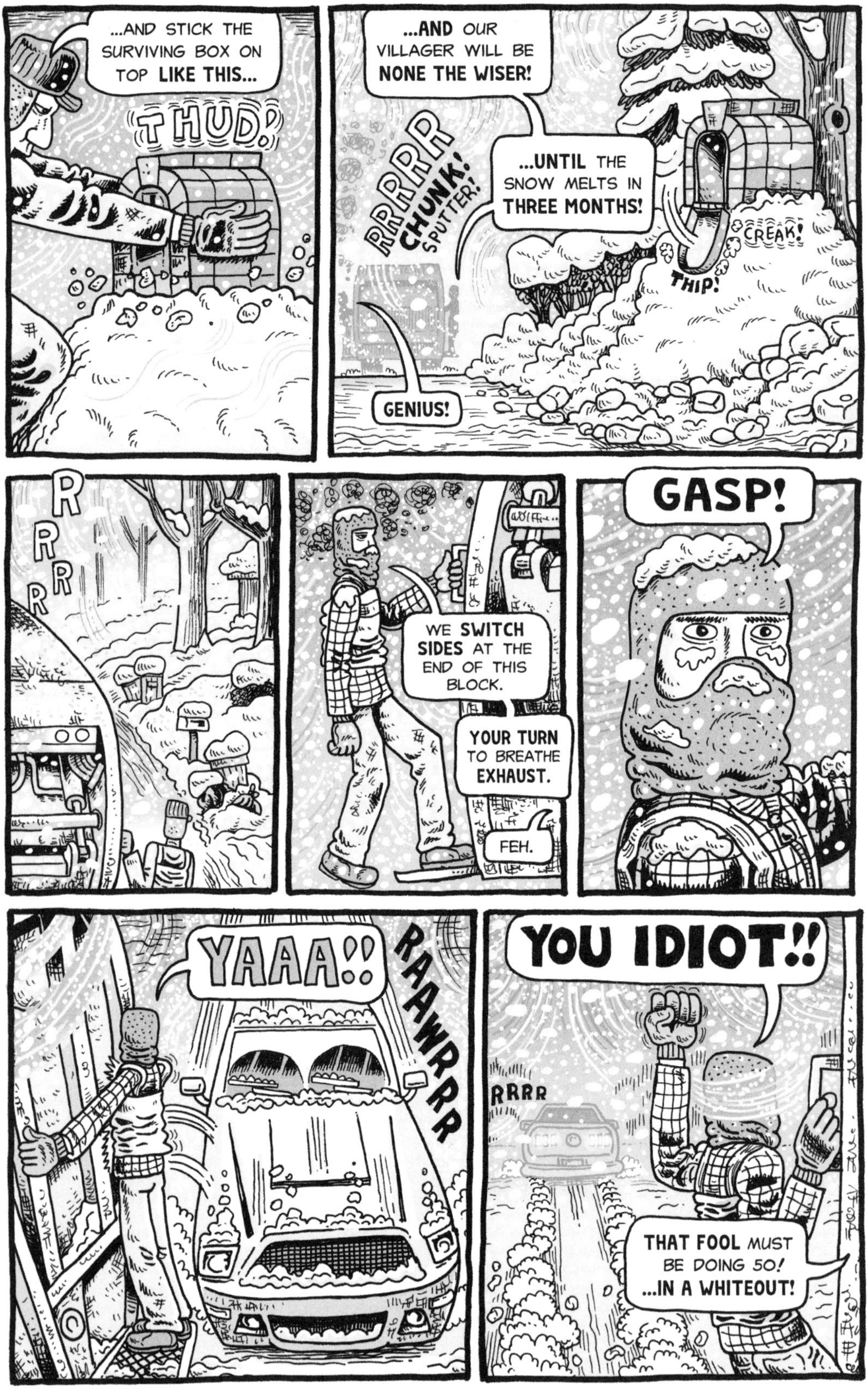
...AND STICK THE SURVIVING BOX ON TOP LIKE THIS...
THUD!
...AND OUR VILLAGER WILL BE NONE THE WISER!
RRRRRR
CHUNK!
SPUTTER!
...UNTIL THE SNOW MELTS IN THREE MONTHS!
CREAK!
THIP!
GENIUS!
RRRR
WE SWITCH SIDES AT THE END OF THIS BLOCK.
YOUR TURN TO BREATHE EXHAUST.
FEH.
GASP!
YAAA!!
RAAWRRR
YOU IDIOT!!
RRRR
THAT FOOL MUST BE DOING 50! ...IN A WHITEOUT!

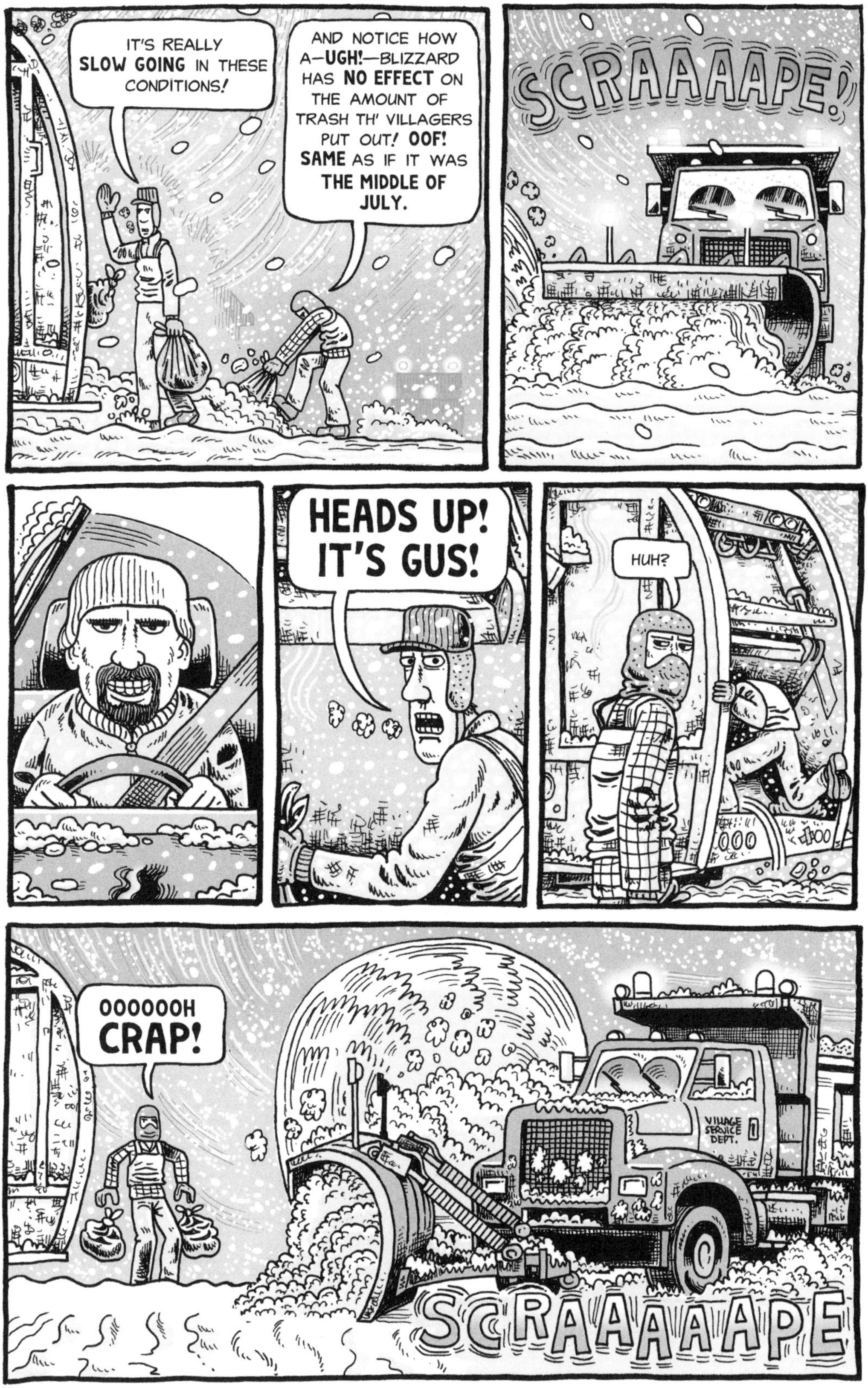
IT'S REALLY SLOW GOING IN THESE CONDITIONS!
AND NOTICE HOW A—UGH!—BLIZZARD HAS NO EFFECT ON THE AMOUNT OF TRASH TH' VILLAGERS PUT OUT! OOF! SAME AS IF IT WAS THE MIDDLE OF JULY.
SCRAAAAPE!
HEADS UP! IT'S GUS!
HUH?
OOOOOOH CRAP!
VILLAGE SERVICE DEPT.
SCRAAAAAPE

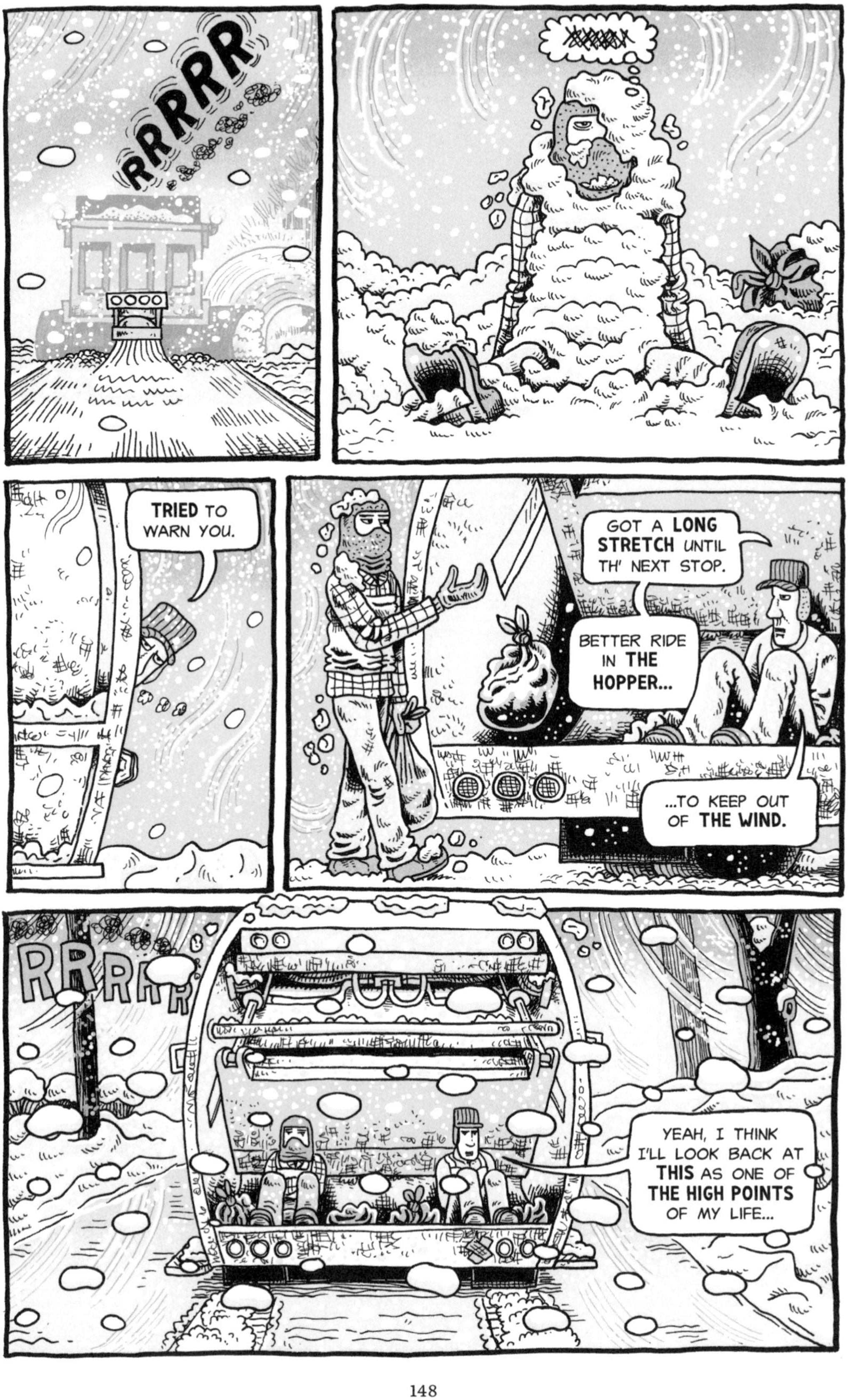
RRRRR
TRIED TO WARN YOU.
GOT A LONG STRETCH UNTIL TH' NEXT STOP.
BETTER RIDE IN THE HOPPER...
...TO KEEP OUT OF THE WIND.
RRRRR
YEAH, I THINK I'LL LOOK BACK AT THIS AS ONE OF THE HIGH POINTS OF MY LIFE...

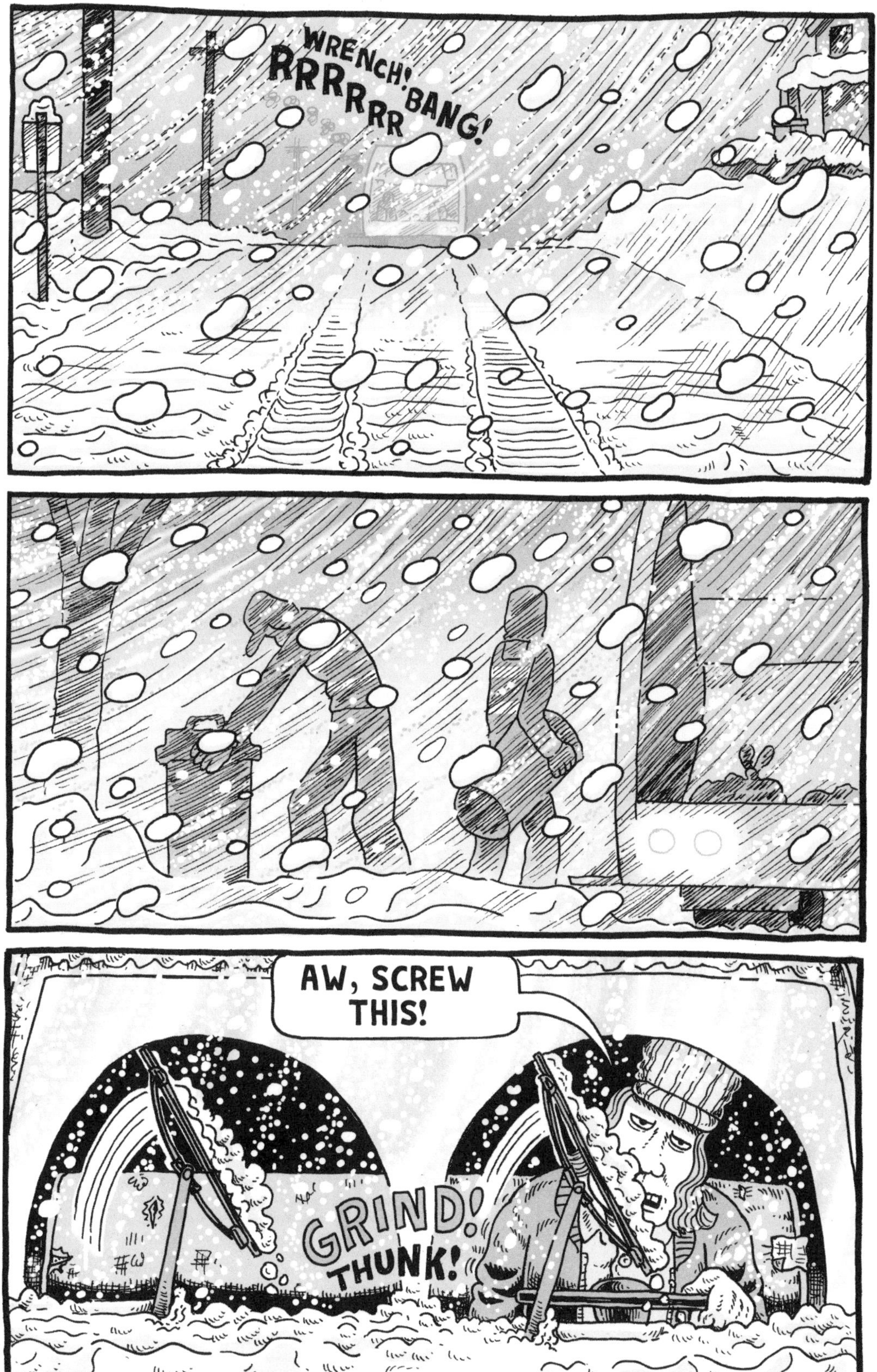
WRENCH!
RRRRRRR
BANG!
AW, SCREW THIS!
GRIND!
THUNK!

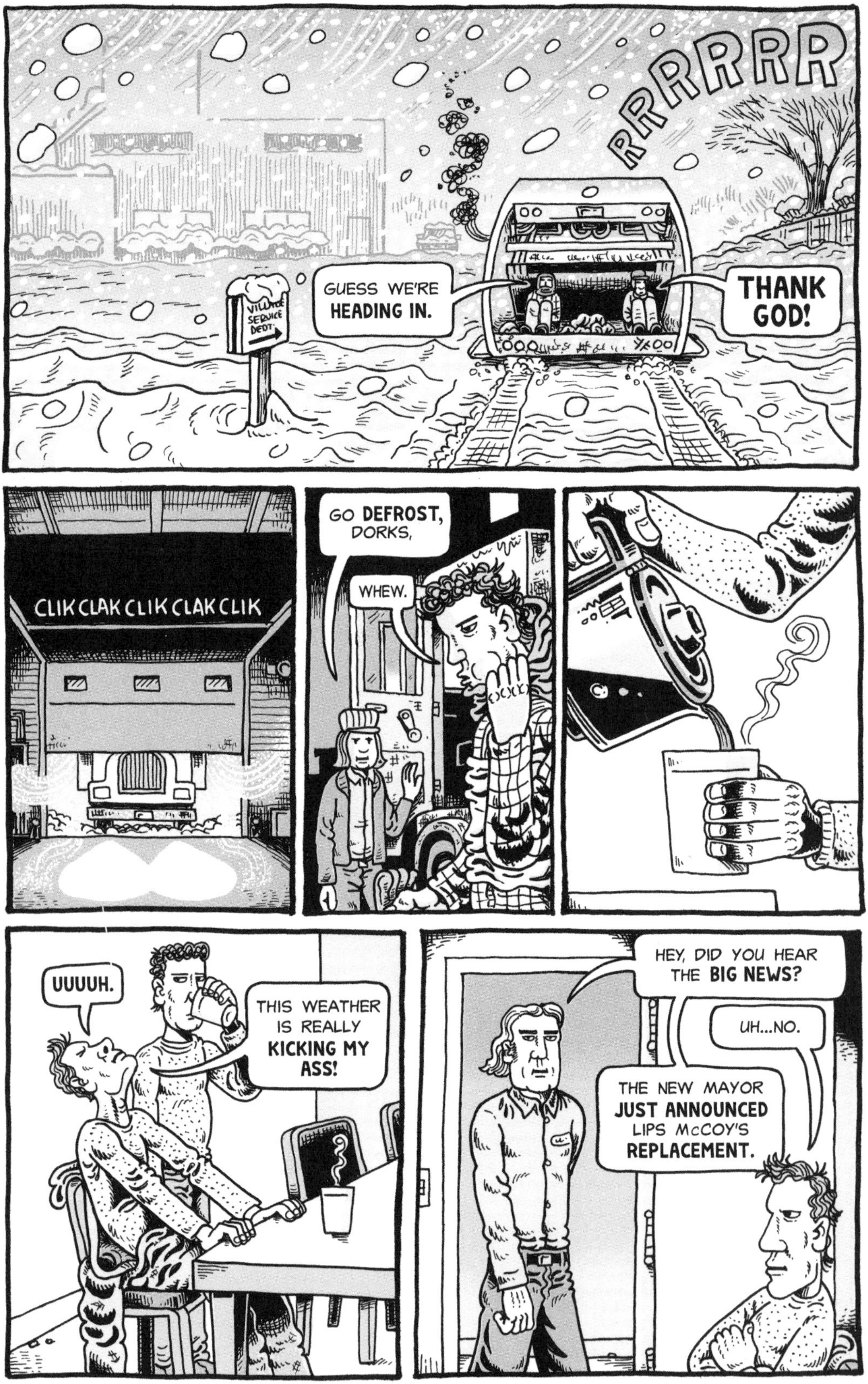
RRRRRR
VILLAGE SERVICE DEPT.
GUESS WE'RE HEADING IN.
THANK GOD!
CLIK CLAK CLIK CLAK CLIK
GO DEFROST, DORKS,
WHEW.
UUUUH.
THIS WEATHER IS REALLY KICKING MY ASS!
HEY, DID YOU HEAR THE BIG NEWS?
UH...NO.
THE NEW MAYOR JUST ANNOUNCED LIPS McCOY'S REPLACEMENT.

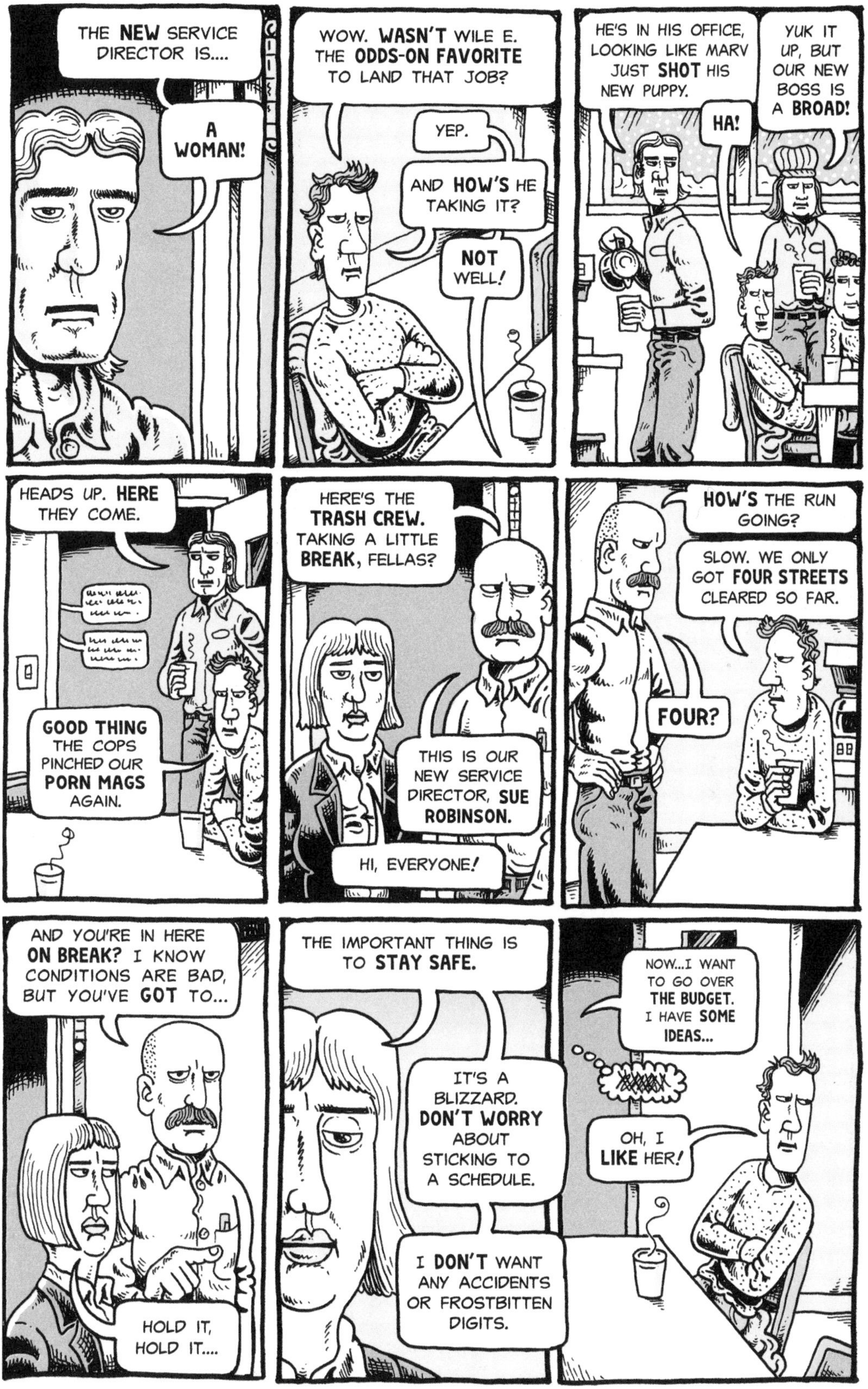
THE NEW SERVICE DIRECTOR IS....
A WOMAN!
WOW. WASN'T WILE E. THE ODDS-ON FAVORITE TO LAND THAT JOB?
YEP.
AND HOW'S HE TAKING IT?
NOT WELL!
HE'S IN HIS OFFICE, LOOKING LIKE MARV JUST SHOT HIS NEW PUPPY.
HA!
YUK IT UP, BUT OUR NEW BOSS IS A BROAD!
HEADS UP. HERE THEY COME.
GOOD THING THE COPS PINCHED OUR PORN MAGS AGAIN.
HERE'S THE TRASH CREW. TAKING A LITTLE BREAK, FELLAS?
THIS IS OUR NEW SERVICE DIRECTOR, SUE ROBINSON.
HI, EVERYONE!
HOW'S THE RUN GOING?
SLOW. WE ONLY GOT FOUR STREETS CLEARED SO FAR.
FOUR?
AND YOU'RE IN HERE ON BREAK? I KNOW CONDITIONS ARE BAD, BUT YOU'VE GOT TO...
HOLD IT, HOLD IT....
THE IMPORTANT THING IS TO STAY SAFE.
IT'S A BLIZZARD. DON'T WORRY ABOUT STICKING TO A SCHEDULE.
I DON'T WANT ANY ACCIDENTS OR FROSTBITTEN DIGITS.
NOW...I WANT TO GO OVER THE BUDGET. I HAVE SOME IDEAS...
OH, I LIKE HER!

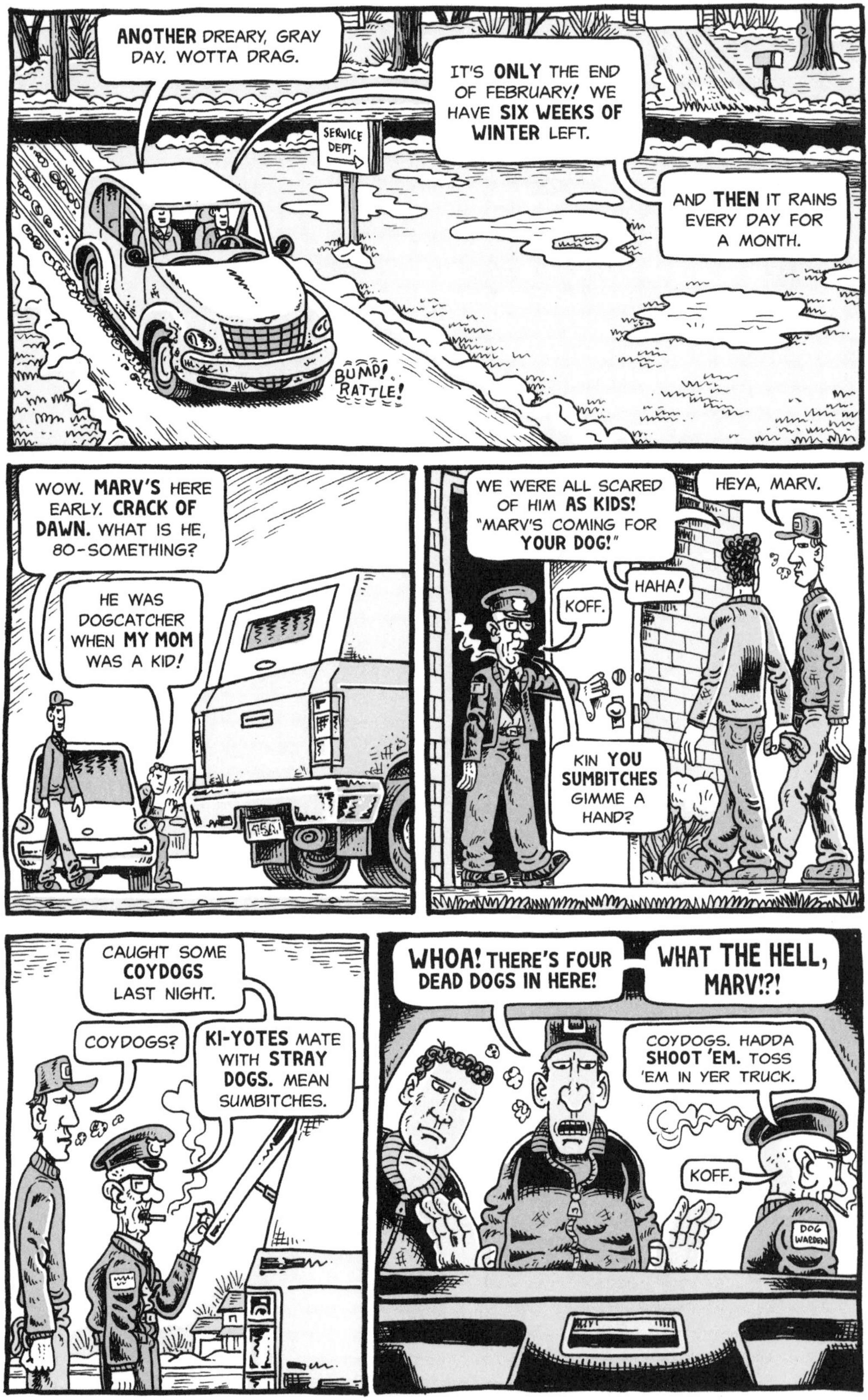
ANOTHER DREARY, GRAY DAY. WOTTA DRAG.
IT'S ONLY THE END OF FEBRUARY! WE HAVE SIX WEEKS OF WINTER LEFT.
AND THEN IT RAINS EVERY DAY FOR A MONTH.
SERVICE DEPT.
BUMP! RATTLE!
WOW. MARV'S HERE EARLY. CRACK OF DAWN. WHAT IS HE, 80-SOMETHING?
HE WAS DOGCATCHER WHEN MY MOM WAS A KID!
WE WERE ALL SCARED OF HIM AS KIDS! "MARV'S COMING FOR YOUR DOG!"
HEYA, MARV.
HAHA!
KOFF.
KIN YOU SUMBITCHES GIMME A HAND?
CAUGHT SOME COYDOGS LAST NIGHT.
COYDOGS?
KI-YOTES MATE WITH STRAY DOGS. MEAN SUMBITCHES.
WHOA! THERE'S FOUR DEAD DOGS IN HERE!
WHAT THE HELL, MARV!?!
COYDOGS. HADDA SHOOT 'EM. TOSS 'EM IN YER TRUCK.
KOFF.
DOG WARDEN

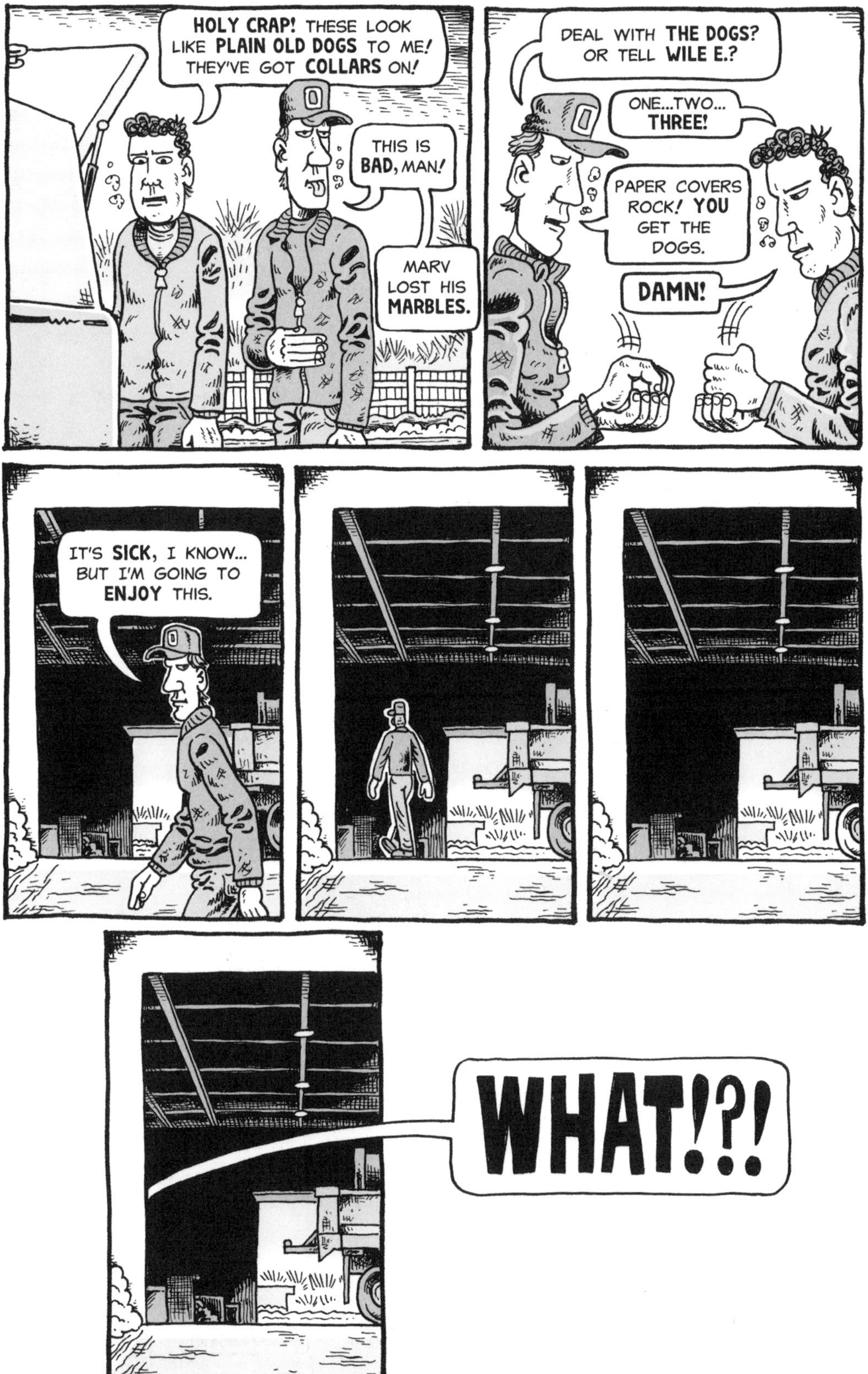
HOLY CRAP! THESE LOOK LIKE PLAIN OLD DOGS TO ME! THEY'VE GOT COLLARS ON!
THIS IS BAD, MAN!
MARV LOST HIS MARBLES.
DEAL WITH THE DOGS? OR TELL WILE E.?
ONE...TWO... THREE!
PAPER COVERS ROCK! YOU GET THE DOGS.
DAMN!
IT'S SICK, I KNOW... BUT I'M GOING TO ENJOY THIS.
WHAT!?!

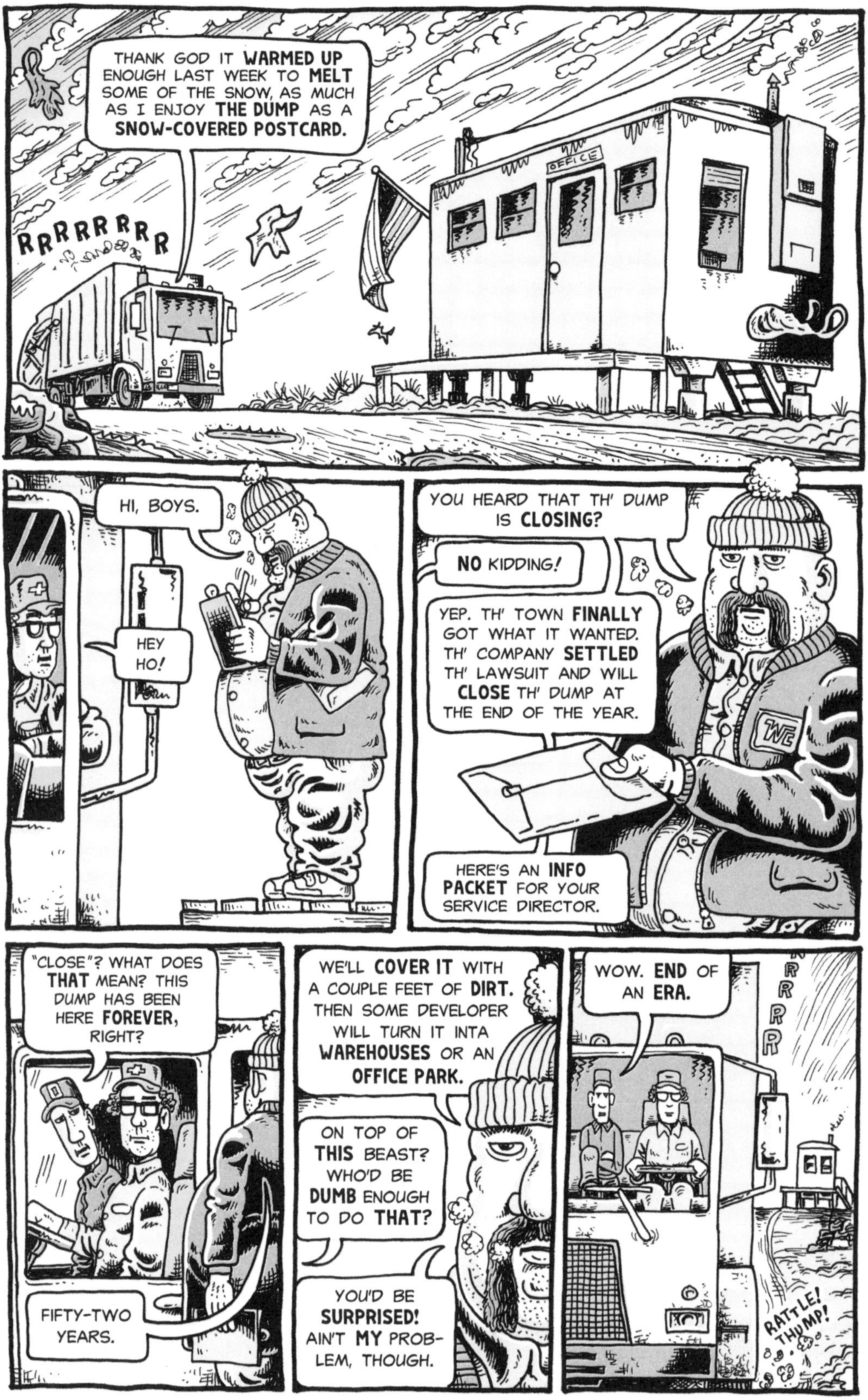
THANK GOD IT WARMED UP ENOUGH LAST WEEK TO MELT SOME OF THE SNOW, AS MUCH AS I ENJOY THE DUMP AS A SNOW-COVERED POSTCARD.
RRRRRRRRR
OFFICE
HI, BOYS.
HEY HO!
YOU HEARD THAT TH' DUMP IS CLOSING?
NO KIDDING!
YEP. TH' TOWN FINALLY GOT WHAT IT WANTED. TH' COMPANY SETTLED TH' LAWSUIT AND WILL CLOSE TH' DUMP AT THE END OF THE YEAR.
HERE'S AN INFO PACKET FOR YOUR SERVICE DIRECTOR.
"CLOSE"? WHAT DOES THAT MEAN? THIS DUMP HAS BEEN HERE FOREVER, RIGHT?
FIFTY-TWO YEARS.
WE'LL COVER IT WITH A COUPLE FEET OF DIRT. THEN SOME DEVELOPER WILL TURN IT INTA WAREHOUSES OR AN OFFICE PARK.
ON TOP OF THIS BEAST? WHO'D BE DUMB ENOUGH TO DO THAT?
YOU'D BE SURPRISED! AIN'T MY PROB-LEM, THOUGH.
WOW. END OF AN ERA.
RRRR
RATTLE! THUMP!

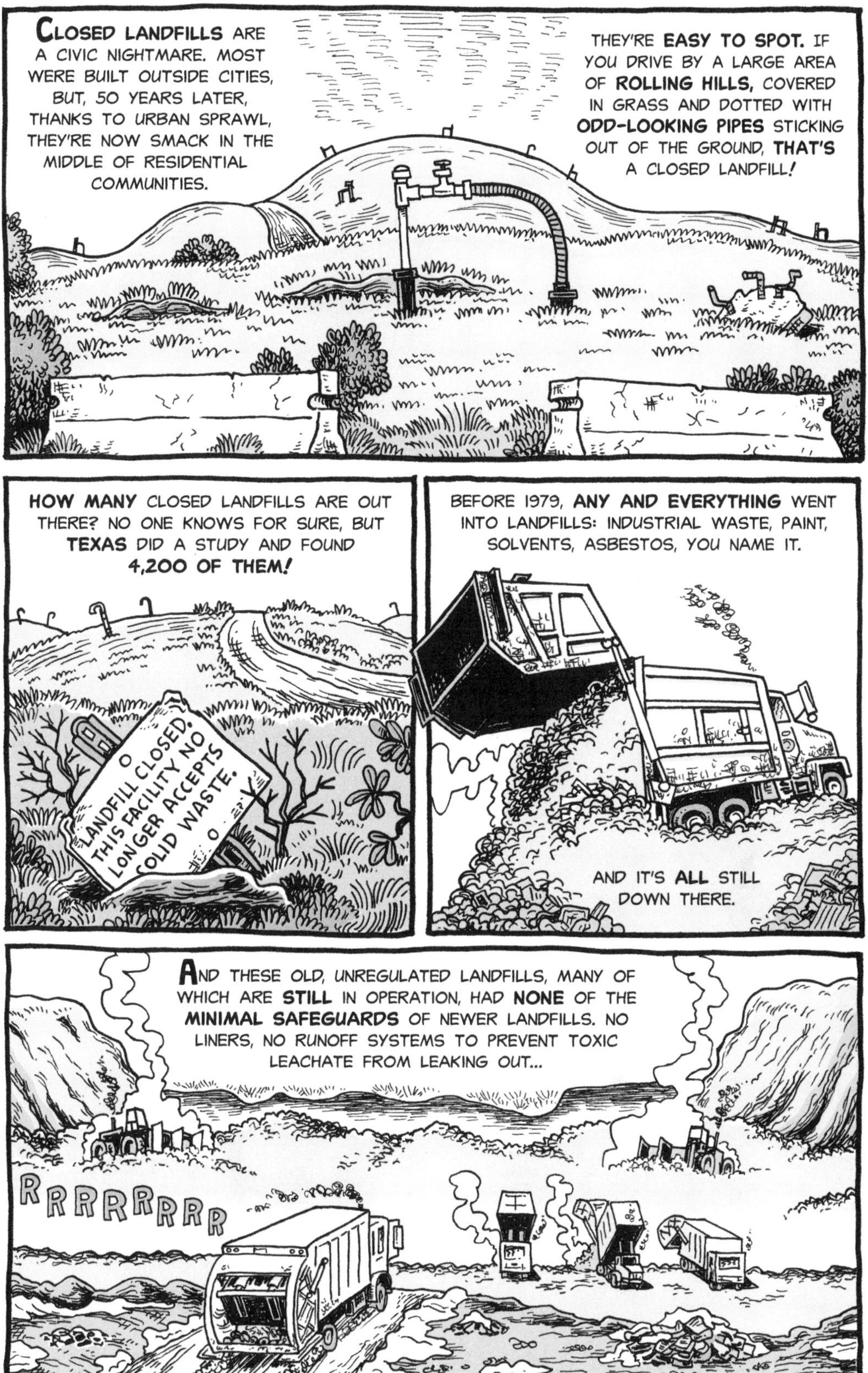
CLOSED LANDFILLS ARE A CIVIC NIGHTMARE. MOST WERE BUILT OUTSIDE CITIES, BUT, 50 YEARS LATER, THANKS TO URBAN SPRAWL, THEY'RE NOW SMACK IN THE MIDDLE OF RESIDENTIAL COMMUNITIES.
THEY'RE **EASY TO SPOT.** IF YOU DRIVE BY A LARGE AREA OF **ROLLING HILLS,** COVERED IN GRASS AND DOTTED WITH **ODD-LOOKING PIPES** STICKING OUT OF THE GROUND, **THAT'S** A CLOSED LANDFILL!
HOW MANY CLOSED LANDFILLS ARE OUT THERE? NO ONE KNOWS FOR SURE, BUT **TEXAS** DID A STUDY AND FOUND **4,200 OF THEM!**
LANDFILL CLOSED. THIS FACILITY NO LONGER ACCEPTS SOLID WASTE.
BEFORE 1979, **ANY AND EVERYTHING** WENT INTO LANDFILLS: INDUSTRIAL WASTE, PAINT, SOLVENTS, ASBESTOS, YOU NAME IT.
AND IT'S **ALL** STILL DOWN THERE.
AND THESE OLD, UNREGULATED LANDFILLS, MANY OF WHICH ARE **STILL IN OPERATION, HAD **NONE** OF THE **MINIMAL SAFEGUARDS** OF NEWER LANDFILLS. NO LINERS, NO RUNOFF SYSTEMS TO PREVENT TOXIC LEACHATE FROM LEAKING OUT...
RRRRRRRRR

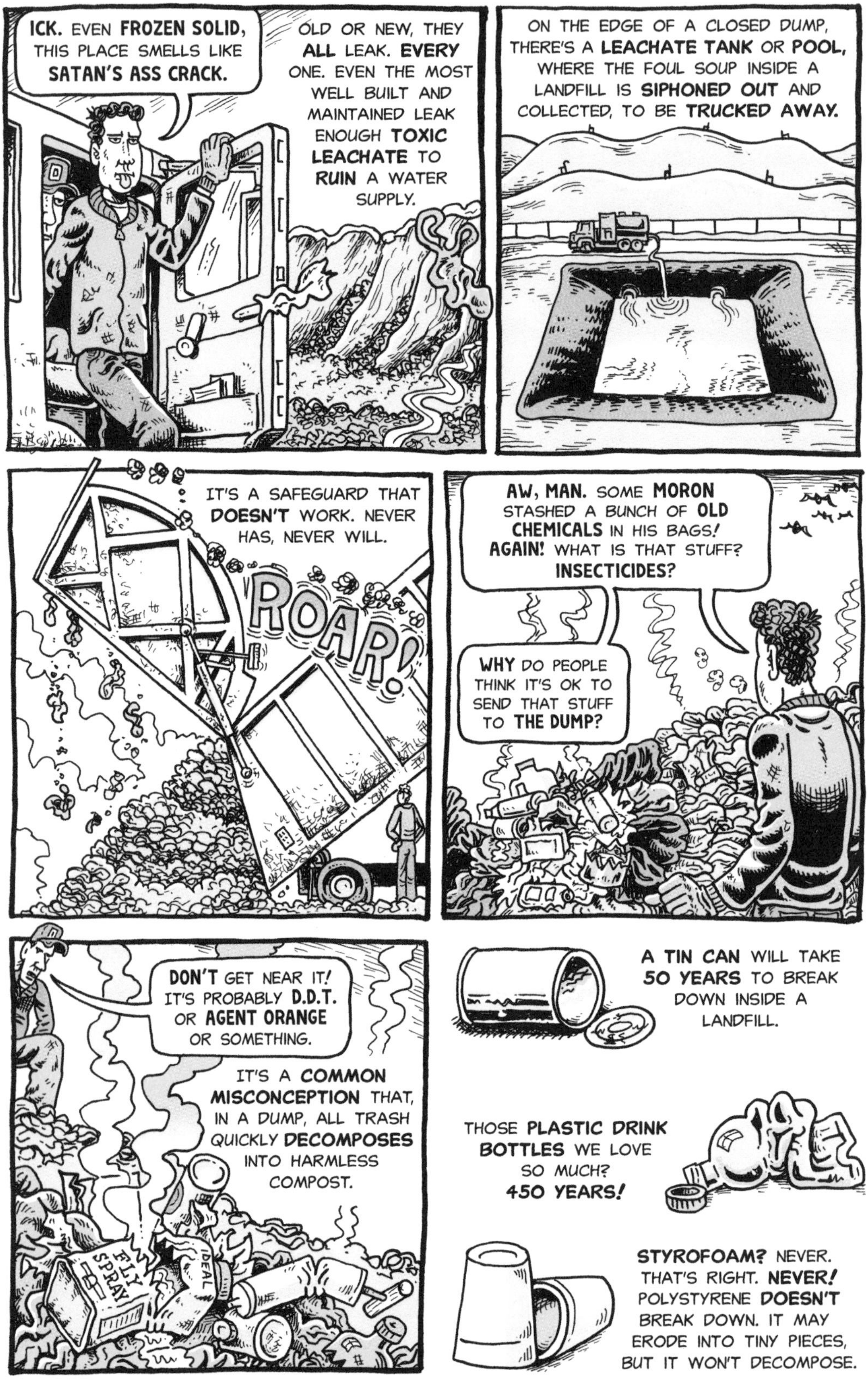

ICK. EVEN FROZEN SOLID, THIS PLACE SMELLS LIKE SATAN'S ASS CRACK.
OLD OR NEW, THEY ALL LEAK. EVERY ONE. EVEN THE MOST WELL BUILT AND MAINTAINED LEAK ENOUGH TOXIC LEACHATE TO RUIN A WATER SUPPLY.
ON THE EDGE OF A CLOSED DUMP, THERE'S A LEACHATE TANK OR POOL, WHERE THE FOUL SOUP INSIDE A LANDFILL IS SIPHONED OUT AND COLLECTED, TO BE TRUCKED AWAY.
IT'S A SAFEGUARD THAT DOESN'T WORK. NEVER HAS, NEVER WILL.
ROAR!
AW, MAN. SOME MORON STASHED A BUNCH OF OLD CHEMICALS IN HIS BAGS! AGAIN! WHAT IS THAT STUFF? INSECTICIDES?
WHY DO PEOPLE THINK IT'S OK TO SEND THAT STUFF TO THE DUMP?
DON'T GET NEAR IT! IT'S PROBABLY D.D.T. OR AGENT ORANGE OR SOMETHING.
IT'S A COMMON MISCONCEPTION THAT, IN A DUMP, ALL TRASH QUICKLY DECOMPOSES INTO HARMLESS COMPOST.
FLY SPRAY
IDEAL
A TIN CAN WILL TAKE 50 YEARS TO BREAK DOWN INSIDE A LANDFILL.
THOSE PLASTIC DRINK BOTTLES WE LOVE SO MUCH? 450 YEARS!
STYROFOAM? NEVER. THAT'S RIGHT. NEVER! POLYSTYRENE DOESN'T BREAK DOWN. IT MAY ERODE INTO TINY PIECES, BUT IT WON'T DECOMPOSE.

THEN THERE'S THE **TOXIC SUBSTANCES.** THOSE WILL REMAIN INSIDE A LANDFILL UNTIL THEY'RE **REMOVED,** OR UNTIL THEY **ESCAPE** INTO THE ENVIRONMENT.

KICK **SOME** PAMPERS ON IT TO **SOAK IT UP!**

THE MOST DEADLY ARE **MERCURY** AND **VINYL CHLORIDE.** MERCURY COMES FROM BATTERIES, ELECTRONICS AND THOSE FLOURESCENT LIGHTBULBS WE USE TO BE MORE GREEN. MERCURY BECOMES **MORE** TOXIC WHEN BURIED.

HERE'S **ANOTHER** BAG WITH CANS OF **SOLVENT!**

PROBABLY THE **SAME** GUY...

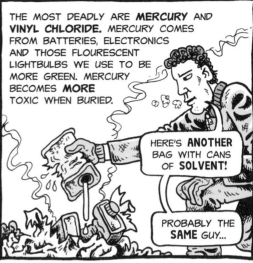

VINYL CHLORIDE IS A GAS THAT FORMS WHEN MANUFACTURED PRODUCTS BREAK DOWN, SUCH AS INDUSTRIAL DE-GREASERS, P.V.C, UPHOLSTERY, AND EVEN KITCHEN WARE...

IT'S HIGHLY **FLAMMABLE** AND EXTREMELY **CARCINOGENIC.**

PEOPLE ARE IDIOTS!

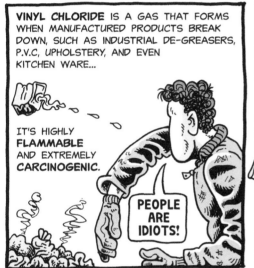

AND THAT'S **ONLY TWO** OF THE MANY TOXIC KILLERS THAT FESTER DEEP INSIDE THOSE MOUNTAINS OF MUMMIFIED GARBAGE.

PING!

BANG!

OLD LANDFILLS ARE ECOLOGICAL TIME BOMBS...**LITERALLY.** BECAUSE, BEYOND THE TOXIC SOUP THAT LEAKS OUT THE SIDES AND BOTTOM OF A DUMP, THE BIGGEST DANGER IS **EXPLOSIVE GAS.** AS GARBAGE DECOMPOSES, IT PRODUCES LARGE UNDERGROUND POCKETS OF CARBONDIOXIDE AND METHANE GASES.

THAT'S WHAT THE PIPES ARE FOR.

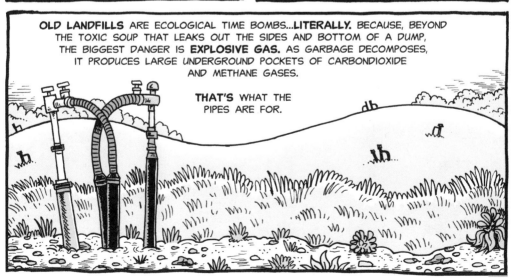

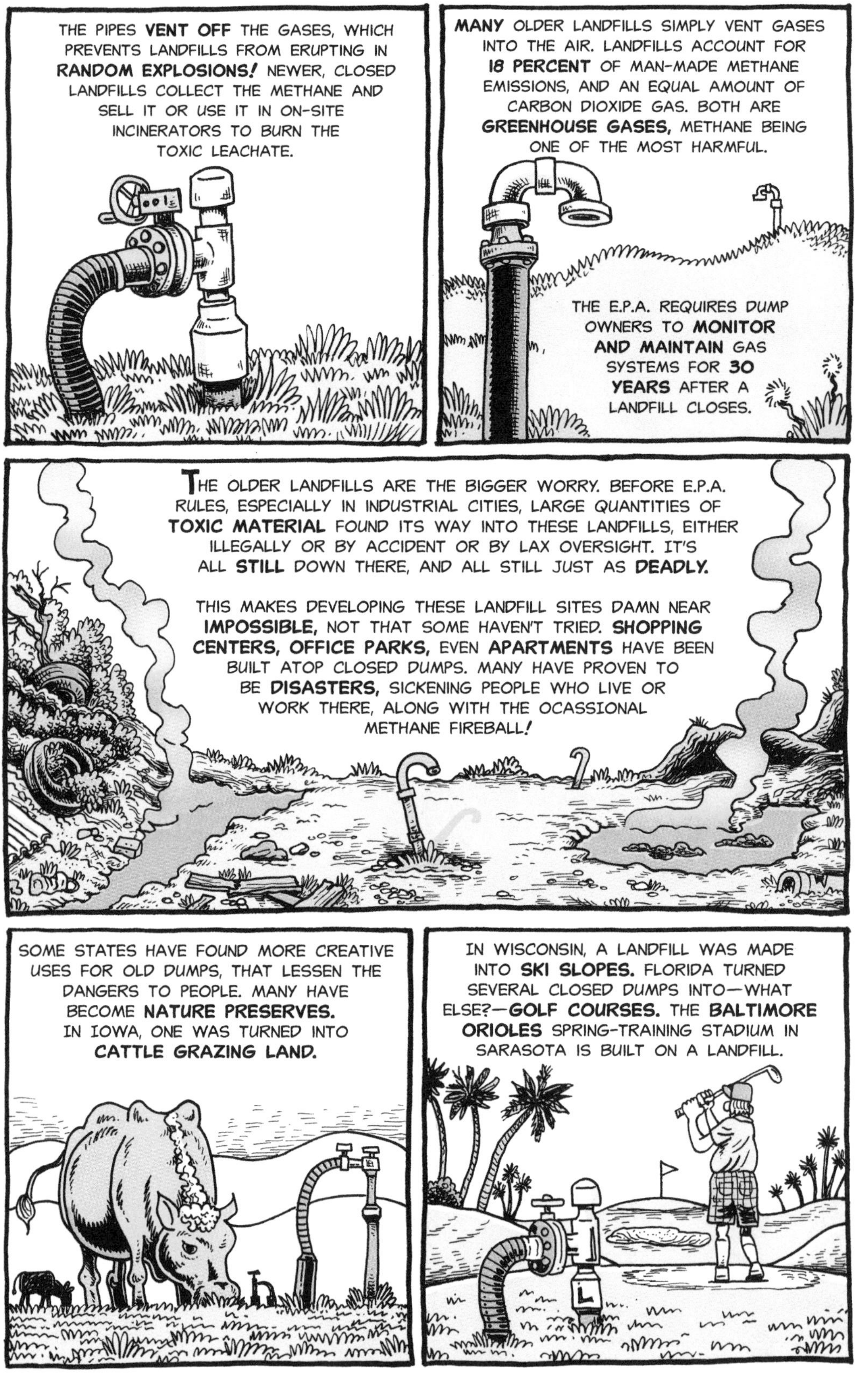
THE PIPES VENT OFF THE GASES, WHICH PREVENTS LANDFILLS FROM ERUPTING IN RANDOM EXPLOSIONS! NEWER, CLOSED LANDFILLS COLLECT THE METHANE AND SELL IT OR USE IT IN ON-SITE INCINERATORS TO BURN THE TOXIC LEACHATE.
MANY OLDER LANDFILLS SIMPLY VENT GASES INTO THE AIR. LANDFILLS ACCOUNT FOR 18 PERCENT OF MAN-MADE METHANE EMISSIONS, AND AN EQUAL AMOUNT OF CARBON DIOXIDE GAS. BOTH ARE GREENHOUSE GASES, METHANE BEING ONE OF THE MOST HARMFUL.
THE E.P.A. REQUIRES DUMP OWNERS TO MONITOR AND MAINTAIN GAS SYSTEMS FOR 30 YEARS AFTER A LANDFILL CLOSES.
THE OLDER LANDFILLS ARE THE BIGGER WORRY. BEFORE E.P.A. RULES, ESPECIALLY IN INDUSTRIAL CITIES, LARGE QUANTITIES OF TOXIC MATERIAL FOUND ITS WAY INTO THESE LANDFILLS, EITHER ILLEGALLY OR BY ACCIDENT OR BY LAX OVERSIGHT. IT'S ALL STILL DOWN THERE, AND ALL STILL JUST AS DEADLY.
THIS MAKES DEVELOPING THESE LANDFILL SITES DAMN NEAR IMPOSSIBLE, NOT THAT SOME HAVEN'T TRIED. SHOPPING CENTERS, OFFICE PARKS, EVEN APARTMENTS HAVE BEEN BUILT ATOP CLOSED DUMPS. MANY HAVE PROVEN TO BE DISASTERS, SICKENING PEOPLE WHO LIVE OR WORK THERE, ALONG WITH THE OCASSIONAL METHANE FIREBALL!
SOME STATES HAVE FOUND MORE CREATIVE USES FOR OLD DUMPS, THAT LESSEN THE DANGERS TO PEOPLE. MANY HAVE BECOME NATURE PRESERVES. IN IOWA, ONE WAS TURNED INTO CATTLE GRAZING LAND.
IN WISCONSIN, A LANDFILL WAS MADE INTO SKI SLOPES. FLORIDA TURNED SEVERAL CLOSED DUMPS INTO—WHAT ELSE?—GOLF COURSES. THE BALTIMORE ORIOLES SPRING-TRAINING STADIUM IN SARASOTA IS BUILT ON A LANDFILL.

LOS ANGELES PLANS TO CONVERT THE JUST-CLOSED **PUENTE HILLS LANDFILL** INTO PARKS AND NATURE PRESERVES. THIS BEHEMOTH OPENED IN 1957 AND, UNTIL RECENTLY, WAS THE **BIGGEST LANDFILL IN THE U.S.** IT'S MORE THAN **500 FEET DEEP** IN PLACES AND COVERS AN AREA THE SIZE OF **CENTRAL PARK.** IT'S SO VAST, IT HAS ITS OWN **MICRO WEATHER SYSTEM!** AND SINCE IT OPERATED **FOR DECADES** BEFORE E.P.A. REGULATIONS, GOD ONLY KNOWS WHAT'S **BURIED** DEEP IN ITS BOWELS. YEAH, **GOOD LUCK** WITH **THAT** PARK.

SEVERAL STATES, SUCH AS MASSACHUSETTS, HAVE PLACED **HUGE SOLAR FARMS** ON TOP OF CLOSED DUMPS. IT'S ALMOST TRITE, RIGHT? TURNING **AN ECOLOGICAL MENACE** INTO A **GREEN ENERGY PLANT!** MAKES PERFECT SENSE, THOUGH. A SOLAR FARM NEEDS A LARGE OPEN AREA AND REQUIRES LITTLE HUMAN PRESENCE.

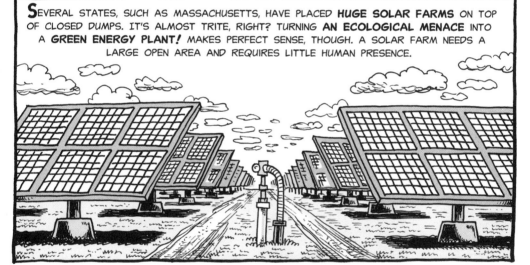

THAT'S IT... **QUITTIN' TIME!**

MAGEE AND I ARE GETTING **PIZZA AND BEER** AT THE PUB.

AGAIN? MAYBE I'LL STOP BY **LATER.**

RRRR

YEAH, I'M GONNA **MISS** THIS PLACE!

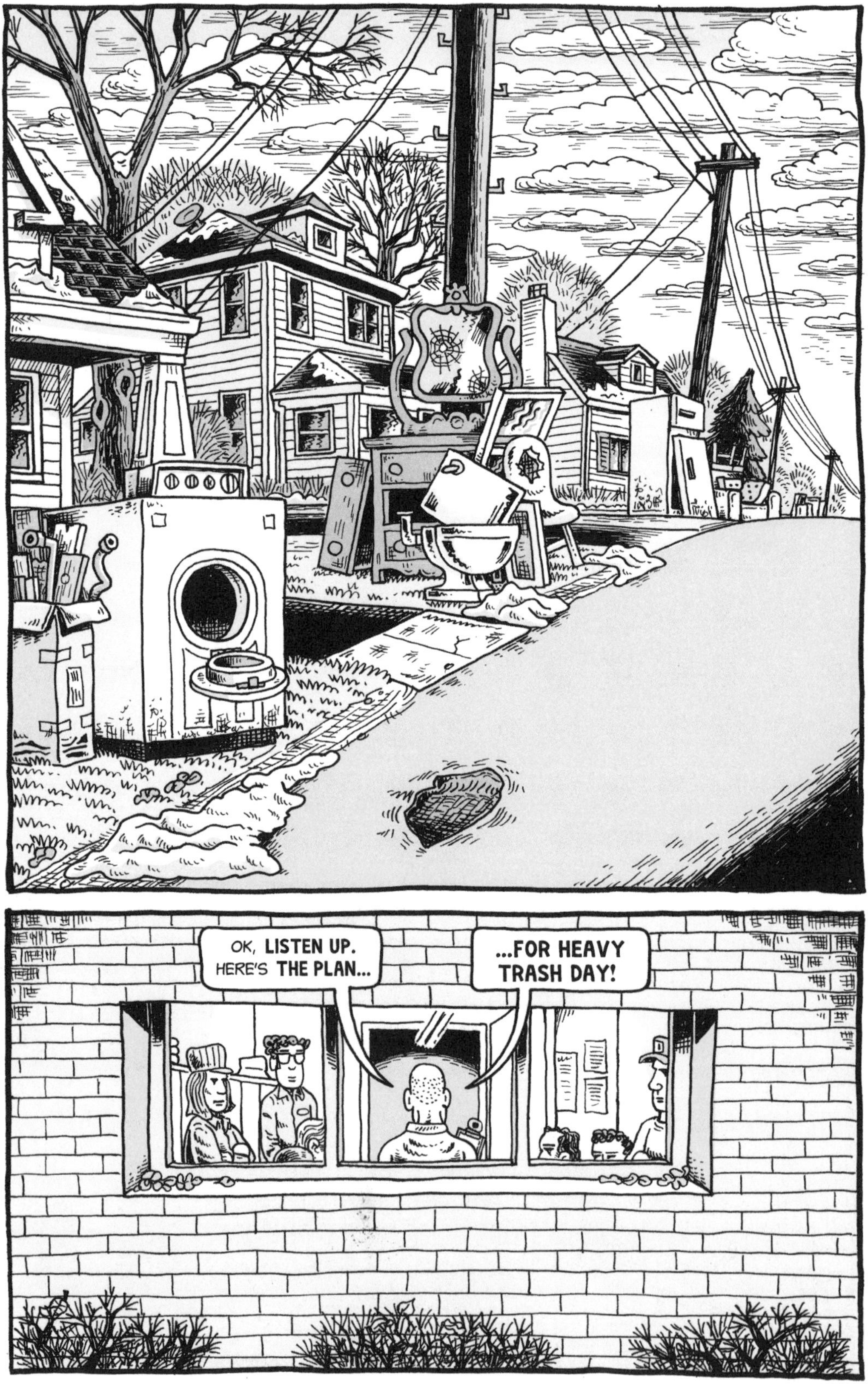
OK, LISTEN UP. HERE'S THE PLAN...
...FOR HEAVY TRASH DAY!

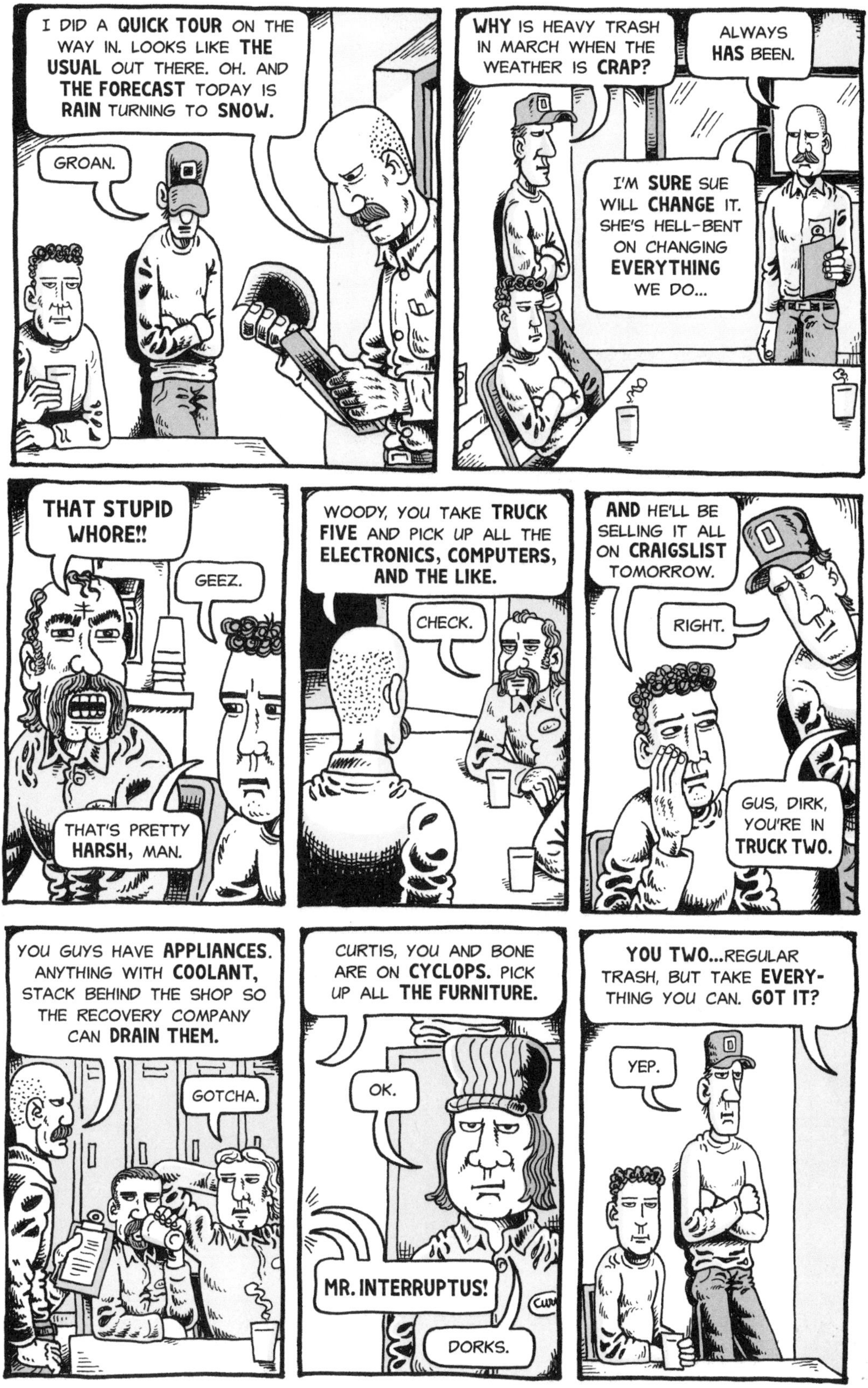
I DID A QUICK TOUR ON THE WAY IN. LOOKS LIKE THE USUAL OUT THERE. OH. AND THE FORECAST TODAY IS RAIN TURNING TO SNOW.
GROAN.
WHY IS HEAVY TRASH IN MARCH WHEN THE WEATHER IS CRAP?
ALWAYS HAS BEEN.
I'M SURE SUE WILL CHANGE IT. SHE'S HELL-BENT ON CHANGING EVERYTHING WE DO...
THAT STUPID WHORE!!
GEEZ.
THAT'S PRETTY HARSH, MAN.
WOODY, YOU TAKE TRUCK FIVE AND PICK UP ALL THE ELECTRONICS, COMPUTERS, AND THE LIKE.
CHECK.
AND HE'LL BE SELLING IT ALL ON CRAIGSLIST TOMORROW.
RIGHT.
GUS, DIRK, YOU'RE IN TRUCK TWO.
YOU GUYS HAVE APPLIANCES. ANYTHING WITH COOLANT, STACK BEHIND THE SHOP SO THE RECOVERY COMPANY CAN DRAIN THEM.
GOTCHA.
CURTIS, YOU AND BONE ARE ON CYCLOPS. PICK UP ALL THE FURNITURE.
OK.
MR. INTERRUPTUS!
DORKS.
YOU TWO...REGULAR TRASH, BUT TAKE EVERYTHING YOU CAN. GOT IT?
YEP.

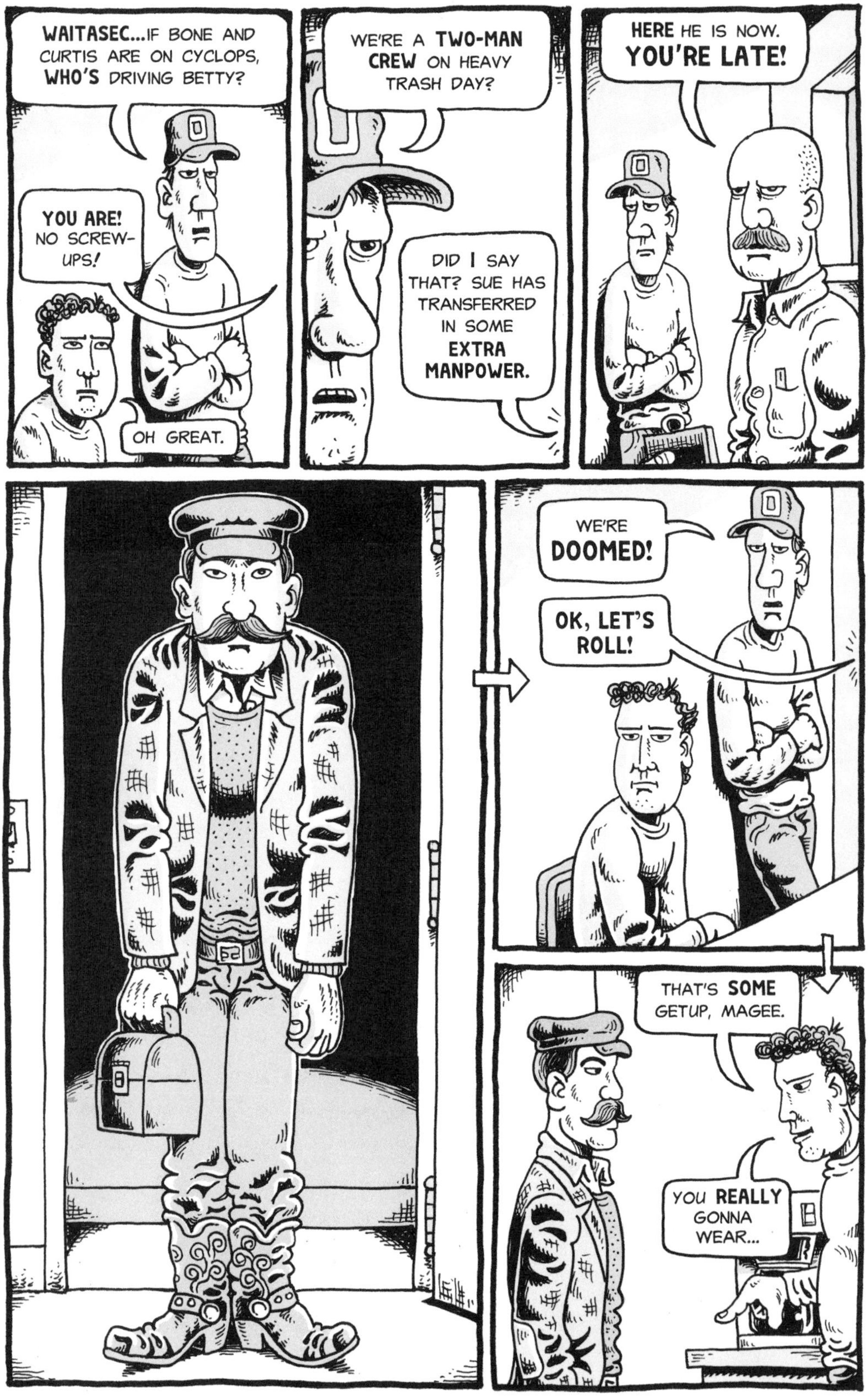
WAITASEC...IF BONE AND CURTIS ARE ON CYCLOPS, WHO'S DRIVING BETTY?
YOU ARE! NO SCREW-UPS!
OH GREAT.
WE'RE A TWO-MAN CREW ON HEAVY TRASH DAY?
DID I SAY THAT? SUE HAS TRANSFERRED IN SOME EXTRA MANPOWER.
HERE HE IS NOW. YOU'RE LATE!
WE'RE DOOMED!
OK, LET'S ROLL!
THAT'S SOME GETUP, MAGEE.
YOU REALLY GONNA WEAR...

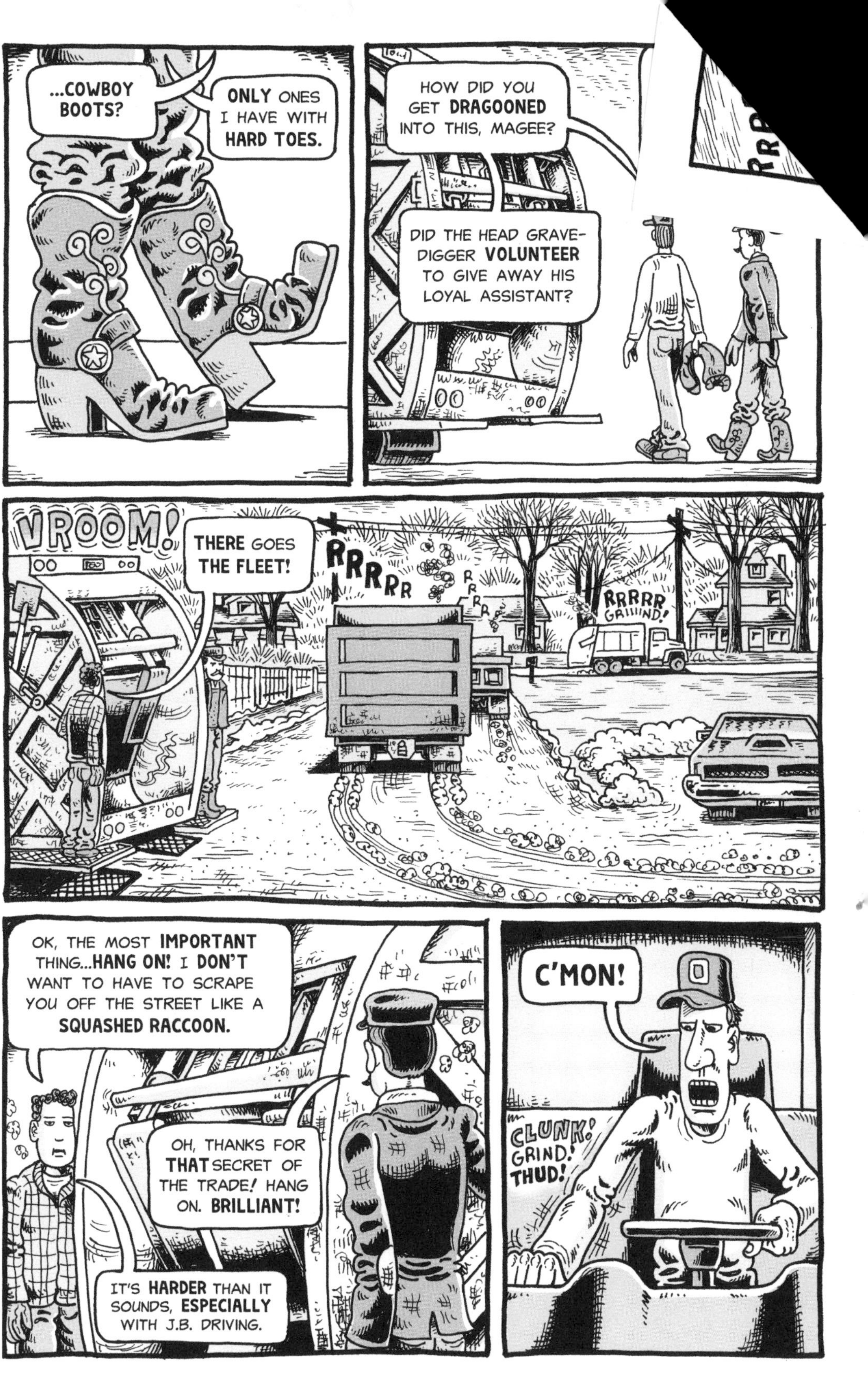
...COWBOY BOOTS?
ONLY ONES I HAVE WITH HARD TOES.
HOW DID YOU GET DRAGOONED INTO THIS, MAGEE?
DID THE HEAD GRAVE-DIGGER VOLUNTEER TO GIVE AWAY HIS LOYAL ASSISTANT?
VROOM!
THERE GOES THE FLEET!
RRRRR
RRRRR
GRIIIND!
OK, THE MOST IMPORTANT THING...HANG ON! I DON'T WANT TO HAVE TO SCRAPE YOU OFF THE STREET LIKE A SQUASHED RACCOON.
OH, THANKS FOR THAT SECRET OF THE TRADE! HANG ON. BRILLIANT!
IT'S HARDER THAN IT SOUNDS, ESPECIALLY WITH J.B. DRIVING.
C'MON!
CLUNK! GRIND! THUD!

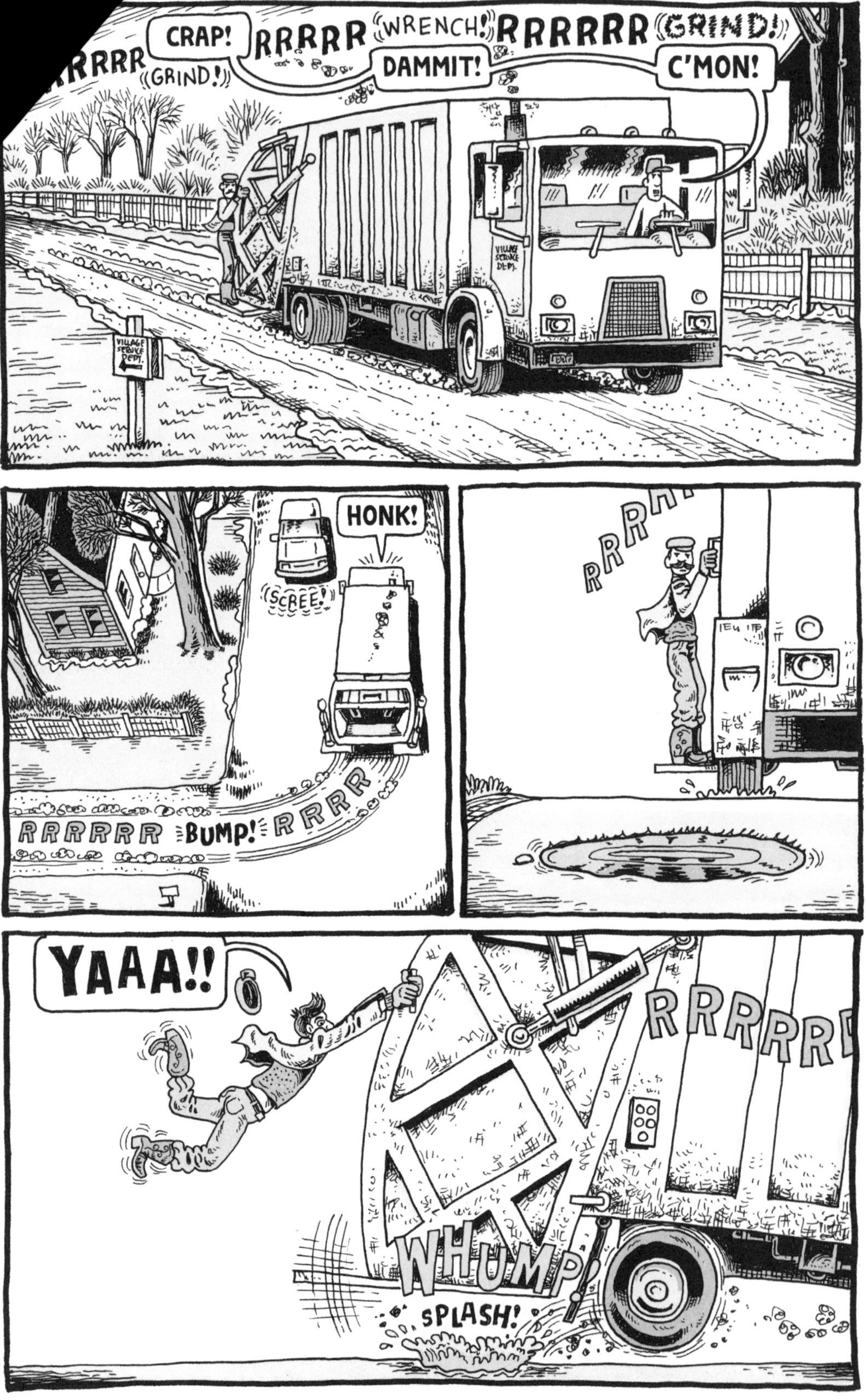
RRRR
GRIND!
CRAP!
RRRRR
WRENCH!
DAMMIT!
RRRRRR
GRIND!
C'MON!
VILLAGE SERVICE DEPT.
VILLAGE SERVICE DEPT.
HONK!
SCREE!
RRRRRR
BUMP!
RRRRR
YAAA!!
RRRRRR
WHUMP!
SPLASH!

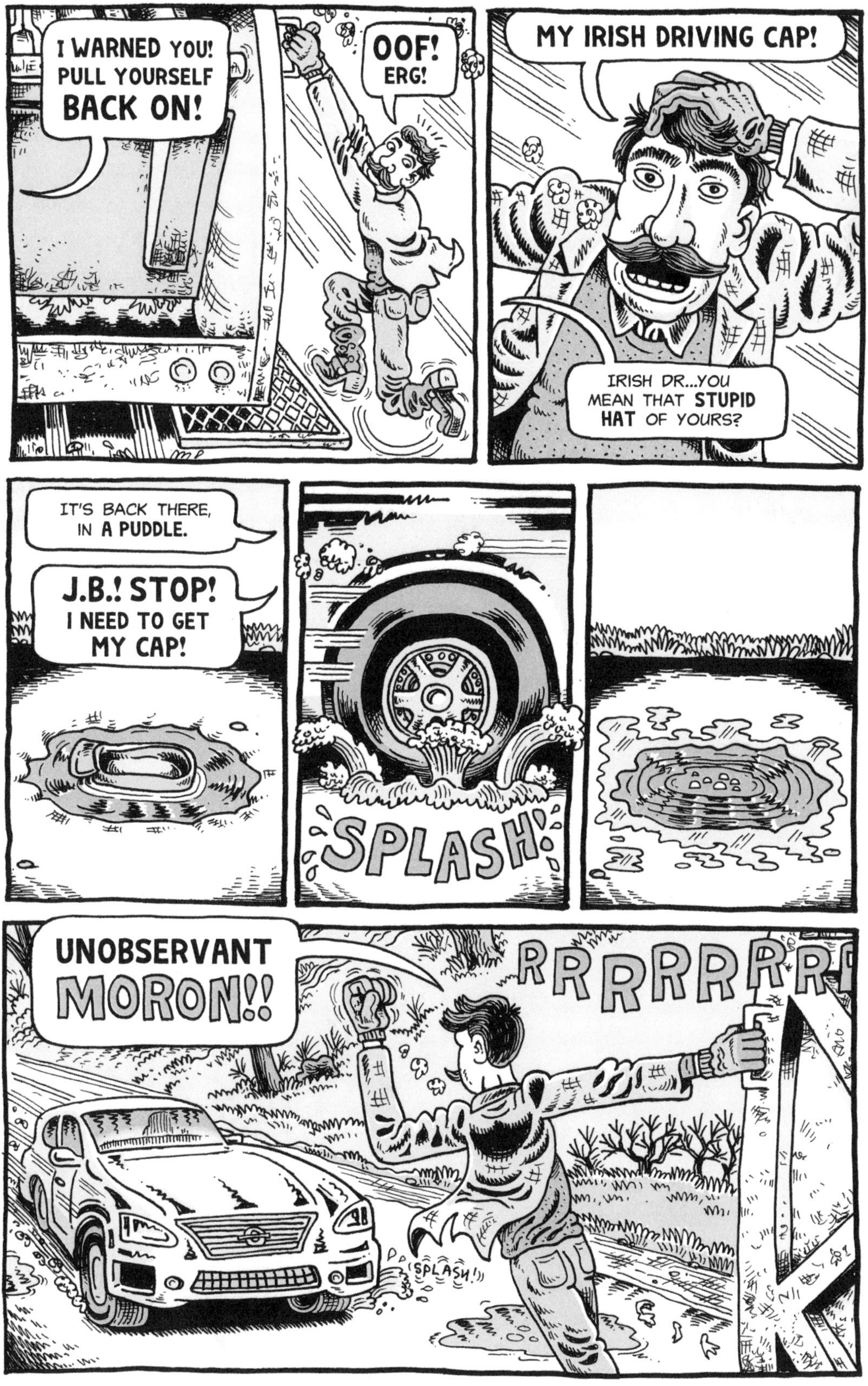
I WARNED YOU! PULL YOURSELF BACK ON!
OOF! ERG!
MY IRISH DRIVING CAP!
IRISH DR...YOU MEAN THAT STUPID HAT OF YOURS?
IT'S BACK THERE, IN A PUDDLE.
J.B.! STOP! I NEED TO GET MY CAP!
SPLASH
UNOBSERVANT MORON!!
RRRRRRR
SPLASH!

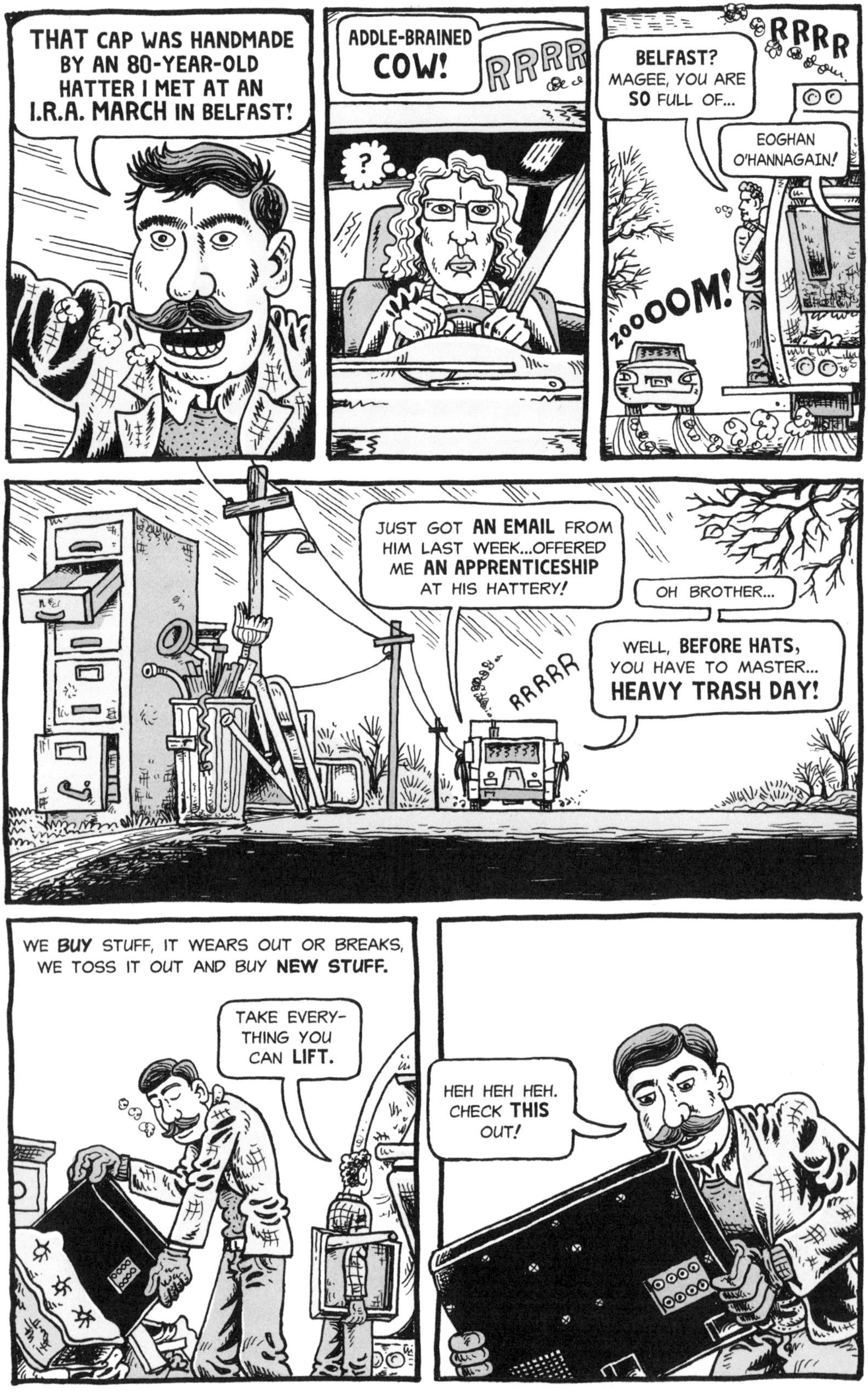
THAT CAP WAS HANDMADE BY AN 80-YEAR-OLD HATTER I MET AT AN I.R.A. MARCH IN BELFAST!
ADDLE-BRAINED COW!
RRRR
?
BELFAST? MAGEE, YOU ARE SO FULL OF...
RRRR
EOGHAN O'HANNAGAIN!
ZOOOOM!
JUST GOT AN EMAIL FROM HIM LAST WEEK...OFFERED ME AN APPRENTICESHIP AT HIS HATTERY!
OH BROTHER...
WELL, BEFORE HATS, YOU HAVE TO MASTER... HEAVY TRASH DAY!
RRRRR
WE BUY STUFF, IT WEARS OUT OR BREAKS, WE TOSS IT OUT AND BUY NEW STUFF.
TAKE EVERY-THING YOU CAN LIFT.
HEH HEH HEH. CHECK THIS OUT!

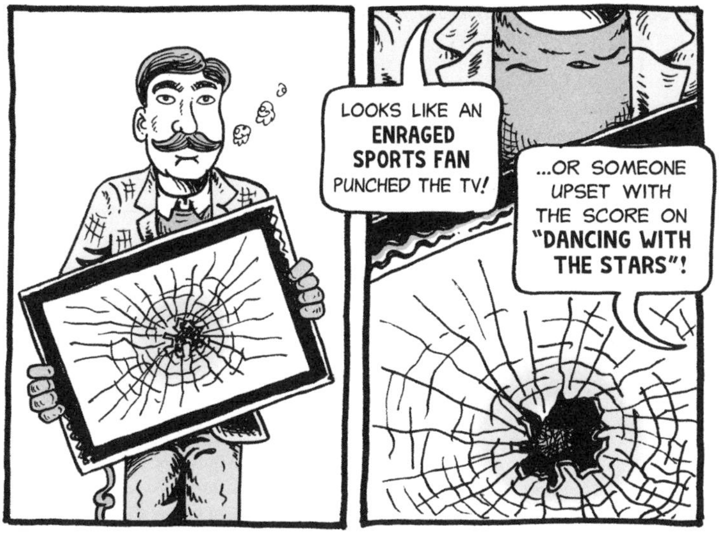
LOOKS LIKE AN ENRAGED SPORTS FAN PUNCHED THE TV!
...OR SOMEONE UPSET WITH THE SCORE ON "DANCING WITH THE STARS"!

MANUFACTURERS DON'T WANT GOODS TO LAST A LONG TIME. **THE AVERAGE LIFE SPAN** OF A REFRIGERATOR IS JUST 13 YEARS. A WASHING MACHINE IS 10 YEARS. A MICROWAVE OVEN IS 9 YEARS. IT'S CALLED **BUILT-IN OBSOLESCENCE.**

AND NOW OUR BELOVED **GADGETS** HAVE BECOME A HUGE DISPOSAL PROBLEM. THERE ARE WELL OVER **300 MILLION** P.C.S IN USE IN THE U.S. AND OVER **1.6 BILLION** WORLDWIDE! THE E.P.A. ESTIMATES THAT THERE ARE ALMOST **THREE TIMES** THAT NUMBER THAT ARE OBSOLETE AND MOTHBALLED.

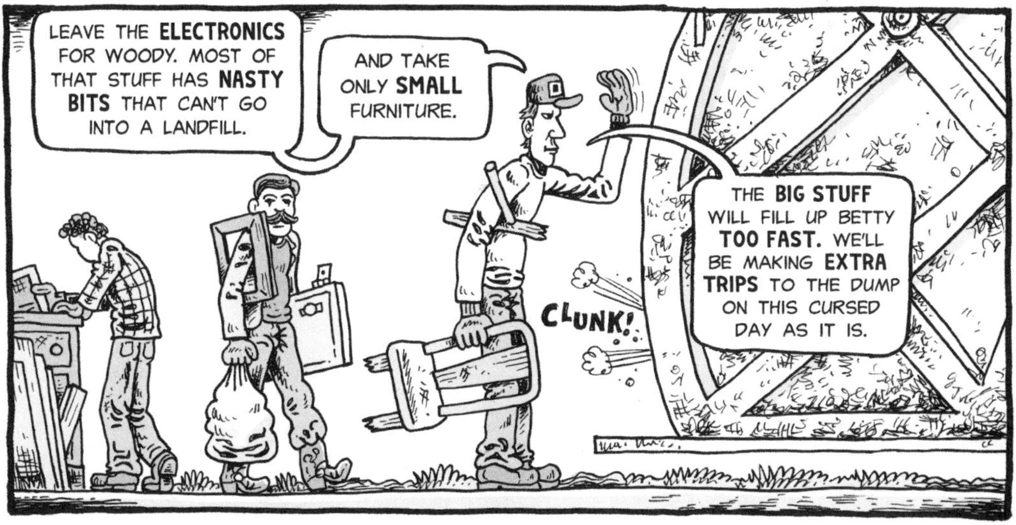
LEAVE THE ELECTRONICS FOR WOODY. MOST OF THAT STUFF HAS NASTY BITS THAT CAN'T GO INTO A LANDFILL.
AND TAKE ONLY SMALL FURNITURE.
THE BIG STUFF WILL FILL UP BETTY TOO FAST. WE'LL BE MAKING EXTRA TRIPS TO THE DUMP ON THIS CURSED DAY AS IT IS.
CLUNK!

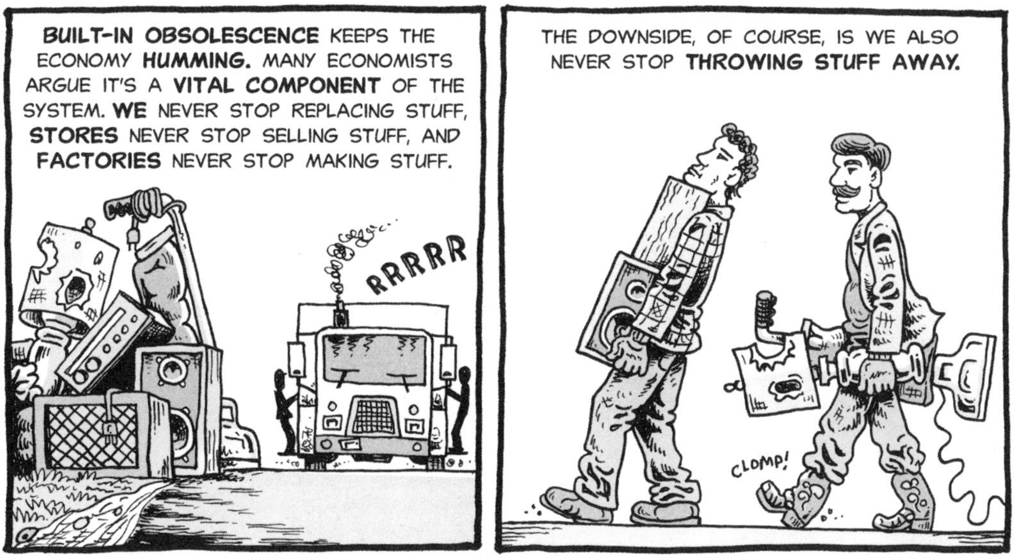
BUILT-IN OBSOLESCENCE KEEPS THE ECONOMY HUMMING. MANY ECONOMISTS ARGUE IT'S A VITAL COMPONENT OF THE SYSTEM. WE NEVER STOP REPLACING STUFF, STORES NEVER STOP SELLING STUFF, AND FACTORIES NEVER STOP MAKING STUFF.
RRRRR
THE DOWNSIDE, OF COURSE, IS WE ALSO NEVER STOP THROWING STUFF AWAY.
CLOMP!

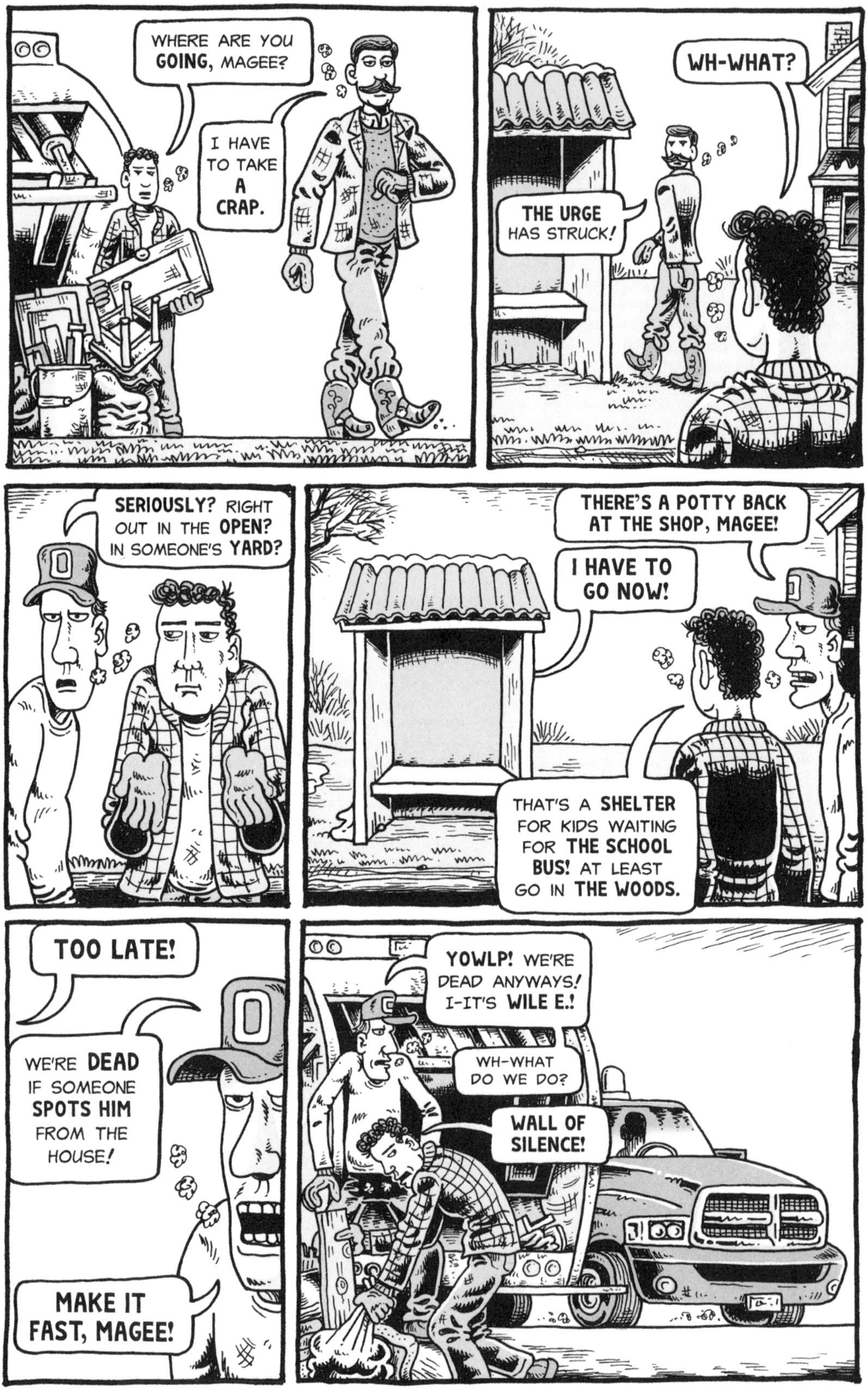
WHERE ARE YOU GOING, MAGEE?
I HAVE TO TAKE A CRAP.
WH-WHAT?
THE URGE HAS STRUCK!
SERIOUSLY? RIGHT OUT IN THE OPEN? IN SOMEONE'S YARD?
THERE'S A POTTY BACK AT THE SHOP, MAGEE!
I HAVE TO GO NOW!
THAT'S A SHELTER FOR KIDS WAITING FOR THE SCHOOL BUS! AT LEAST GO IN THE WOODS.
TOO LATE!
WE'RE DEAD IF SOMEONE SPOTS HIM FROM THE HOUSE!
MAKE IT FAST, MAGEE!
YOWLP! WE'RE DEAD ANYWAYS! I-IT'S WILE E.!
WH-WHAT DO WE DO?
WALL OF SILENCE!

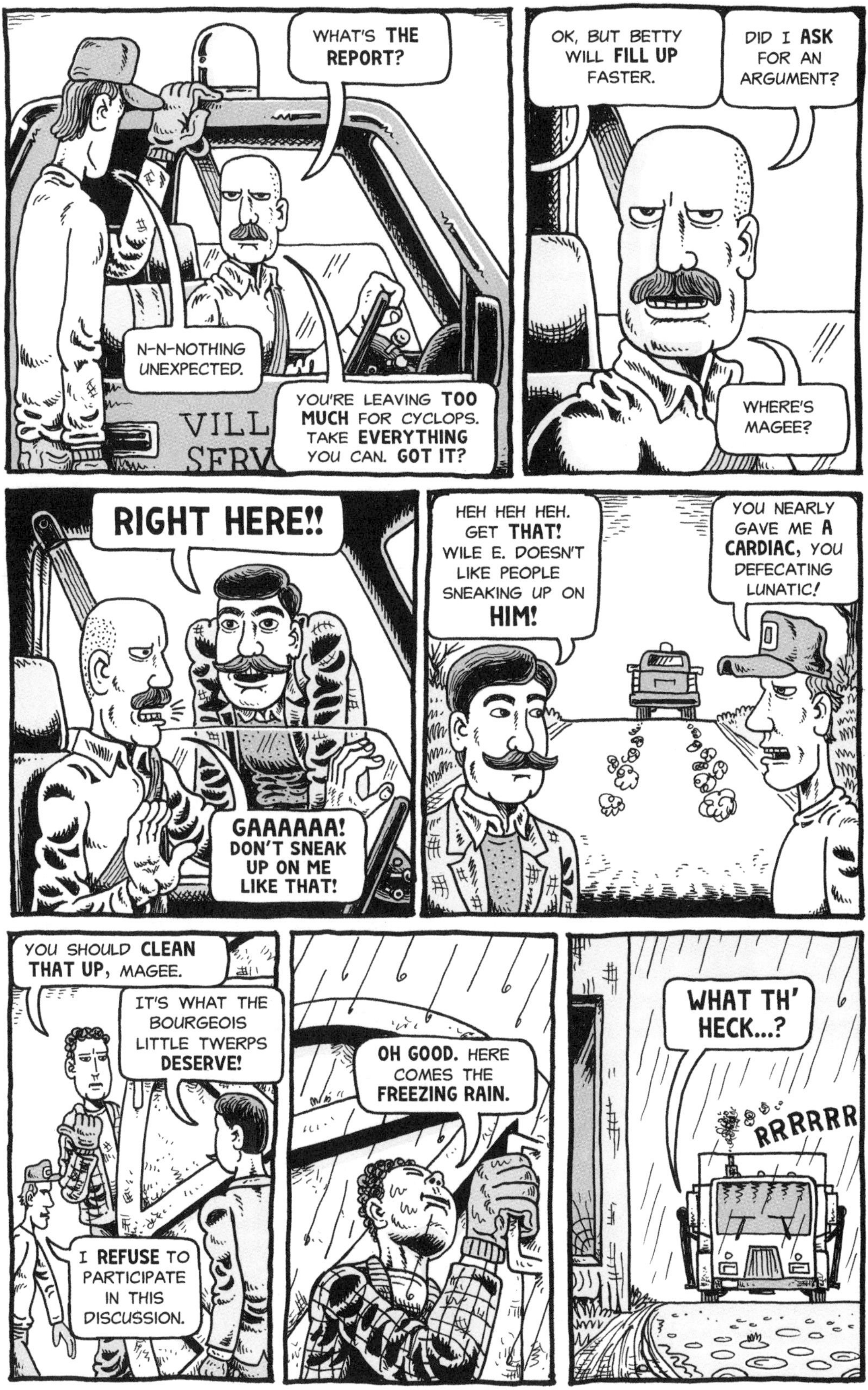
WHAT'S THE REPORT?
N-N-NOTHING UNEXPECTED.
YOU'RE LEAVING TOO MUCH FOR CYCLOPS. TAKE EVERYTHING YOU CAN. GOT IT?
VILL
SERV
OK, BUT BETTY WILL FILL UP FASTER.
DID I ASK FOR AN ARGUMENT?
WHERE'S MAGEE?
RIGHT HERE!!
GAAAAAA! DON'T SNEAK UP ON ME LIKE THAT!
HEH HEH HEH. GET THAT! WILE E. DOESN'T LIKE PEOPLE SNEAKING UP ON HIM!
YOU NEARLY GAVE ME A CARDIAC, YOU DEFECATING LUNATIC!
YOU SHOULD CLEAN THAT UP, MAGEE.
IT'S WHAT THE BOURGEOIS LITTLE TWERPS DESERVE!
I REFUSE TO PARTICIPATE IN THIS DISCUSSION.
OH GOOD. HERE COMES THE FREEZING RAIN.
WHAT TH' HECK...?
RRRRRR

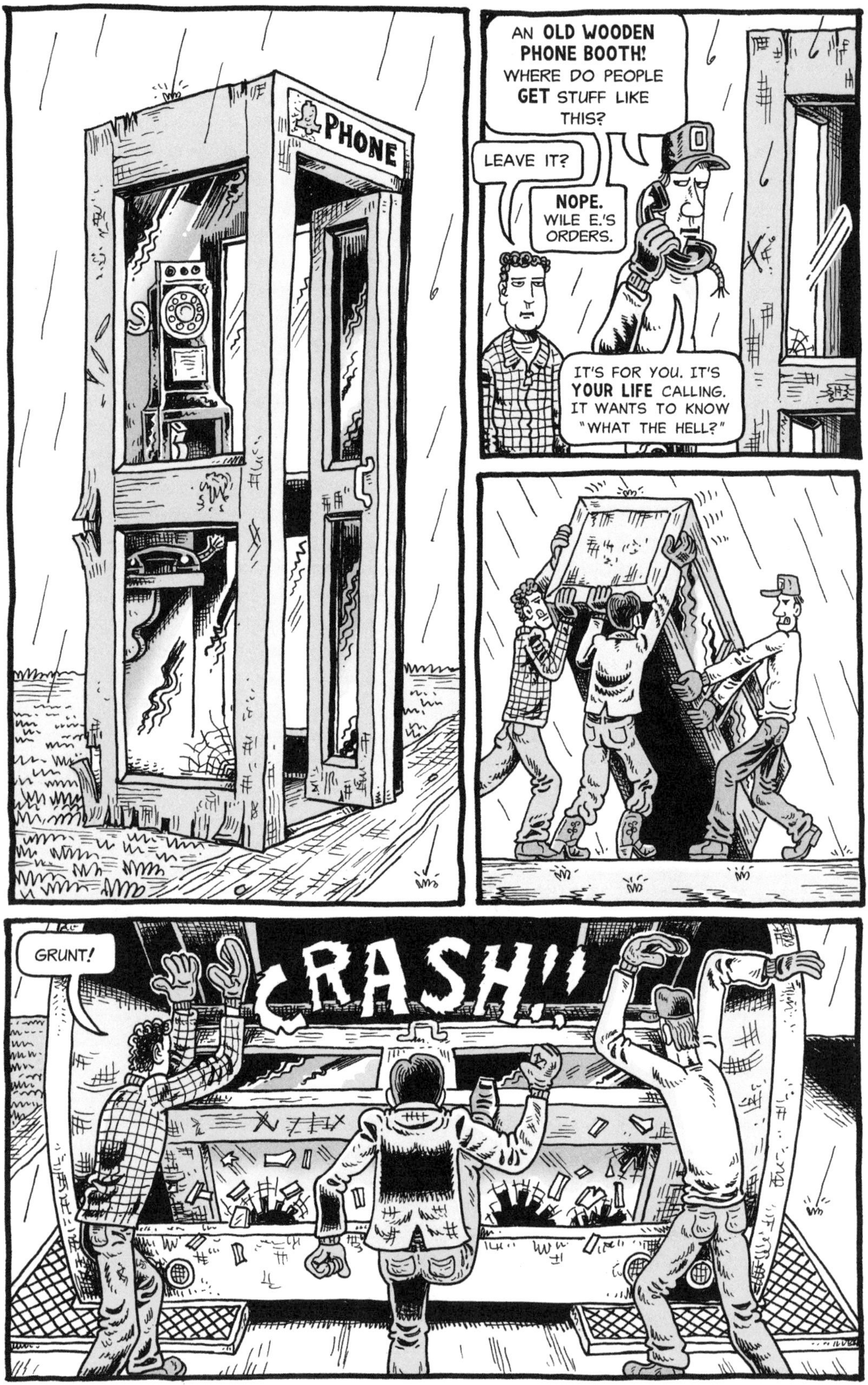
PHONE
AN OLD WOODEN PHONE BOOTH! WHERE DO PEOPLE GET STUFF LIKE THIS?
LEAVE IT?
NOPE. WILE E.'S ORDERS.
IT'S FOR YOU. IT'S YOUR LIFE CALLING. IT WANTS TO KNOW "WHAT THE HELL?"
GRUNT!
CRASH!!

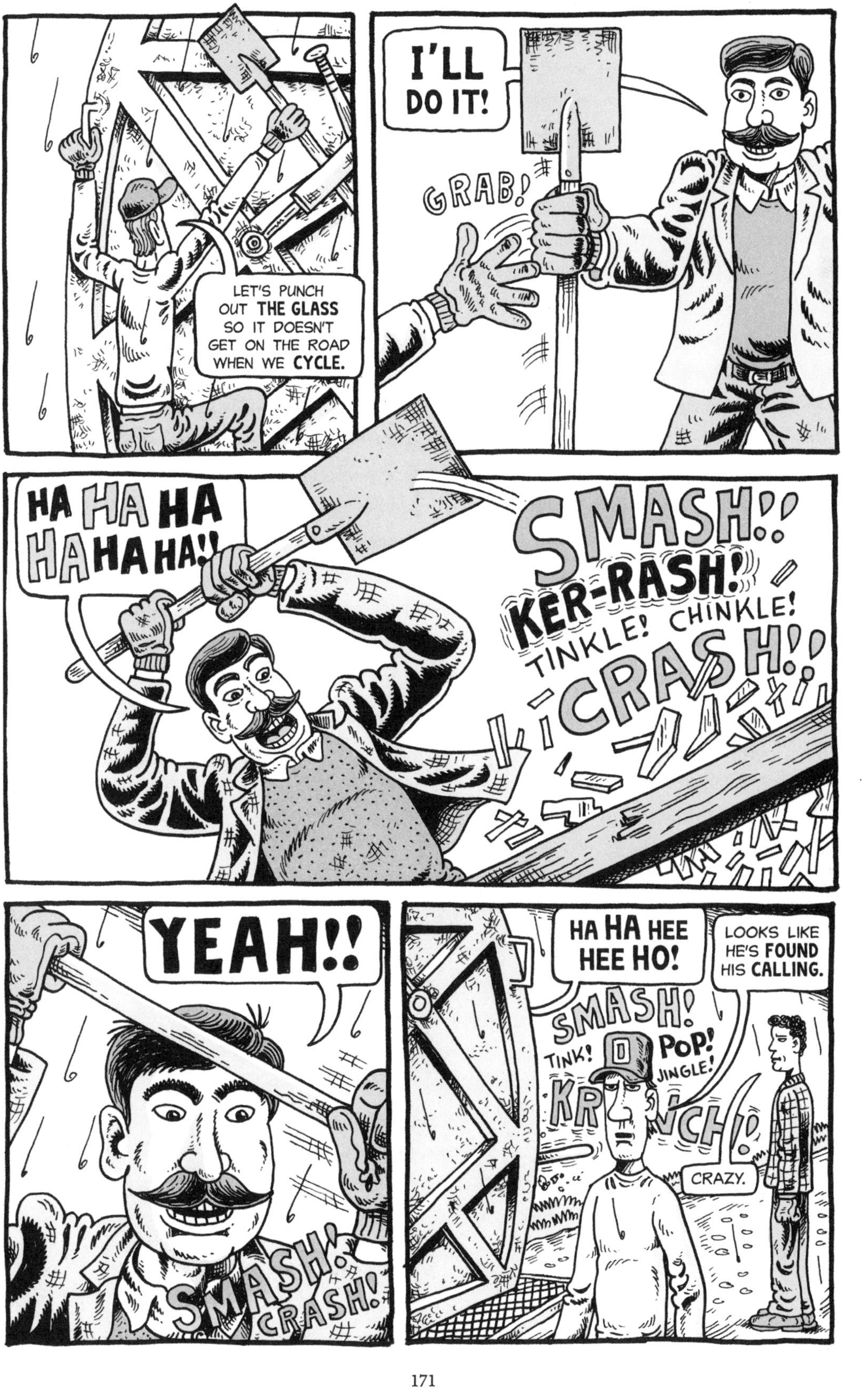
LET'S PUNCH OUT THE GLASS SO IT DOESN'T GET ON THE ROAD WHEN WE CYCLE.
I'LL DO IT!
GRAB!
HA HA HA HA HA HA HA!!
SMASH!!
KER-RASH!
TINKLE! CHINKLE!
CRASH!!
YEAH!!
SMASH!
CRASH!
HA HA HEE HEE HO!
LOOKS LIKE HE'S FOUND HIS CALLING.
SMASH!
TINK!
POP!
JINGLE!
CRAZY.

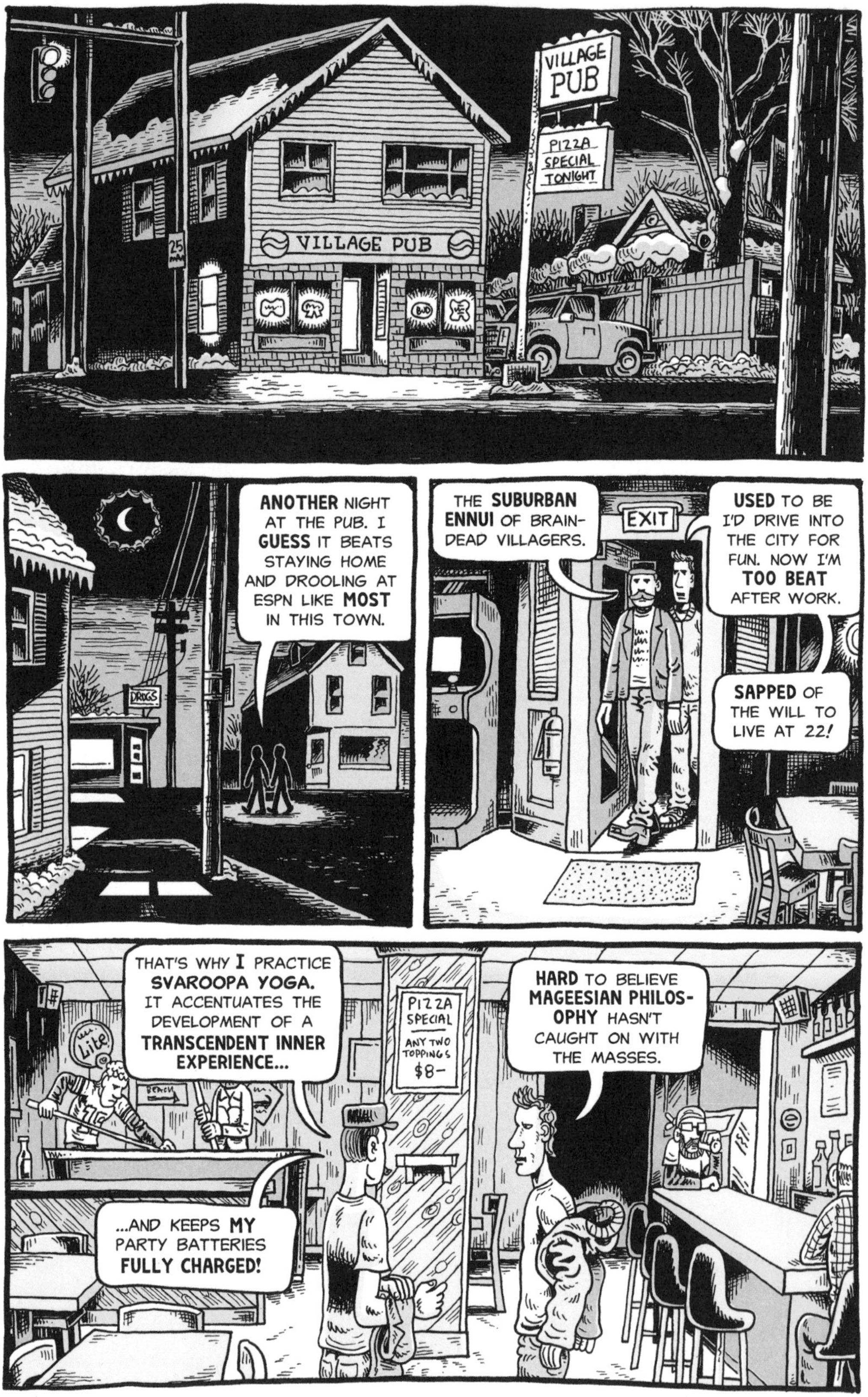
VILLAGE PUB
PIZZA SPECIAL TONIGHT
VILLAGE PUB
25
ANOTHER NIGHT AT THE PUB. I GUESS IT BEATS STAYING HOME AND DROOLING AT ESPN LIKE MOST IN THIS TOWN.
DRUGS
THE SUBURBAN ENNUI OF BRAIN-DEAD VILLAGERS.
EXIT
USED TO BE I'D DRIVE INTO THE CITY FOR FUN. NOW I'M TOO BEAT AFTER WORK.
SAPPED OF THE WILL TO LIVE AT 22!
THAT'S WHY I PRACTICE SVAROOPA YOGA. IT ACCENTUATES THE DEVELOPMENT OF A TRANSCENDENT INNER EXPERIENCE...
Lite
PIZZA SPECIAL ANY TWO TOPPINGS $8-
HARD TO BELIEVE MAGEESIAN PHILOSOPHY HASN'T CAUGHT ON WITH THE MASSES.
...AND KEEPS MY PARTY BATTERIES FULLY CHARGED!

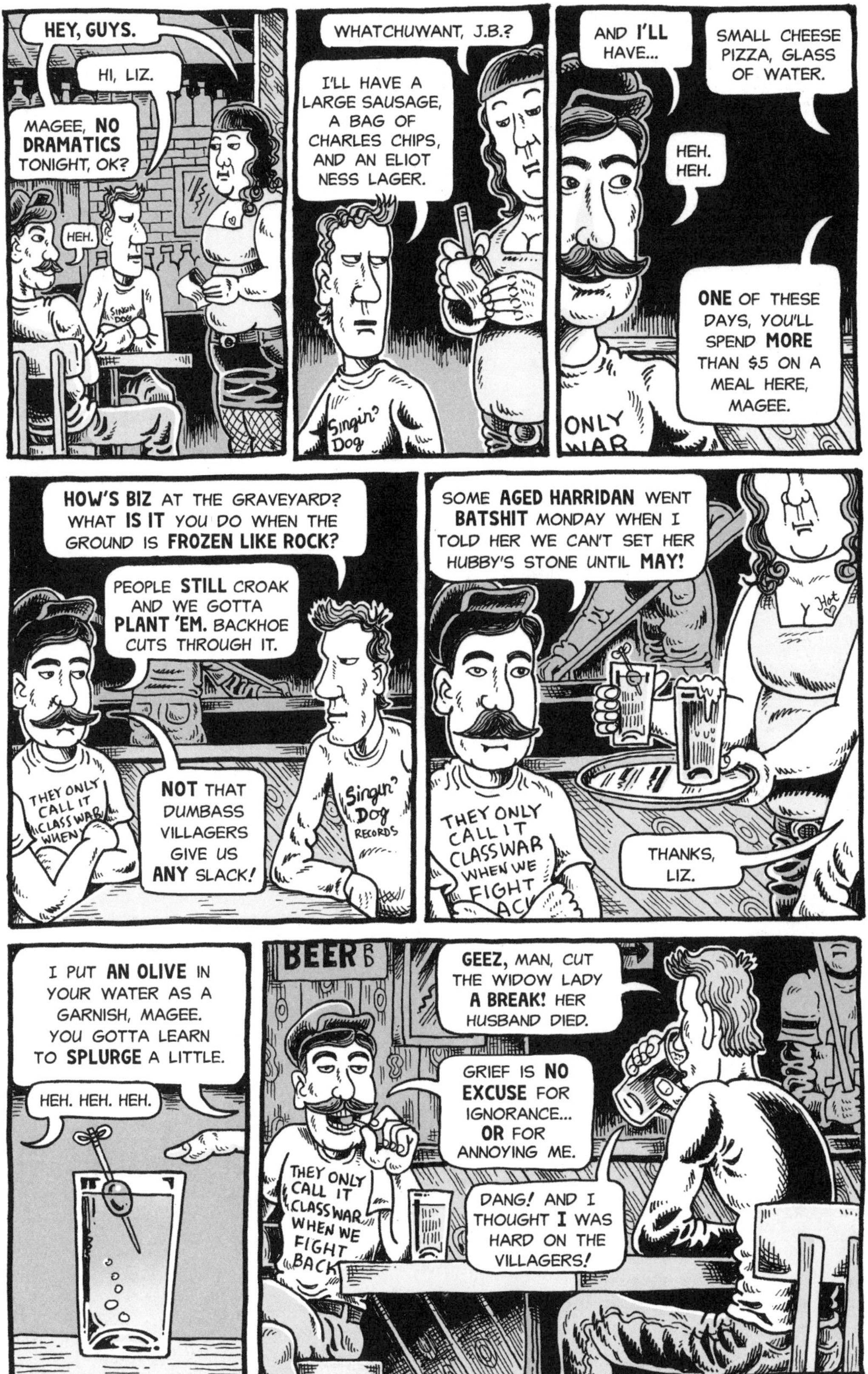
HEY, GUYS.
HI, LIZ.
MAGEE, NO DRAMATICS TONIGHT, OK?
HEH.
SINGIN' DOG
WHATCHUWANT, J.B.?
I'LL HAVE A LARGE SAUSAGE, A BAG OF CHARLES CHIPS, AND AN ELIOT NESS LAGER.
Singin' Dog
AND I'LL HAVE...
SMALL CHEESE PIZZA, GLASS OF WATER.
HEH. HEH.
ONE OF THESE DAYS, YOU'LL SPEND MORE THAN $5 ON A MEAL HERE, MAGEE.
ONLY WAR
HOW'S BIZ AT THE GRAVEYARD? WHAT IS IT YOU DO WHEN THE GROUND IS FROZEN LIKE ROCK?
PEOPLE STILL CROAK AND WE GOTTA PLANT 'EM. BACKHOE CUTS THROUGH IT.
NOT THAT DUMBASS VILLAGERS GIVE US ANY SLACK!
THEY ONLY CALL IT CLASS WAR WHEN
Singin' Dog RECORDS
SOME AGED HARRIDAN WENT BATSHIT MONDAY WHEN I TOLD HER WE CAN'T SET HER HUBBY'S STONE UNTIL MAY!
Hot
THEY ONLY CALL IT CLASS WAR WHEN WE FIGHT
THANKS, LIZ.
I PUT AN OLIVE IN YOUR WATER AS A GARNISH, MAGEE. YOU GOTTA LEARN TO SPLURGE A LITTLE.
HEH. HEH. HEH.
BEER
GEEZ, MAN, CUT THE WIDOW LADY A BREAK! HER HUSBAND DIED.
GRIEF IS NO EXCUSE FOR IGNORANCE... OR FOR ANNOYING ME.
DANG! AND I THOUGHT I WAS HARD ON THE VILLAGERS!
THEY ONLY CALL IT CLASS WAR WHEN WE FIGHT BACK

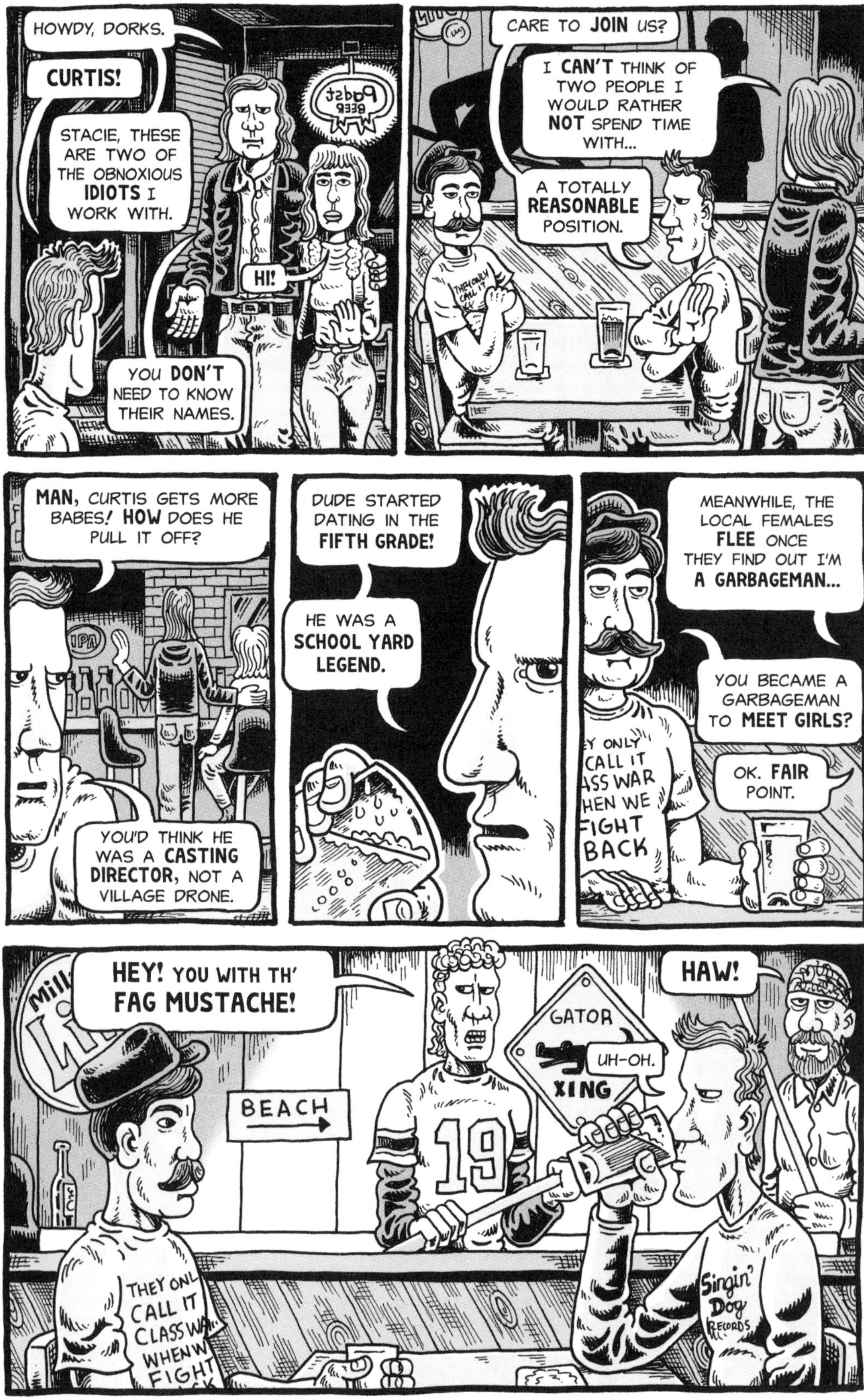
HOWDY, DORKS.
CURTIS!
STACIE, THESE ARE TWO OF THE OBNOXIOUS IDIOTS I WORK WITH.
HI!
YOU DON'T NEED TO KNOW THEIR NAMES.
CARE TO JOIN US?
I CAN'T THINK OF TWO PEOPLE I WOULD RATHER NOT SPEND TIME WITH...
A TOTALLY REASONABLE POSITION.
MAN, CURTIS GETS MORE BABES! HOW DOES HE PULL IT OFF?
YOU'D THINK HE WAS A CASTING DIRECTOR, NOT A VILLAGE DRONE.
DUDE STARTED DATING IN THE FIFTH GRADE!
HE WAS A SCHOOL YARD LEGEND.
MEANWHILE, THE LOCAL FEMALES FLEE ONCE THEY FIND OUT I'M A GARBAGEMAN...
YOU BECAME A GARBAGEMAN TO MEET GIRLS?
OK. FAIR POINT.
HEY! YOU WITH TH' FAG MUSTACHE!
HAW!
UH-OH.
GATOR XING
BEACH
19
Singin' Dog Records

WHAT'S TH' DEAL WITH THAT SHIRT?

WHAT ABOUT IT?

THEY ONLY CALL IT CLASS WAR WHEN WE FIGHT BACK

WHAT ARE YOU, A LIBTARD OR SOMETHIN'...?

HAW!

BEA

YOU GOT THAT RIGHT! I'M YOUR GAY, COMMUNIST BOGEYMAN... AND I'M HERE TO CONFISCATE YOUR GUNS AND HAVE SWEATY HOMO SEX WITH YOUR DAD!

THEY ONLY CALL IT CLASS WAR WHEN WE FIGHT BACK

NOT WORTH IT, MAGEE.

WHAAAAT? YER A REAL SICKO, YOU KNOW THAT?

19

I REMEMBER YOU FROM HIGH SCHOOL...

YOU WERE A SMARTASS DICK BACK THEN, TOO.

19

AND WHY DO I CARE WHAT AN INBRED, SMALL-TOWN SPIDER MONKEY THINKS!?!

HERE WE GO AGAIN.

IS THERE A PROBLEM HERE?

JUST THE USUAL ONE.

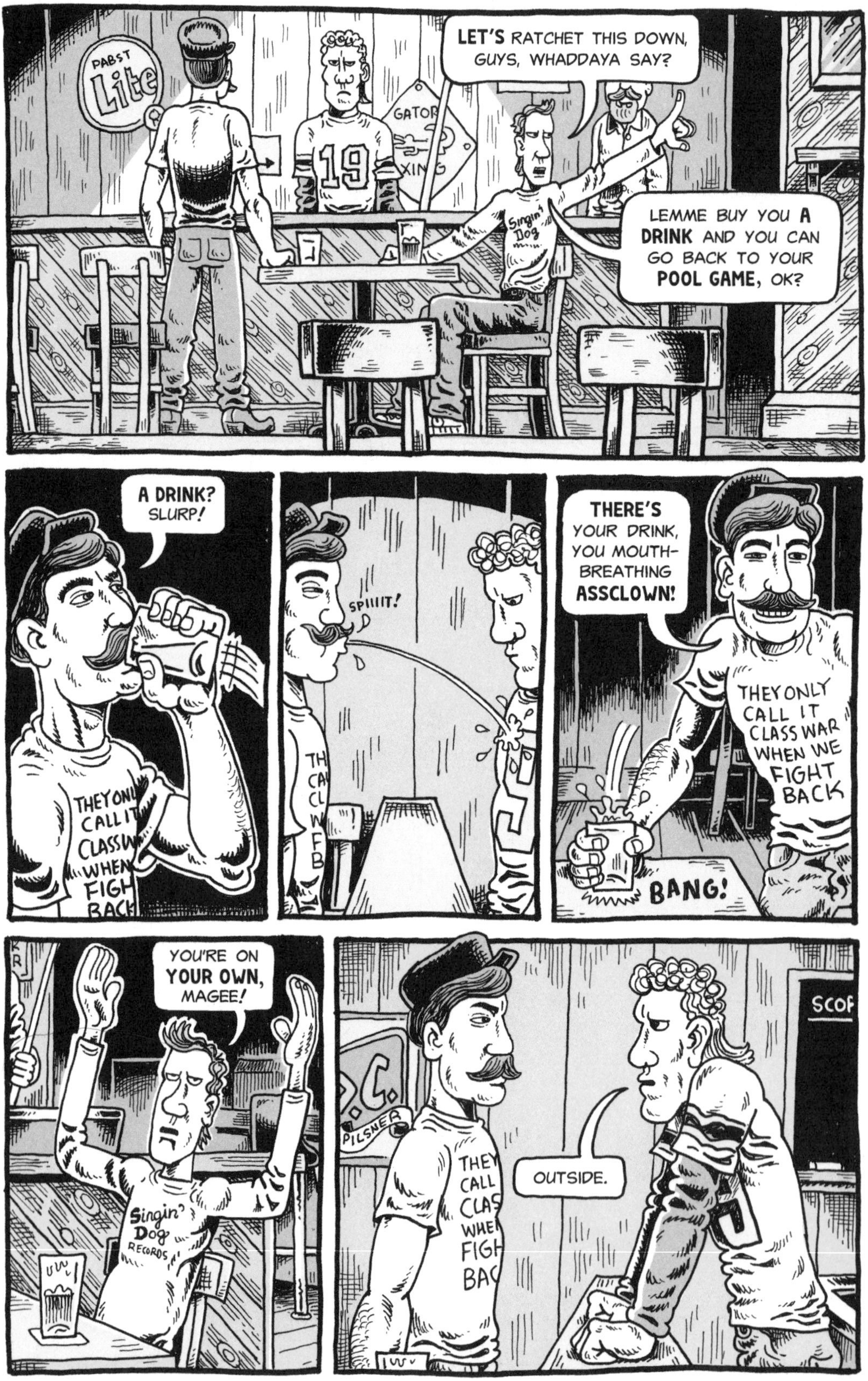
PABST Lite
GATOR XING
LET'S RATCHET THIS DOWN, GUYS, WHADDAYA SAY?
LEMME BUY YOU A DRINK AND YOU CAN GO BACK TO YOUR POOL GAME, OK?
A DRINK? SLURP!
SPIIIIT!
THERE'S YOUR DRINK, YOU MOUTH-BREATHING ASSCLOWN!
THEY ONLY CALL IT CLASS WAR WHEN WE FIGHT BACK
BANG!
YOU'RE ON YOUR OWN, MAGEE!
Singin' Dog RECORDS
PILSNER
OUTSIDE.

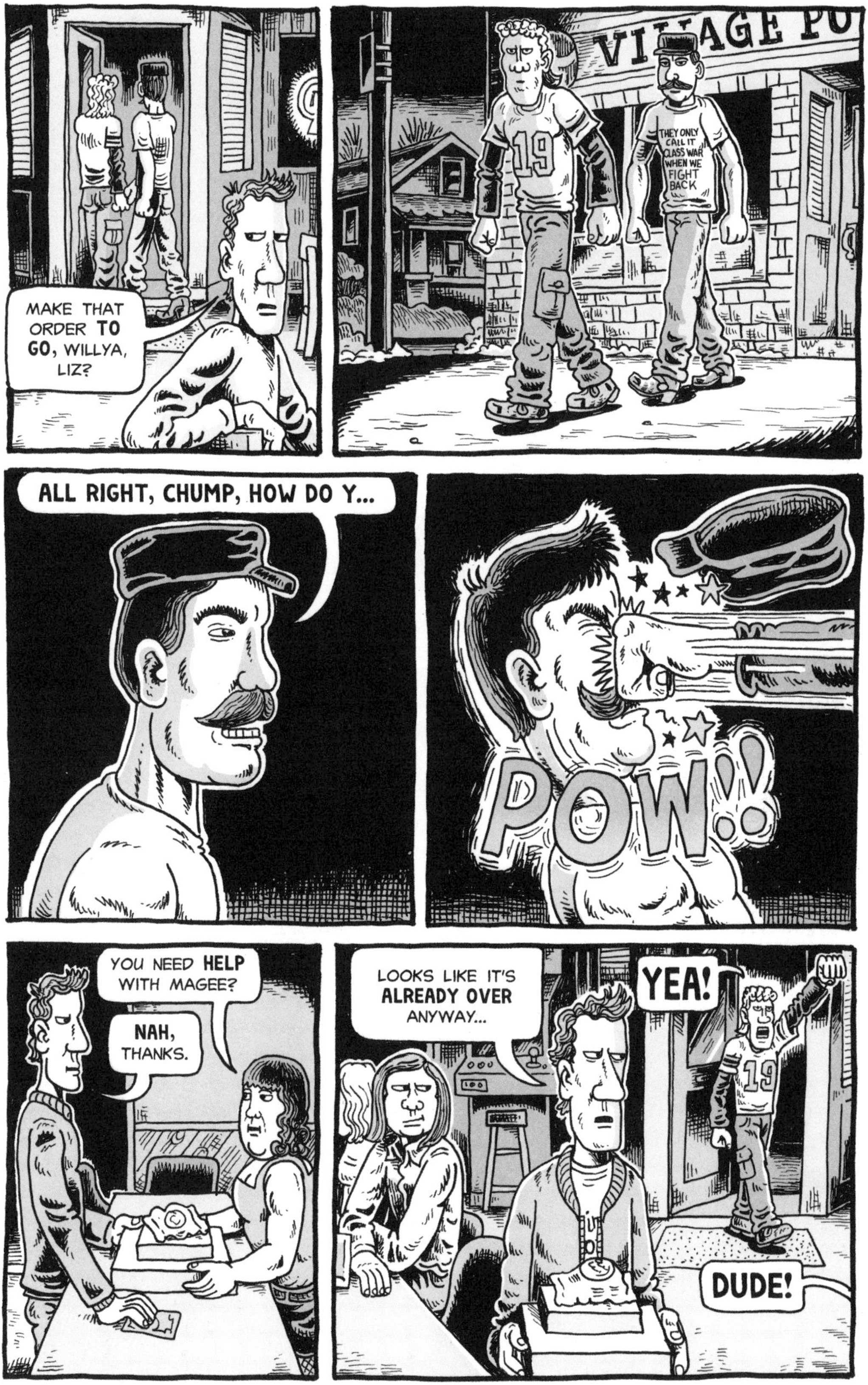
MAKE THAT ORDER TO GO, WILLYA, LIZ?
THEY ONLY CALL IT CLASS WAR WHEN WE FIGHT BACK
ALL RIGHT, CHUMP, HOW DO Y...
POW!!
YOU NEED HELP WITH MAGEE?
NAH, THANKS.
LOOKS LIKE IT'S ALREADY OVER ANYWAY...
YEA!
DUDE!

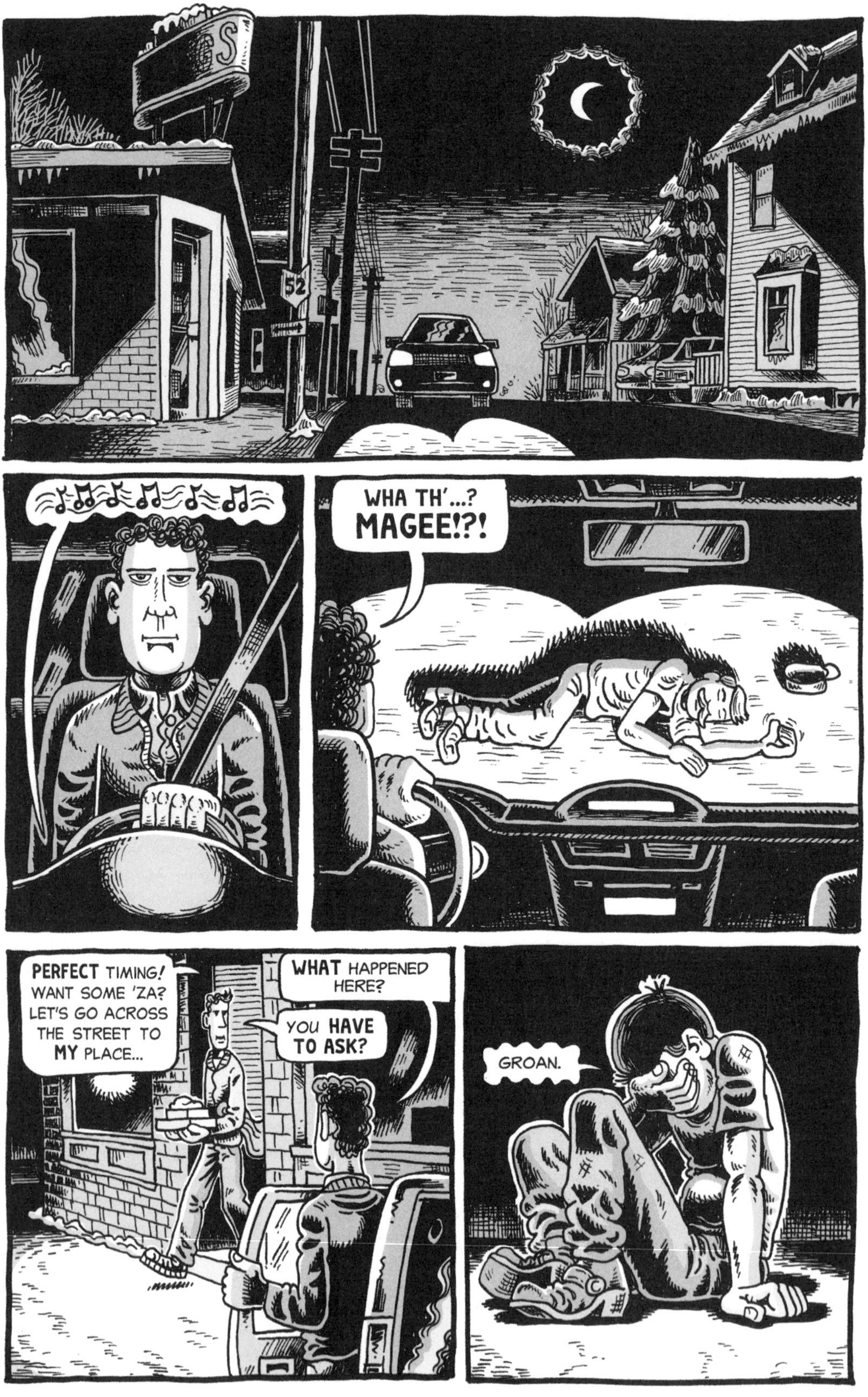
52
WHA TH'...? MAGEE!?!
PERFECT TIMING! WANT SOME 'ZA? LET'S GO ACROSS THE STREET TO MY PLACE...
WHAT HAPPENED HERE?
YOU HAVE TO ASK?
GROAN.

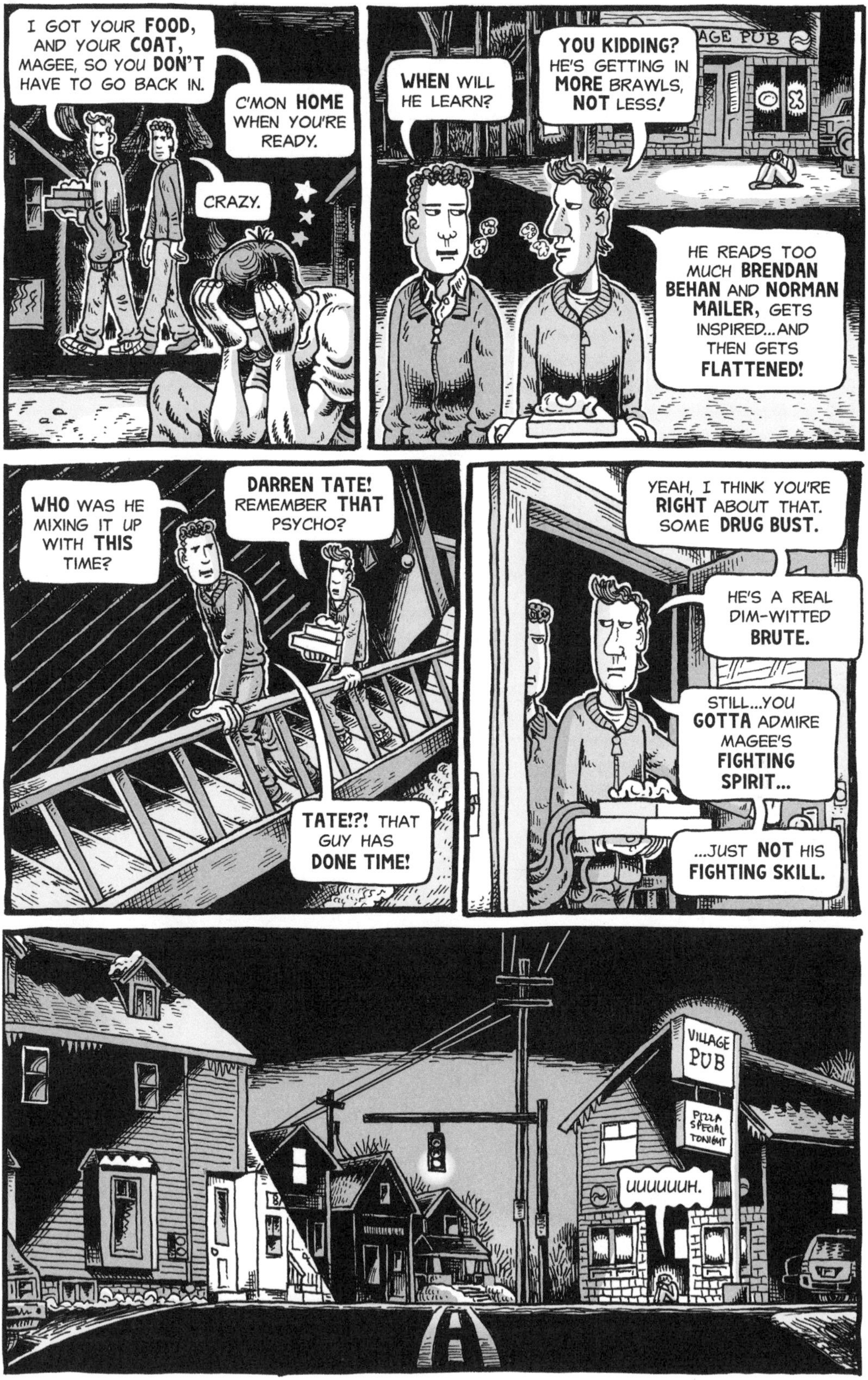
I GOT YOUR FOOD, AND YOUR COAT, MAGEE, SO YOU DON'T HAVE TO GO BACK IN.
C'MON HOME WHEN YOU'RE READY.
CRAZY.
WHEN WILL HE LEARN?
YOU KIDDING? HE'S GETTING IN MORE BRAWLS, NOT LESS!
HE READS TOO MUCH BRENDAN BEHAN AND NORMAN MAILER, GETS INSPIRED...AND THEN GETS FLATTENED!
WHO WAS HE MIXING IT UP WITH THIS TIME?
DARREN TATE! REMEMBER THAT PSYCHO?
TATE!?! THAT GUY HAS DONE TIME!
YEAH, I THINK YOU'RE RIGHT ABOUT THAT. SOME DRUG BUST.
HE'S A REAL DIM-WITTED BRUTE.
STILL...YOU GOTTA ADMIRE MAGEE'S FIGHTING SPIRIT...
...JUST NOT HIS FIGHTING SKILL.
VILLAGE PUB
PIZZA SPECIAL TONIGHT
UUUUUUH.

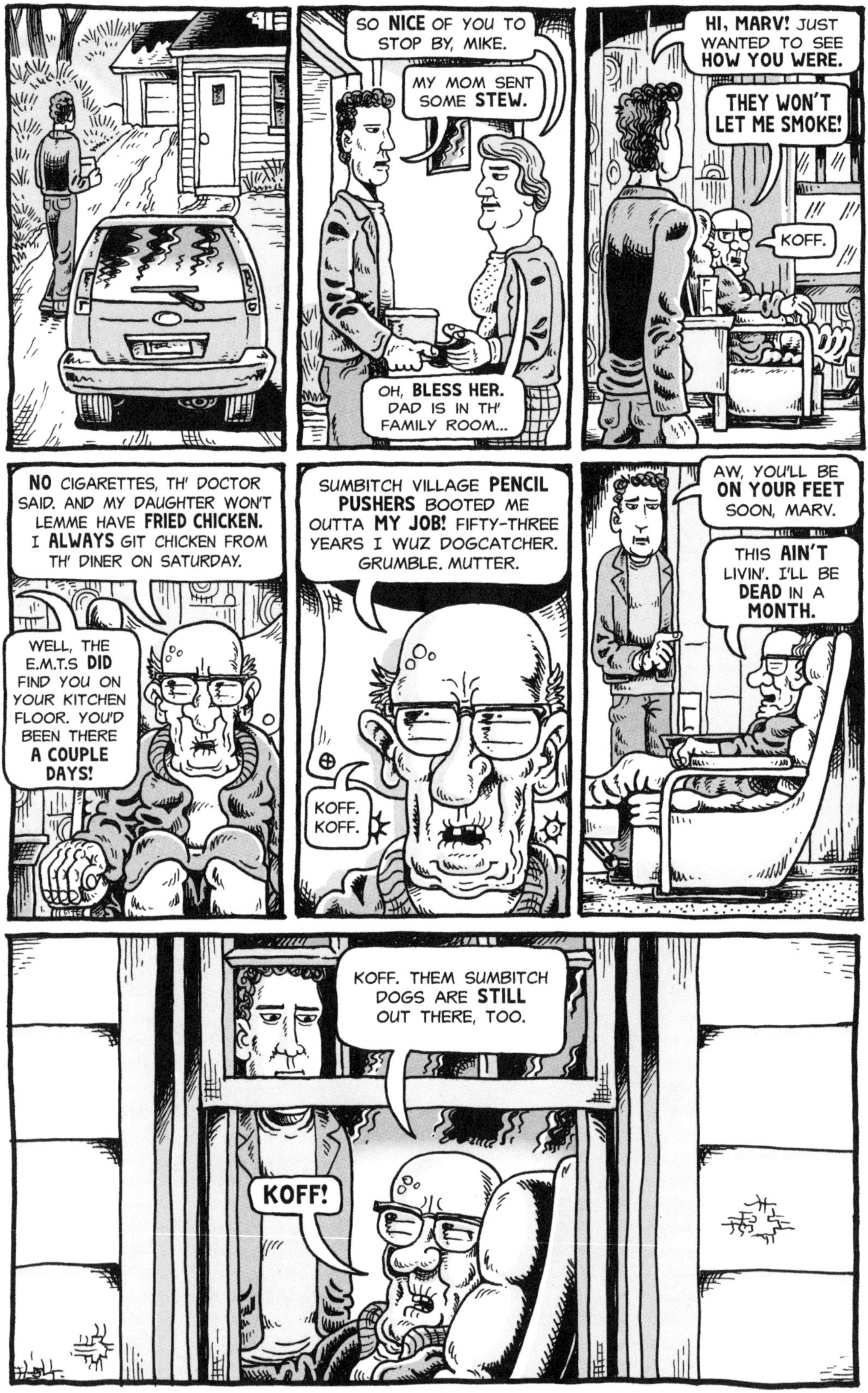
SO NICE OF YOU TO STOP BY, MIKE.
MY MOM SENT SOME STEW.
OH, BLESS HER. DAD IS IN TH' FAMILY ROOM...
HI, MARV! JUST WANTED TO SEE HOW YOU WERE.
THEY WON'T LET ME SMOKE!
KOFF.
NO CIGARETTES, TH' DOCTOR SAID. AND MY DAUGHTER WON'T LEMME HAVE FRIED CHICKEN. I ALWAYS GIT CHICKEN FROM TH' DINER ON SATURDAY.
WELL, THE E.M.T.S DID FIND YOU ON YOUR KITCHEN FLOOR. YOU'D BEEN THERE A COUPLE DAYS!
SUMBITCH VILLAGE PENCIL PUSHERS BOOTED ME OUTTA MY JOB! FIFTY-THREE YEARS I WUZ DOGCATCHER. GRUMBLE. MUTTER.
KOFF. KOFF.
AW, YOU'LL BE ON YOUR FEET SOON, MARV.
THIS AIN'T LIVIN'. I'LL BE DEAD IN A MONTH.
KOFF. THEM SUMBITCH DOGS ARE STILL OUT THERE, TOO.
KOFF!

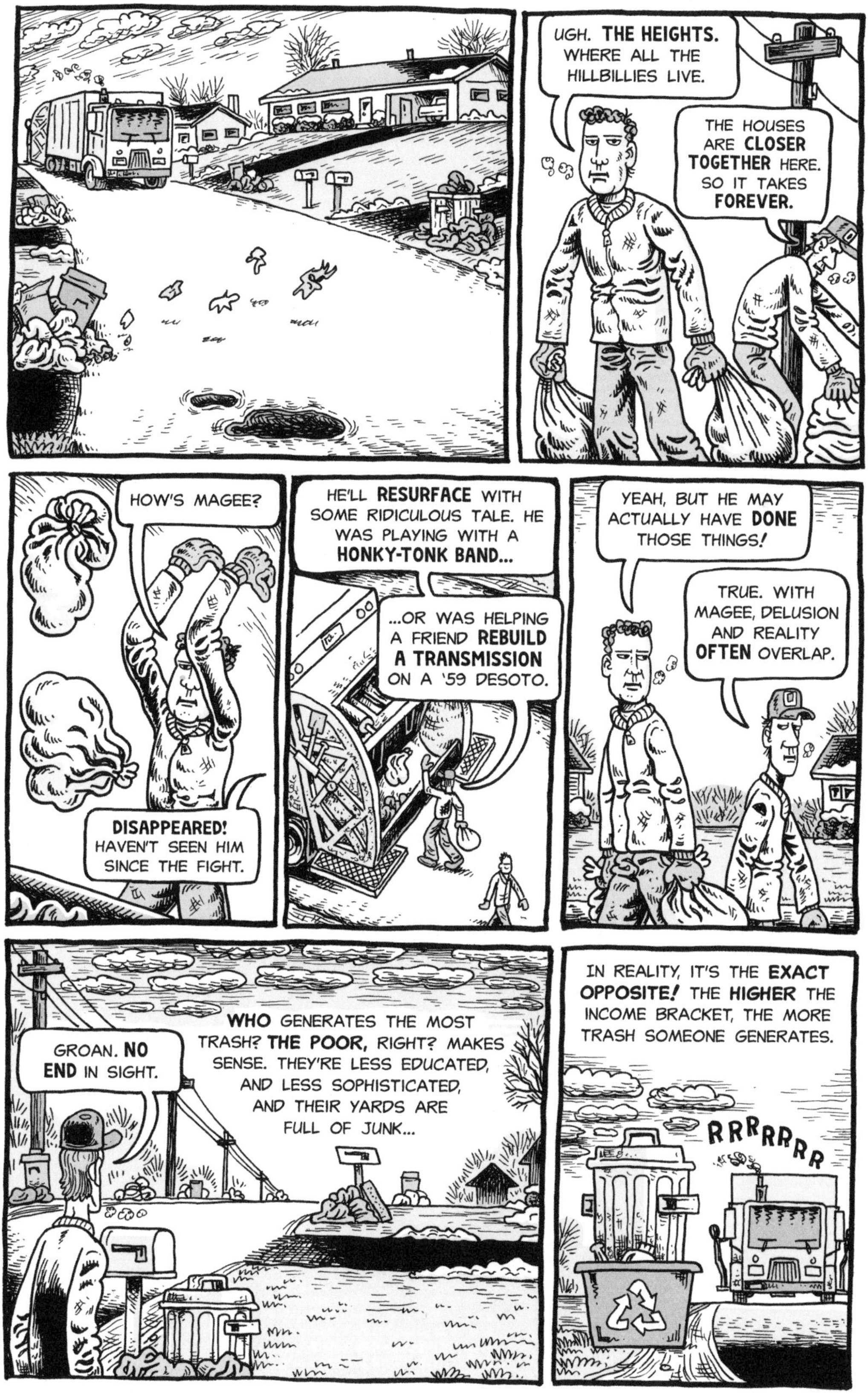
UGH. THE HEIGHTS. WHERE ALL THE HILLBILLIES LIVE.
THE HOUSES ARE CLOSER TOGETHER HERE. SO IT TAKES FOREVER.
HOW'S MAGEE?
DISAPPEARED! HAVEN'T SEEN HIM SINCE THE FIGHT.
HE'LL RESURFACE WITH SOME RIDICULOUS TALE. HE WAS PLAYING WITH A HONKY-TONK BAND...
...OR WAS HELPING A FRIEND REBUILD A TRANSMISSION ON A '59 DESOTO.
YEAH, BUT HE MAY ACTUALLY HAVE DONE THOSE THINGS!
TRUE. WITH MAGEE, DELUSION AND REALITY OFTEN OVERLAP.
WHO GENERATES THE MOST TRASH? THE POOR, RIGHT? MAKES SENSE. THEY'RE LESS EDUCATED, AND LESS SOPHISTICATED, AND THEIR YARDS ARE FULL OF JUNK...
GROAN. NO END IN SIGHT.
IN REALITY, IT'S THE EXACT OPPOSITE! THE HIGHER THE INCOME BRACKET, THE MORE TRASH SOMEONE GENERATES.
RRRRRRRR

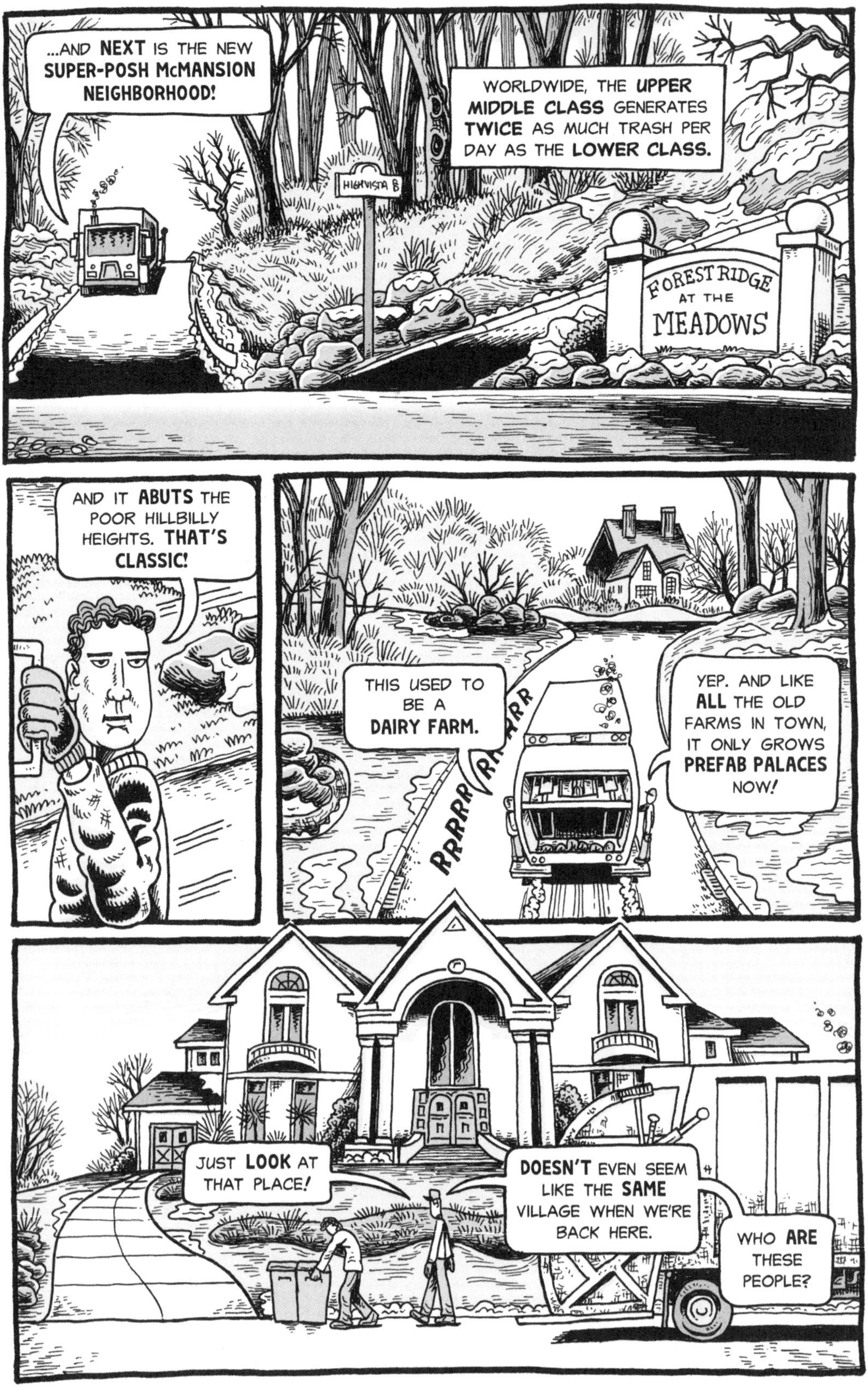

...AND NEXT IS THE NEW SUPER-POSH McMANSION NEIGHBORHOOD!
WORLDWIDE, THE UPPER MIDDLE CLASS GENERATES TWICE AS MUCH TRASH PER DAY AS THE LOWER CLASS.
HIGHVISTA B
FOREST RIDGE AT THE MEADOWS
AND IT ABUTS THE POOR HILLBILLY HEIGHTS. THAT'S CLASSIC!
THIS USED TO BE A DAIRY FARM.
RRRRR
RRRRR
YEP. AND LIKE ALL THE OLD FARMS IN TOWN, IT ONLY GROWS PREFAB PALACES NOW!
JUST LOOK AT THAT PLACE!
DOESN'T EVEN SEEM LIKE THE SAME VILLAGE WHEN WE'RE BACK HERE.
WHO ARE THESE PEOPLE?

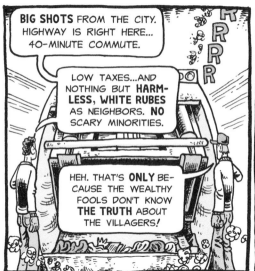

BIG SHOTS FROM THE CITY. HIGHWAY IS RIGHT HERE... 40-MINUTE COMMUTE.

LOW TAXES...AND NOTHING BUT **HARM-LESS, WHITE RUBES** AS NEIGHBORS. **NO** SCARY MINORITIES.

HEH. THAT'S **ONLY** BE-CAUSE THE WEALTHY FOOLS DON'T KNOW **THE TRUTH** ABOUT THE VILLAGERS!

HEY!

WAIT!!

BONE **HASN'T SPOTTED** HIM YET. SHOULD WE SIGNAL HIM TO STOP?

I DUNNO...I KINDA **ENJOY** HAVING SOME RICH GUY FUTILELY CHASE AFTER US.

PANT. PANT.

RRRRRRRRR

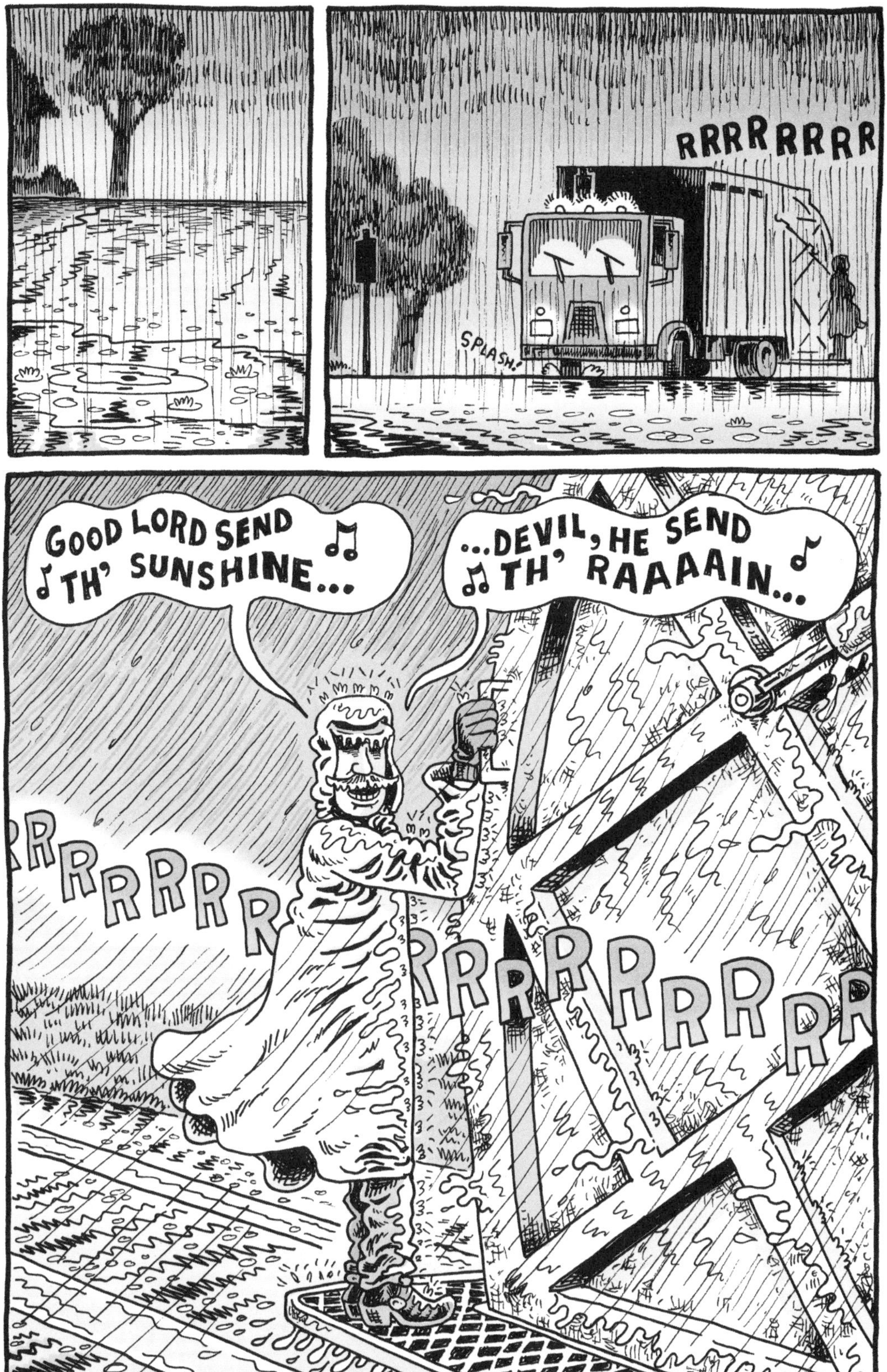
SPLASH!
RRRRRRRR
GOOD LORD SEND TH' SUNSHINE...
...DEVIL, HE SEND TH' RAAAAIN...
RRRRRRR
RRRRRRRRR

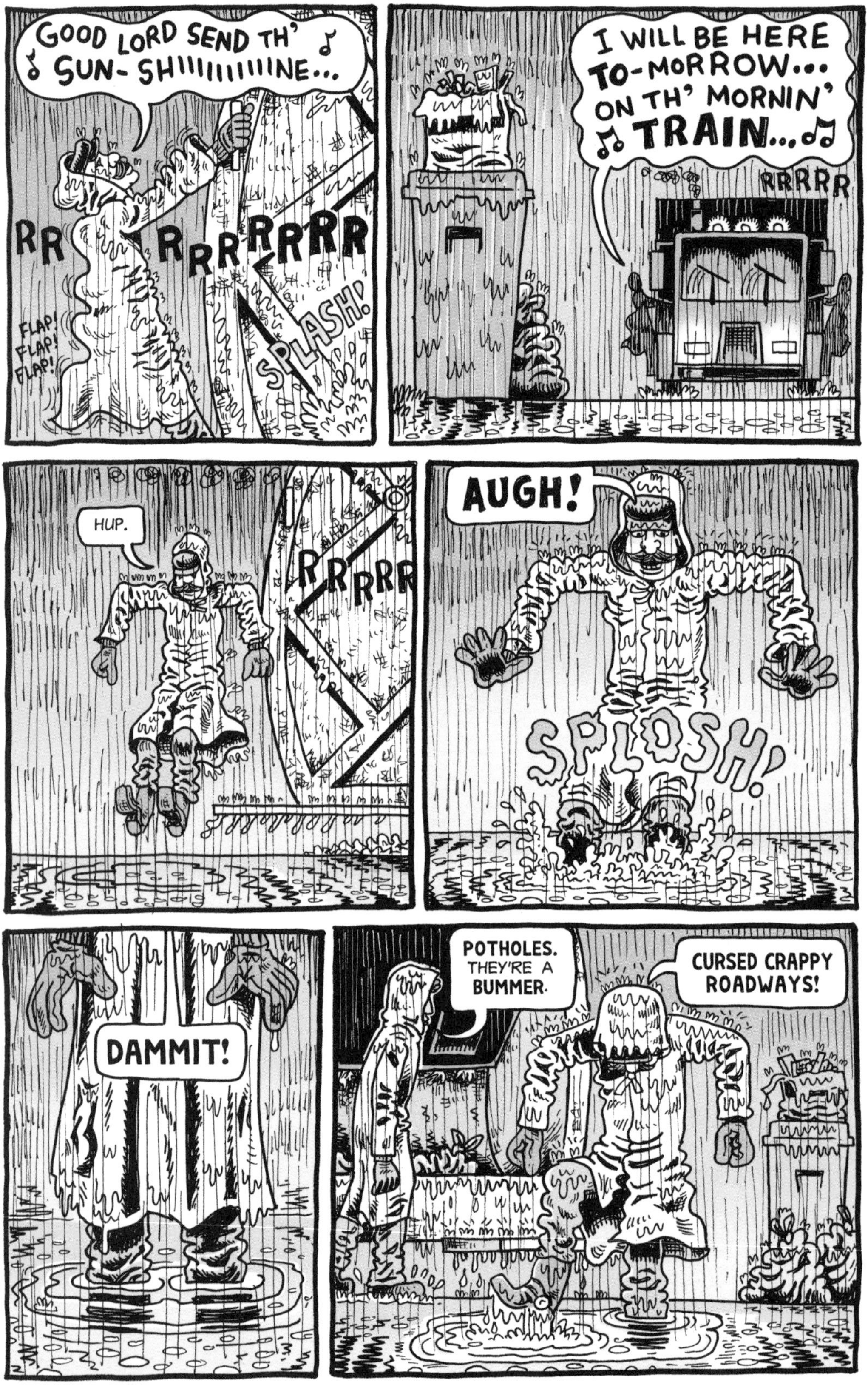
GOOD LORD SEND TH' SUN-SHIIIIIIIIINE...
RR RRRRRRR
FLAP! FLAP! FLAP!
SPLASH!
I WILL BE HERE TO-MORROW... ON TH' MORNIN' TRAIN...
RRRRR
HUP.
RRRRR
AUGH!
SPLOSH!
DAMMIT!
POTHOLES. THEY'RE A BUMMER.
CURSED CRAPPY ROADWAYS!

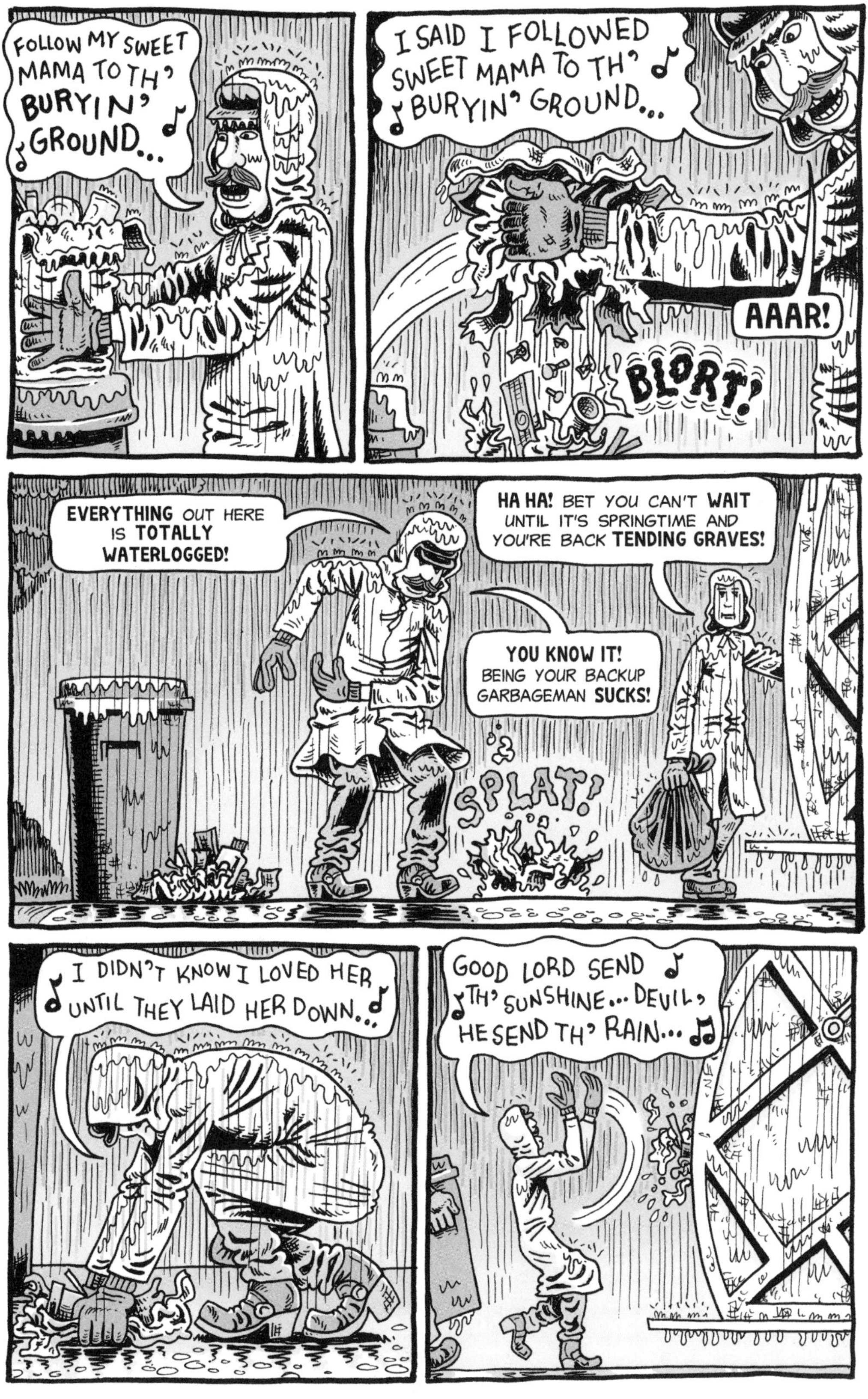
FOLLOW MY SWEET MAMA TO TH' BURYIN' GROUND...
I SAID I FOLLOWED SWEET MAMA TO TH' BURYIN' GROUND...
AAAR!
BLORT!
EVERYTHING OUT HERE IS TOTALLY WATERLOGGED!
HA HA! BET YOU CAN'T WAIT UNTIL IT'S SPRINGTIME AND YOU'RE BACK TENDING GRAVES!
YOU KNOW IT! BEING YOUR BACKUP GARBAGEMAN SUCKS!
SPLAT!
I DIDN'T KNOW I LOVED HER UNTIL THEY LAID HER DOWN...
GOOD LORD SEND TH' SUNSHINE... DEVIL, HE SEND TH' RAIN...

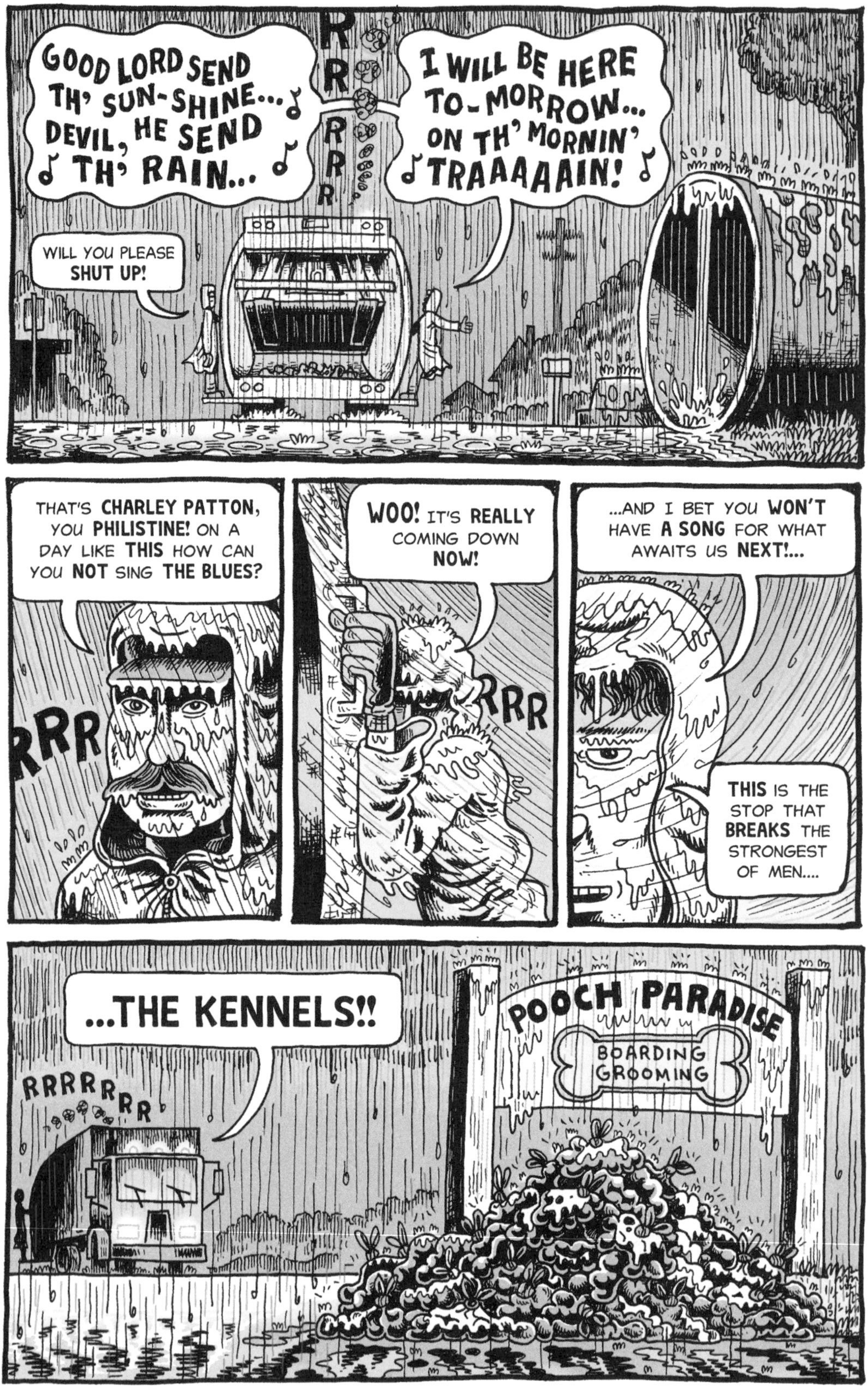
GOOD LORD SEND TH' SUN-SHINE... ♪ DEVIL, HE SEND ♪ TH' RAIN... ♪
RRRRRR
I WILL BE HERE TO-MORROW... ON TH' MORNIN' ♪ TRAAAAAIN! ♪
WILL YOU PLEASE SHUT UP!
THAT'S CHARLEY PATTON, YOU PHILISTINE! ON A DAY LIKE THIS HOW CAN YOU NOT SING THE BLUES?
RRR
WOO! IT'S REALLY COMING DOWN NOW!
RRR
...AND I BET YOU WON'T HAVE A SONG FOR WHAT AWAITS US NEXT!...
THIS IS THE STOP THAT BREAKS THE STRONGEST OF MEN....
...THE KENNELS!!
RRRRRRRR
POOCH PARADISE
BOARDING GROOMING

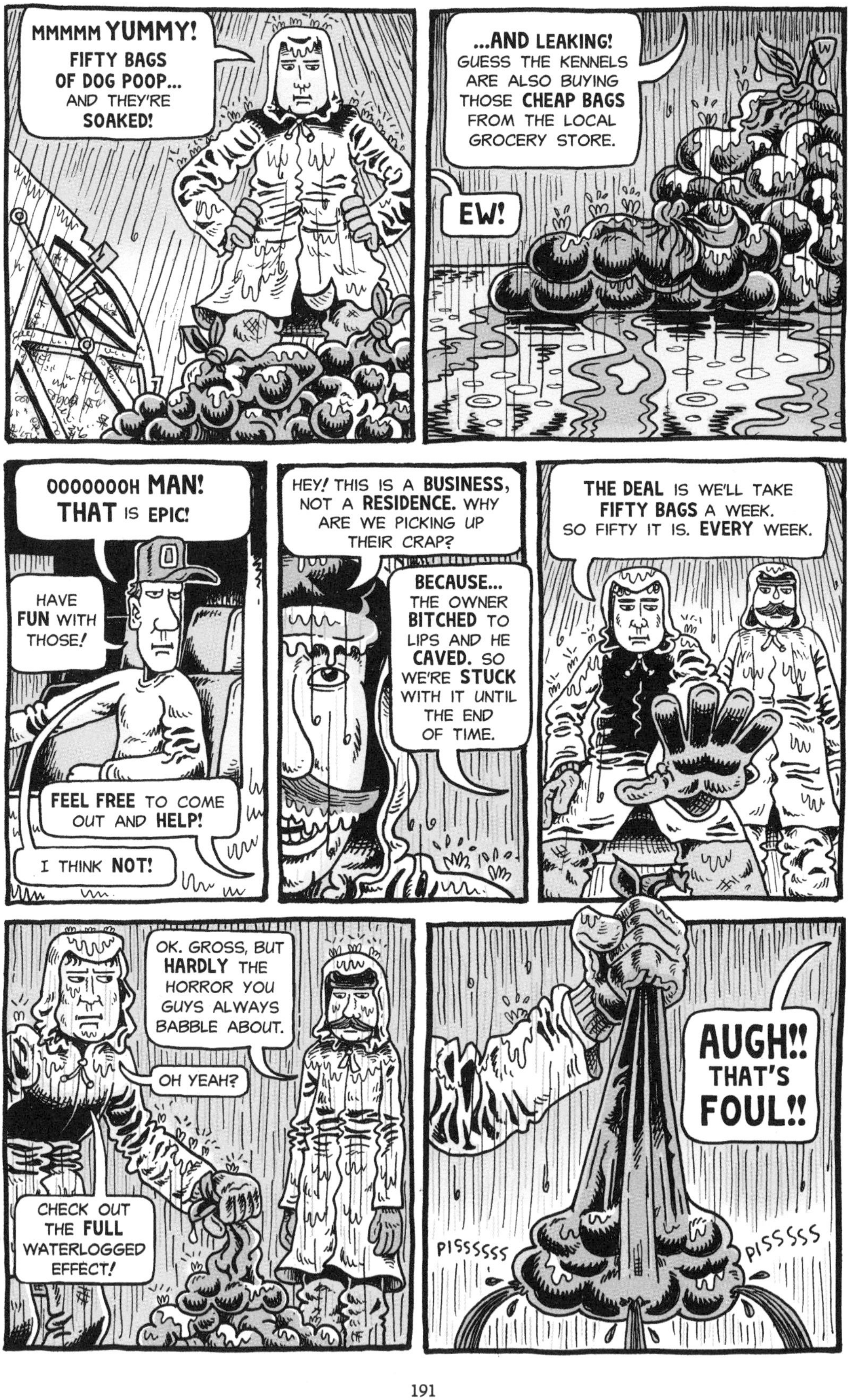
MMMMM YUMMY! FIFTY BAGS OF DOG POOP... AND THEY'RE SOAKED!
...AND LEAKING! GUESS THE KENNELS ARE ALSO BUYING THOSE CHEAP BAGS FROM THE LOCAL GROCERY STORE.
EW!
OOOOOOOH MAN! THAT IS EPIC!
HAVE FUN WITH THOSE!
FEEL FREE TO COME OUT AND HELP!
I THINK NOT!
HEY! THIS IS A BUSINESS, NOT A RESIDENCE. WHY ARE WE PICKING UP THEIR CRAP?
BECAUSE... THE OWNER BITCHED TO LIPS AND HE CAVED. SO WE'RE STUCK WITH IT UNTIL THE END OF TIME.
THE DEAL IS WE'LL TAKE FIFTY BAGS A WEEK. SO FIFTY IT IS. EVERY WEEK.
OK. GROSS, BUT HARDLY THE HORROR YOU GUYS ALWAYS BABBLE ABOUT.
OH YEAH?
CHECK OUT THE FULL WATERLOGGED EFFECT!
AUGH!! THAT'S FOUL!!
PISSSSSS
PISSSSS

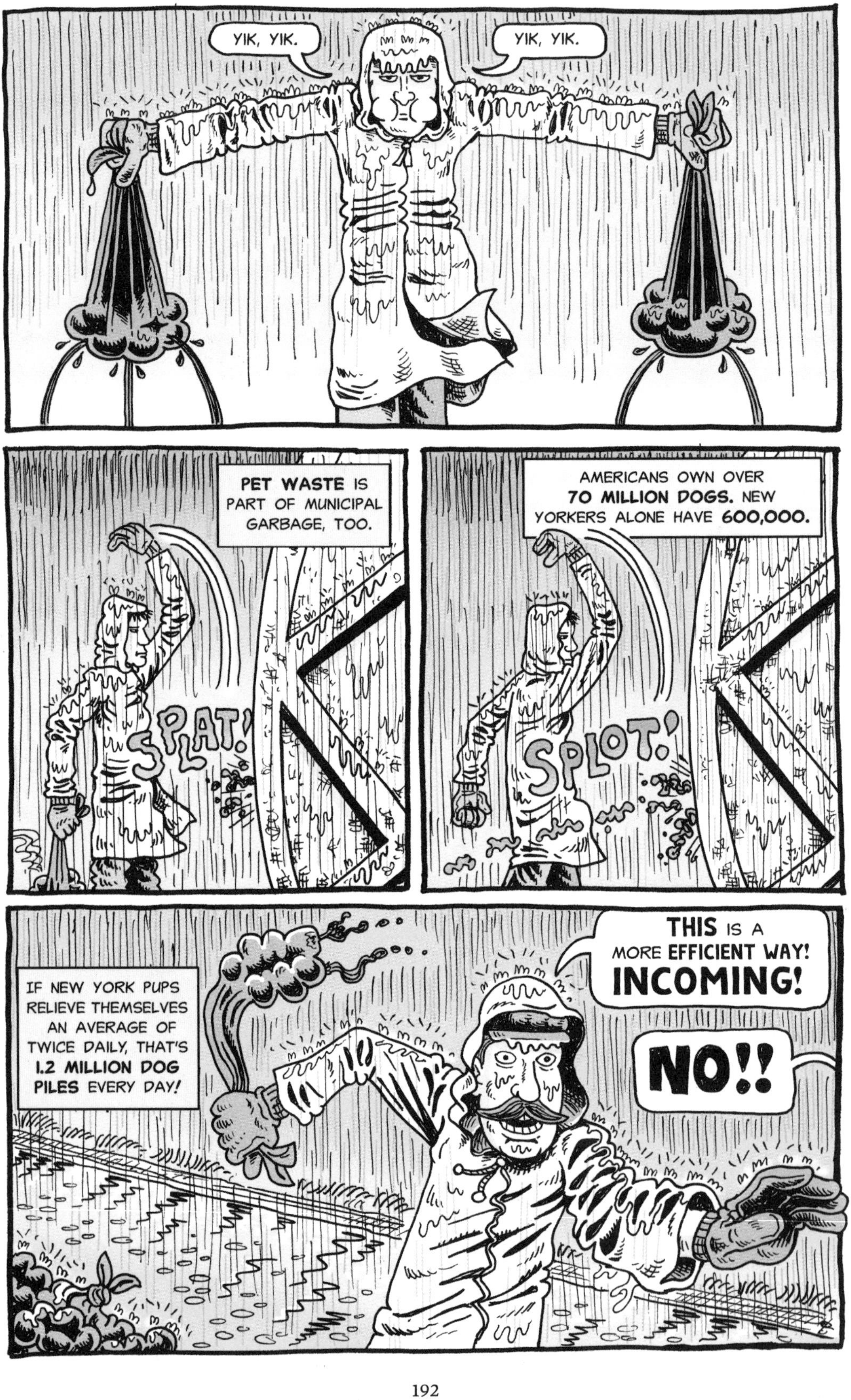
YIK, YIK.
YIK, YIK.
PET WASTE IS PART OF MUNICIPAL GARBAGE, TOO.
SPLAT!
AMERICANS OWN OVER 70 MILLION DOGS. NEW YORKERS ALONE HAVE 600,000.
SPLOT!
IF NEW YORK PUPS RELIEVE THEMSELVES AN AVERAGE OF TWICE DAILY, THAT'S 1.2 MILLION DOG PILES EVERY DAY!
THIS IS A MORE EFFICIENT WAY! INCOMING!
NO!!

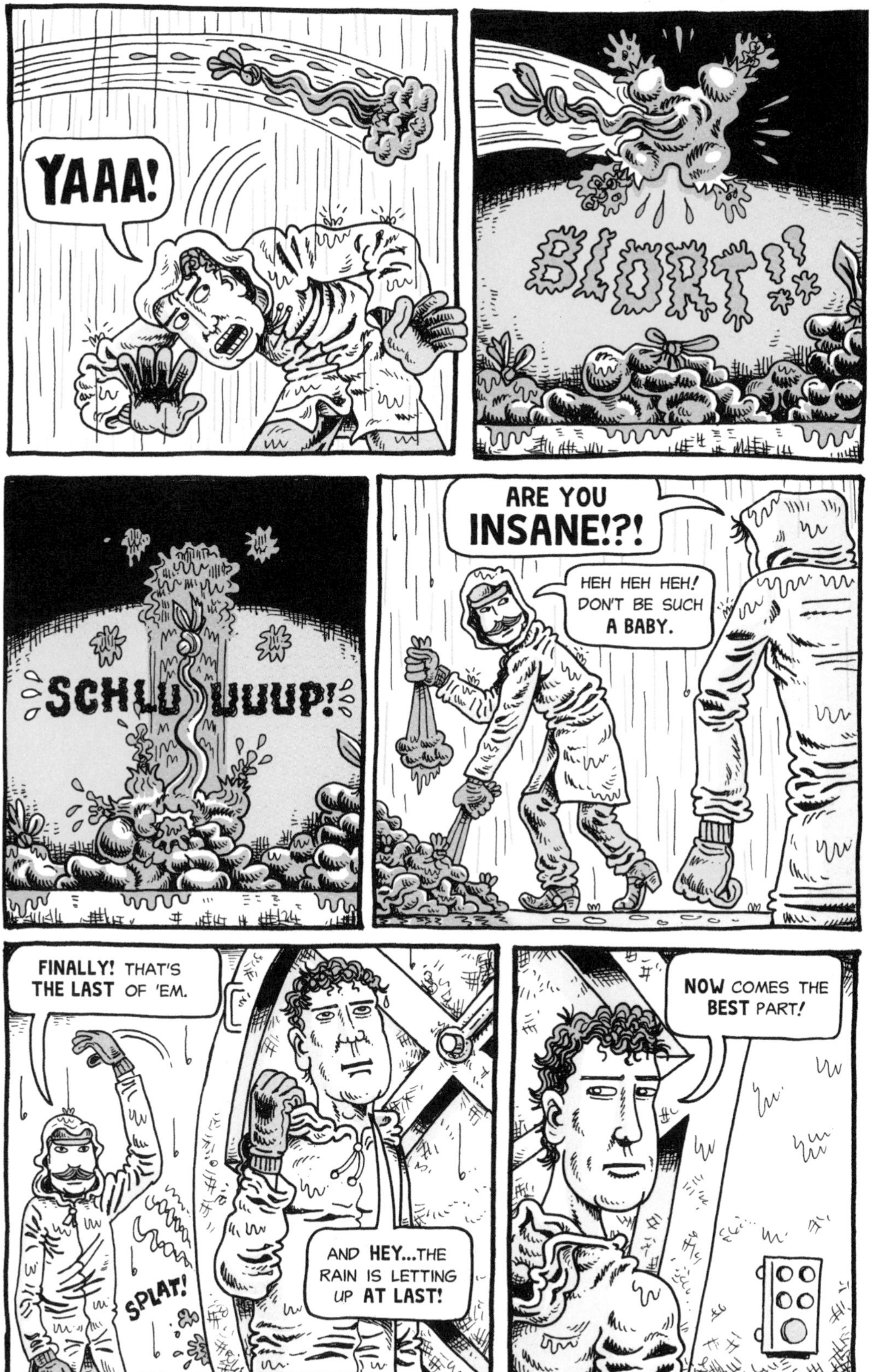
YAAA!
BLORT!
SCHLU UUUP!
ARE YOU INSANE!?!
HEH HEH HEH! DON'T BE SUCH A BABY.
FINALLY! THAT'S THE LAST OF 'EM.
SPLAT!
AND HEY...THE RAIN IS LETTING UP AT LAST!
NOW COMES THE BEST PART!

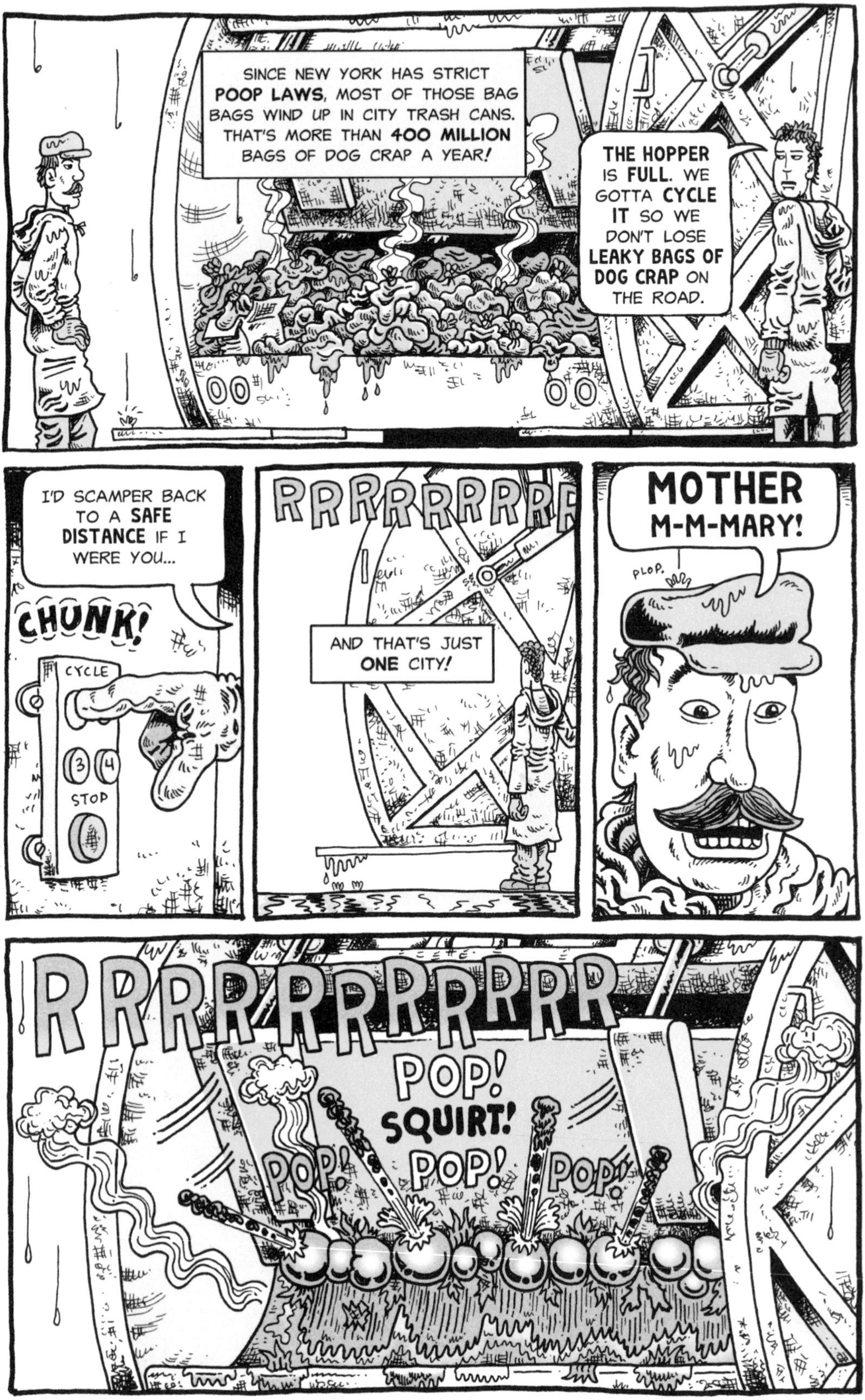
SINCE NEW YORK HAS STRICT POOP LAWS, MOST OF THOSE BAG BAGS WIND UP IN CITY TRASH CANS. THAT'S MORE THAN 400 MILLION BAGS OF DOG CRAP A YEAR!
THE HOPPER IS FULL. WE GOTTA CYCLE IT SO WE DON'T LOSE LEAKY BAGS OF DOG CRAP ON THE ROAD.
I'D SCAMPER BACK TO A SAFE DISTANCE IF I WERE YOU...
CHUNK!
CYCLE
3
4
STOP
RRRRRRRR
AND THAT'S JUST ONE CITY!
MOTHER M-M-MARY!
PLOP.
RRRRRRRRRR
POP!
SQUIRT!
POP!
POP!
POP!

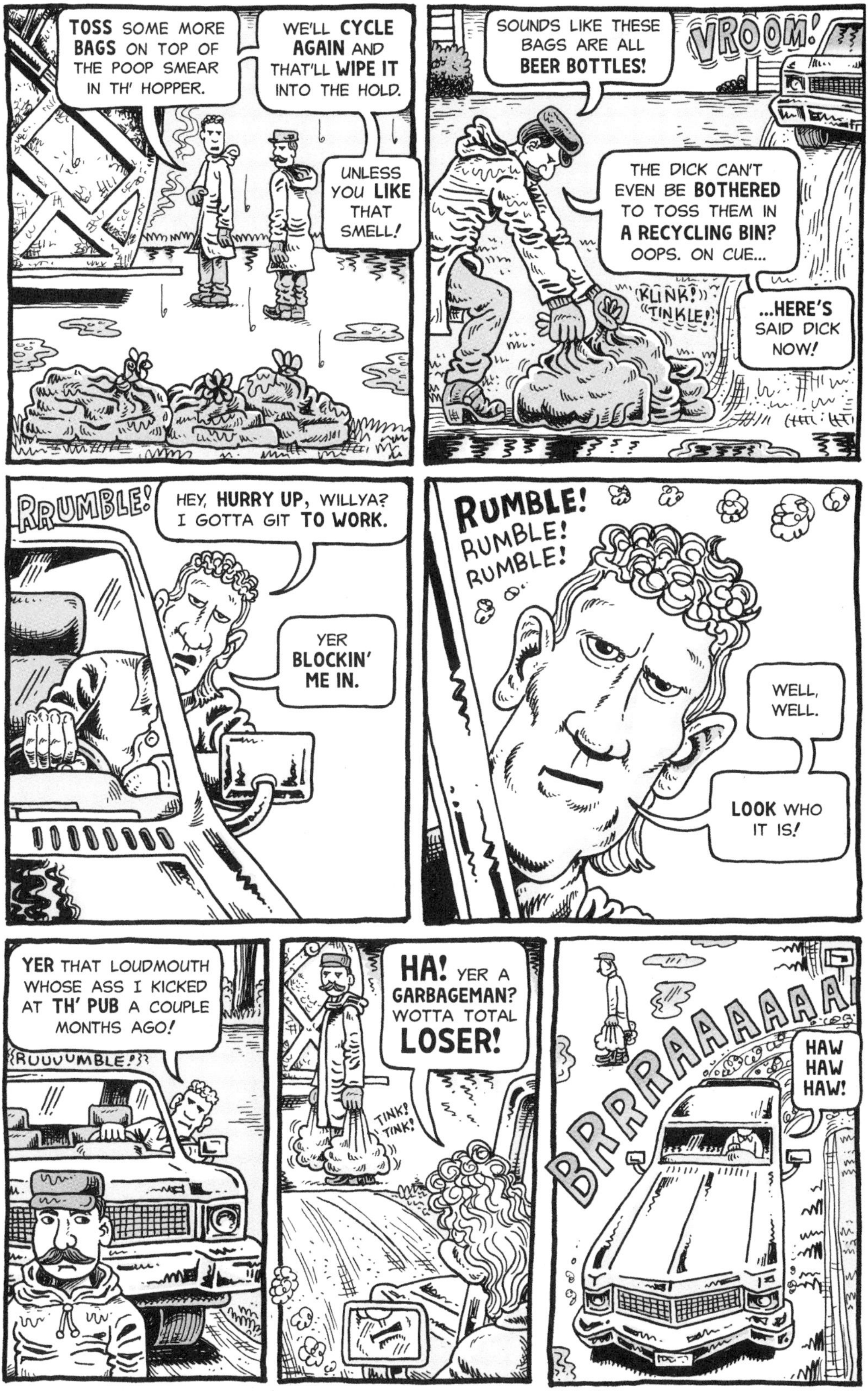
TOSS SOME MORE BAGS ON TOP OF THE POOP SMEAR IN TH' HOPPER.
WE'LL CYCLE AGAIN AND THAT'LL WIPE IT INTO THE HOLD.
UNLESS YOU LIKE THAT SMELL!
SOUNDS LIKE THESE BAGS ARE ALL BEER BOTTLES!
VROOM!
THE DICK CAN'T EVEN BE BOTHERED TO TOSS THEM IN A RECYCLING BIN? OOPS. ON CUE...
KLINK! TINKLE!
...HERE'S SAID DICK NOW!
RRUMBLE!
HEY, HURRY UP, WILLYA? I GOTTA GIT TO WORK.
YER BLOCKIN' ME IN.
RUMBLE! RUMBLE! RUMBLE!
WELL, WELL.
LOOK WHO IT IS!
YER THAT LOUDMOUTH WHOSE ASS I KICKED AT TH' PUB A COUPLE MONTHS AGO!
RUUUUMBLE!
HA! YER A GARBAGEMAN? WOTTA TOTAL LOSER!
TINK! TINK!
BRRRAAAAAAA
HAW HAW HAW!

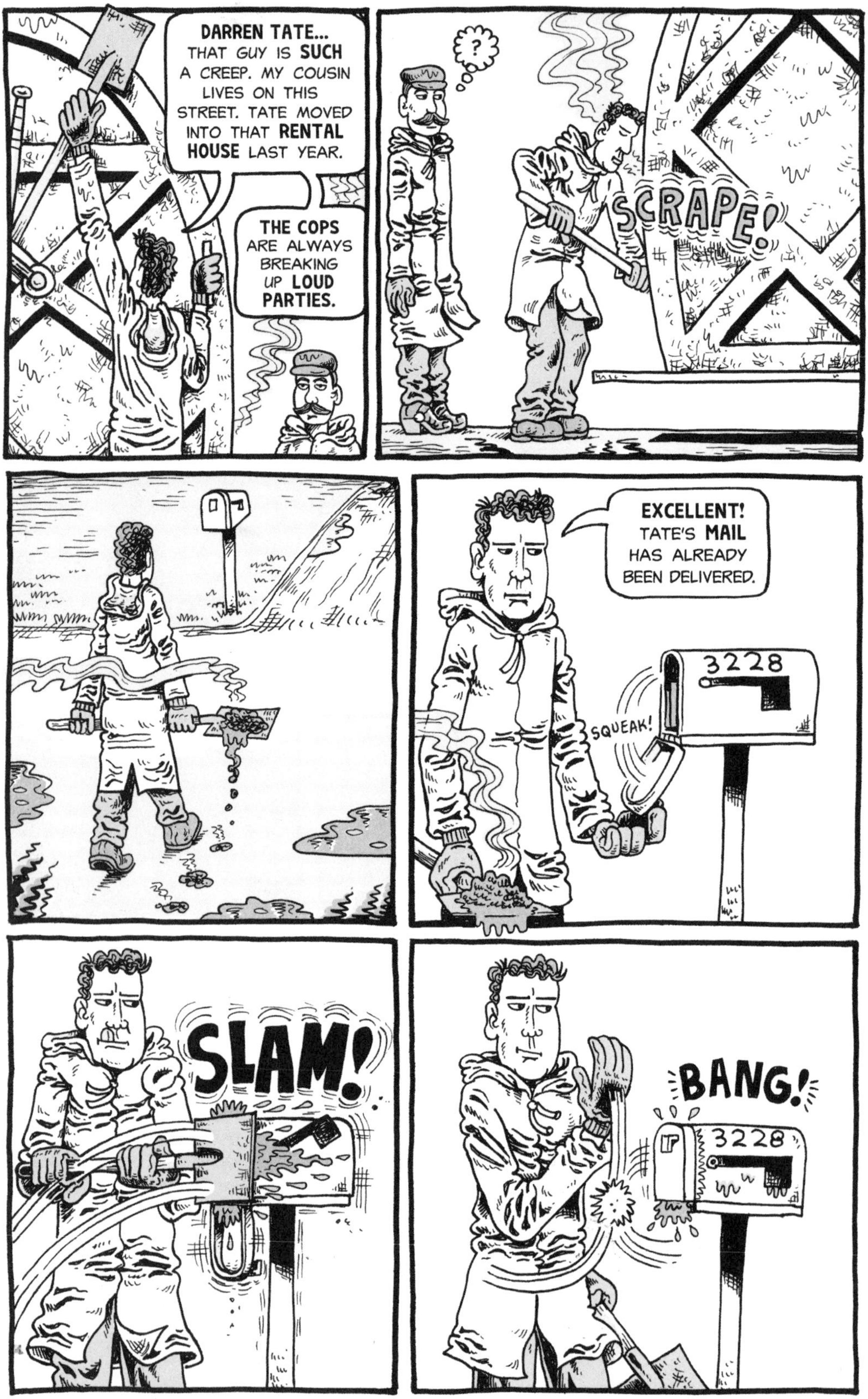
DARREN TATE... THAT GUY IS SUCH A CREEP. MY COUSIN LIVES ON THIS STREET. TATE MOVED INTO THAT RENTAL HOUSE LAST YEAR.
THE COPS ARE ALWAYS BREAKING UP LOUD PARTIES.
?
SCRAPE!
EXCELLENT! TATE'S MAIL HAS ALREADY BEEN DELIVERED.
3228
SQUEAK!
SLAM!
BANG!
3228

DON'T PISS OFF A GARBAGEMAN, MR. TATE!
HAHAHAHA!
HA
HA
HA
HEE
HAHA
HOHO
HAHA
HA
HO
RRRRR
3228
RRRRRRRRRR

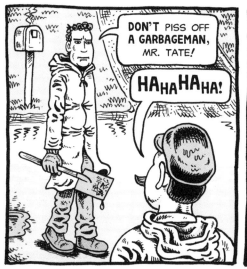

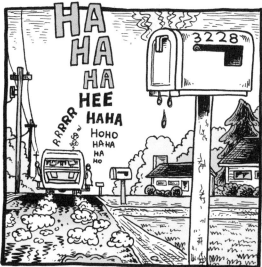

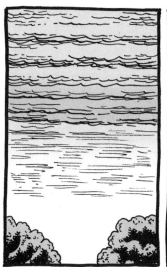

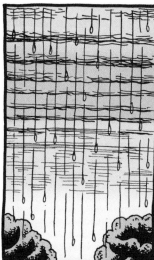

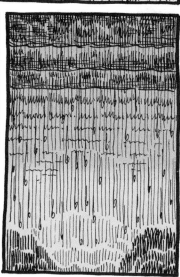

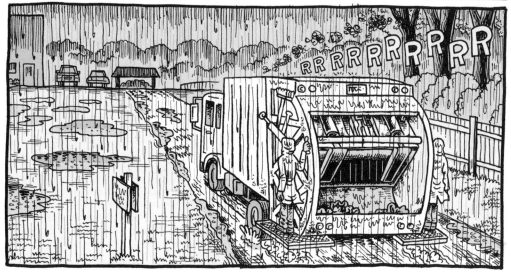

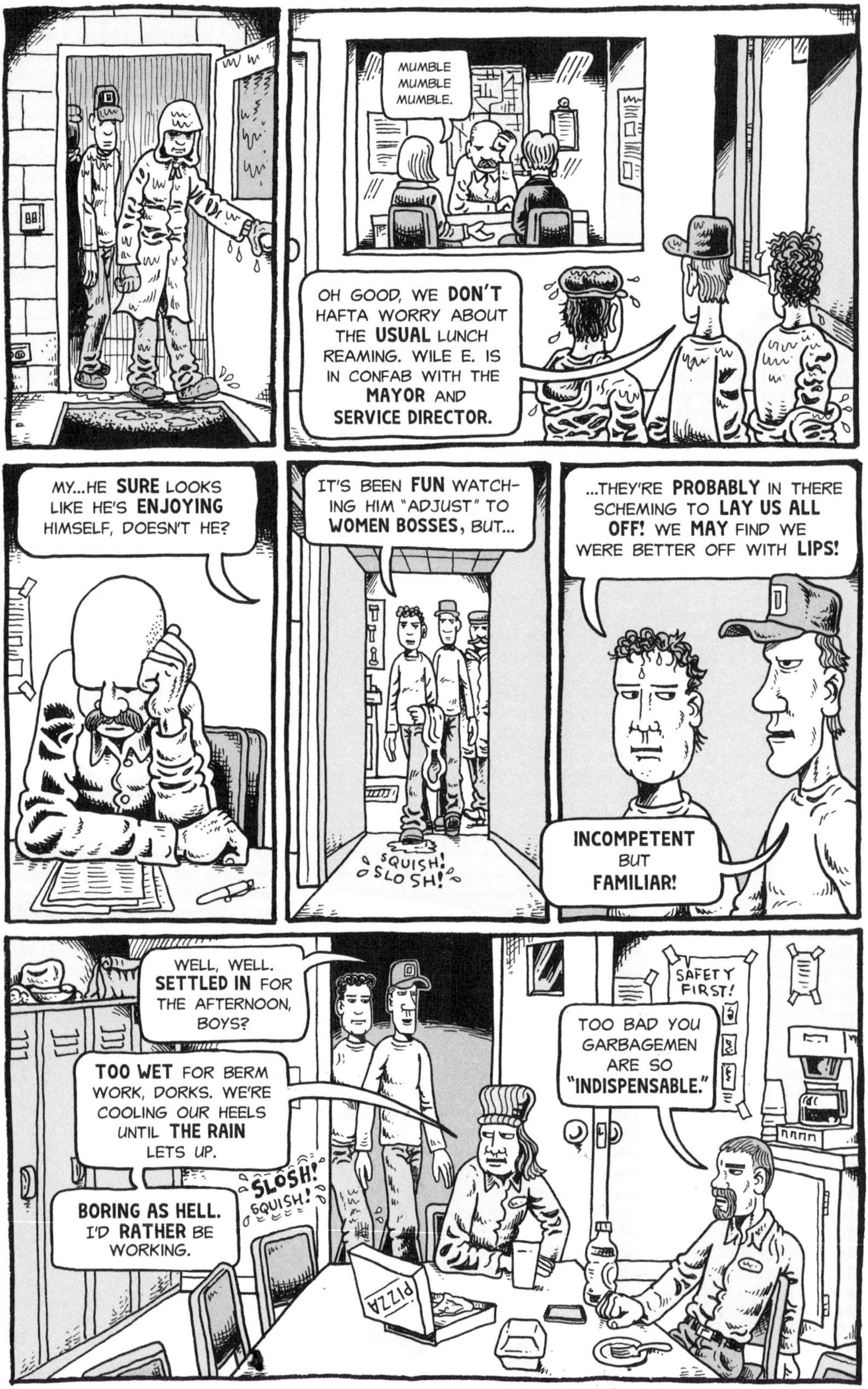
MUMBLE MUMBLE MUMBLE.
OH GOOD, WE DON'T HAFTA WORRY ABOUT THE USUAL LUNCH REAMING. WILE E. IS IN CONFAB WITH THE MAYOR AND SERVICE DIRECTOR.
MY...HE SURE LOOKS LIKE HE'S ENJOYING HIMSELF, DOESN'T HE?
IT'S BEEN FUN WATCHING HIM "ADJUST" TO WOMEN BOSSES, BUT...
SQUISH! SLOSH!
...THEY'RE PROBABLY IN THERE SCHEMING TO LAY US ALL OFF! WE MAY FIND WE WERE BETTER OFF WITH LIPS!
INCOMPETENT BUT FAMILIAR!
WELL, WELL. SETTLED IN FOR THE AFTERNOON, BOYS?
TOO WET FOR BERM WORK, DORKS. WE'RE COOLING OUR HEELS UNTIL THE RAIN LETS UP.
BORING AS HELL. I'D RATHER BE WORKING.
SLOSH! SQUISH!
SAFETY FIRST!
TOO BAD YOU GARBAGEMEN ARE SO "INDISPENSABLE."
PIZZA

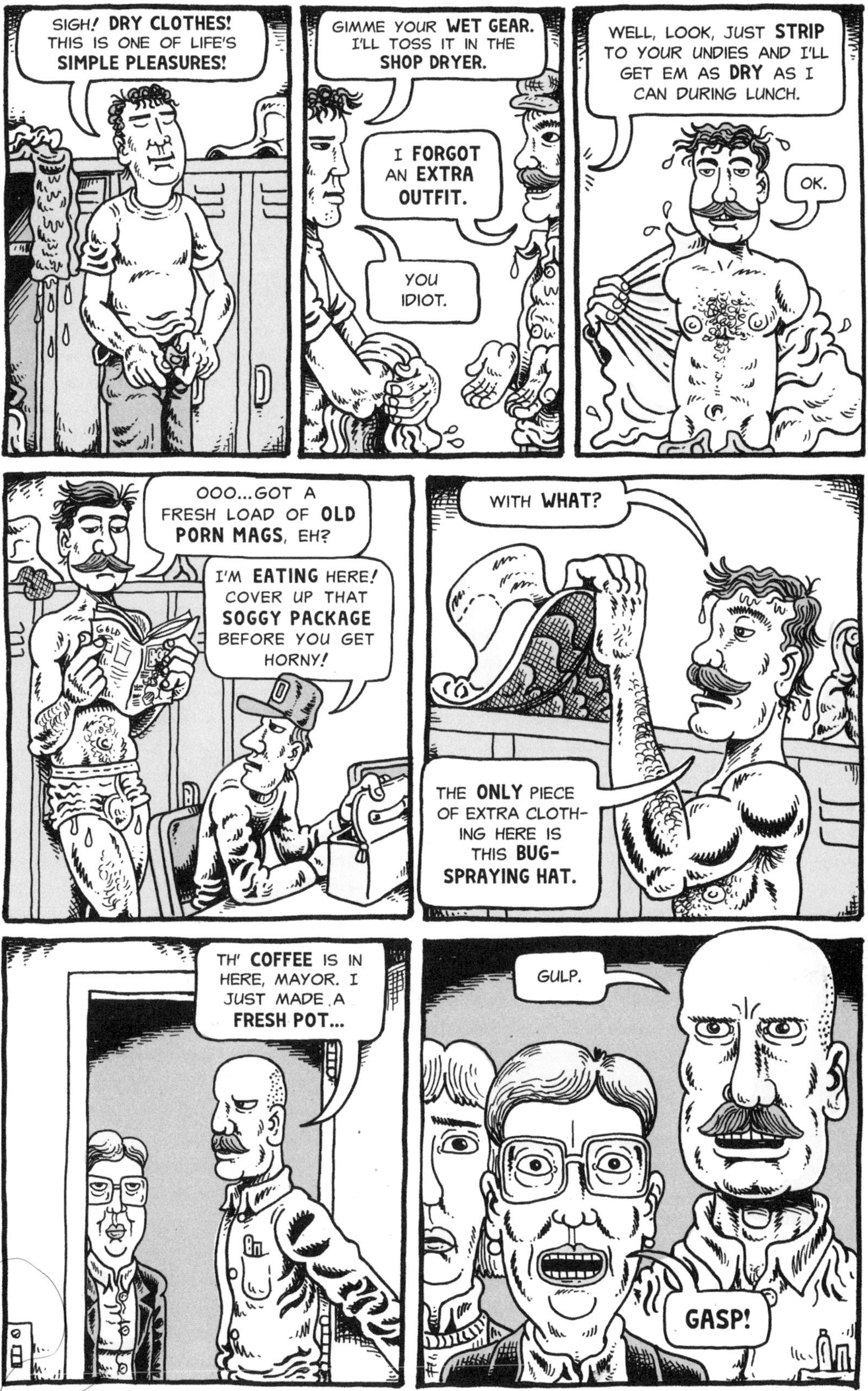
SIGH! DRY CLOTHES! THIS IS ONE OF LIFE'S SIMPLE PLEASURES!
GIMME YOUR WET GEAR. I'LL TOSS IT IN THE SHOP DRYER.
I FORGOT AN EXTRA OUTFIT.
YOU IDIOT.
WELL, LOOK, JUST STRIP TO YOUR UNDIES AND I'LL GET EM AS DRY AS I CAN DURING LUNCH.
OK.
OOO...GOT A FRESH LOAD OF OLD PORN MAGS, EH?
I'M EATING HERE! COVER UP THAT SOGGY PACKAGE BEFORE YOU GET HORNY!
WITH WHAT?
THE ONLY PIECE OF EXTRA CLOTHING HERE IS THIS BUG-SPRAYING HAT.
TH' COFFEE IS IN HERE, MAYOR. I JUST MADE A FRESH POT...
GULP.
GASP!

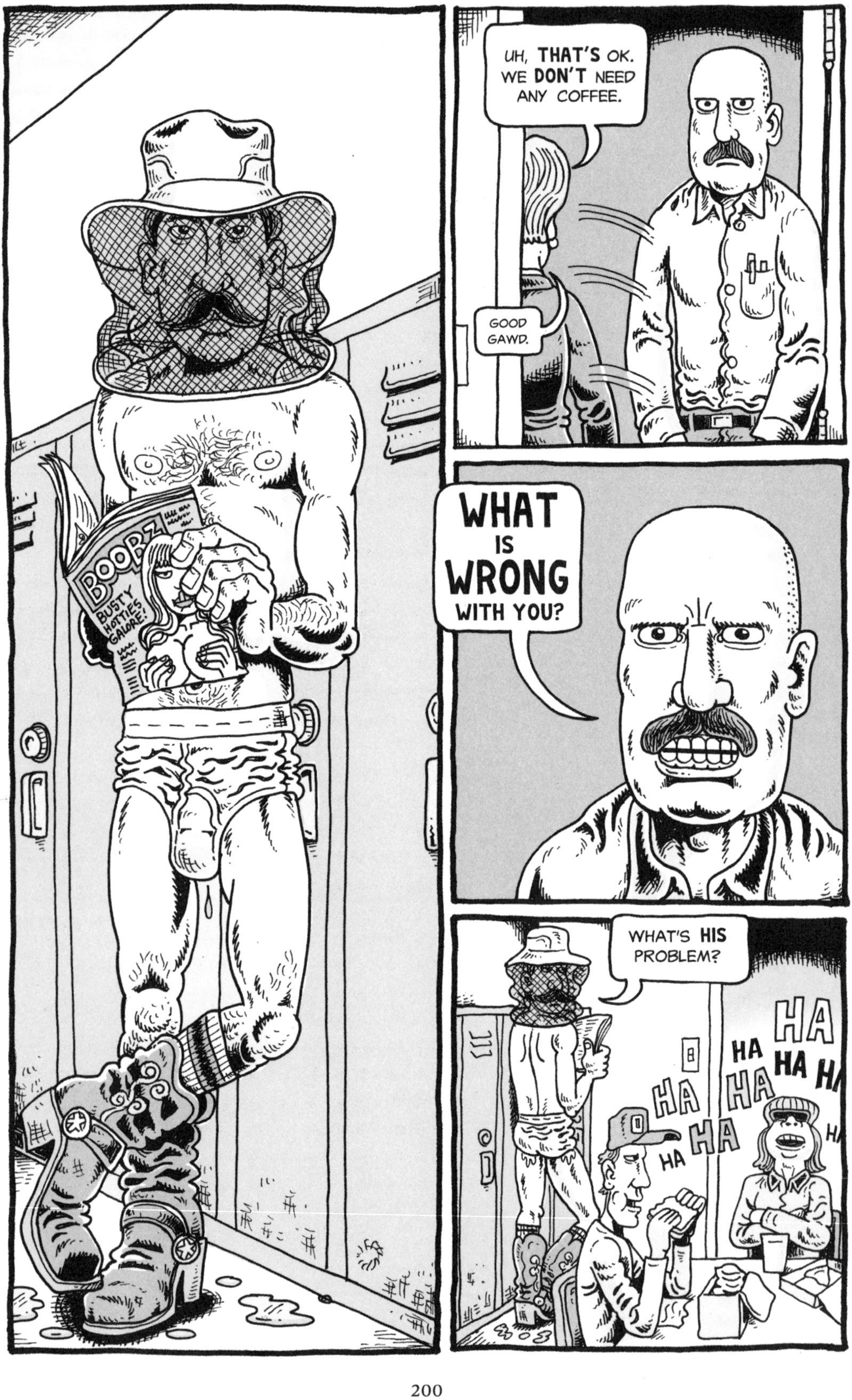
BOOBZ
BUSTY HOTTIES GALORE!
UH, THAT'S OK. WE DON'T NEED ANY COFFEE.
GOOD GAWD.
WHAT IS WRONG WITH YOU?
WHAT'S HIS PROBLEM?
HA HA HA HA HA HA HA HA HA HA

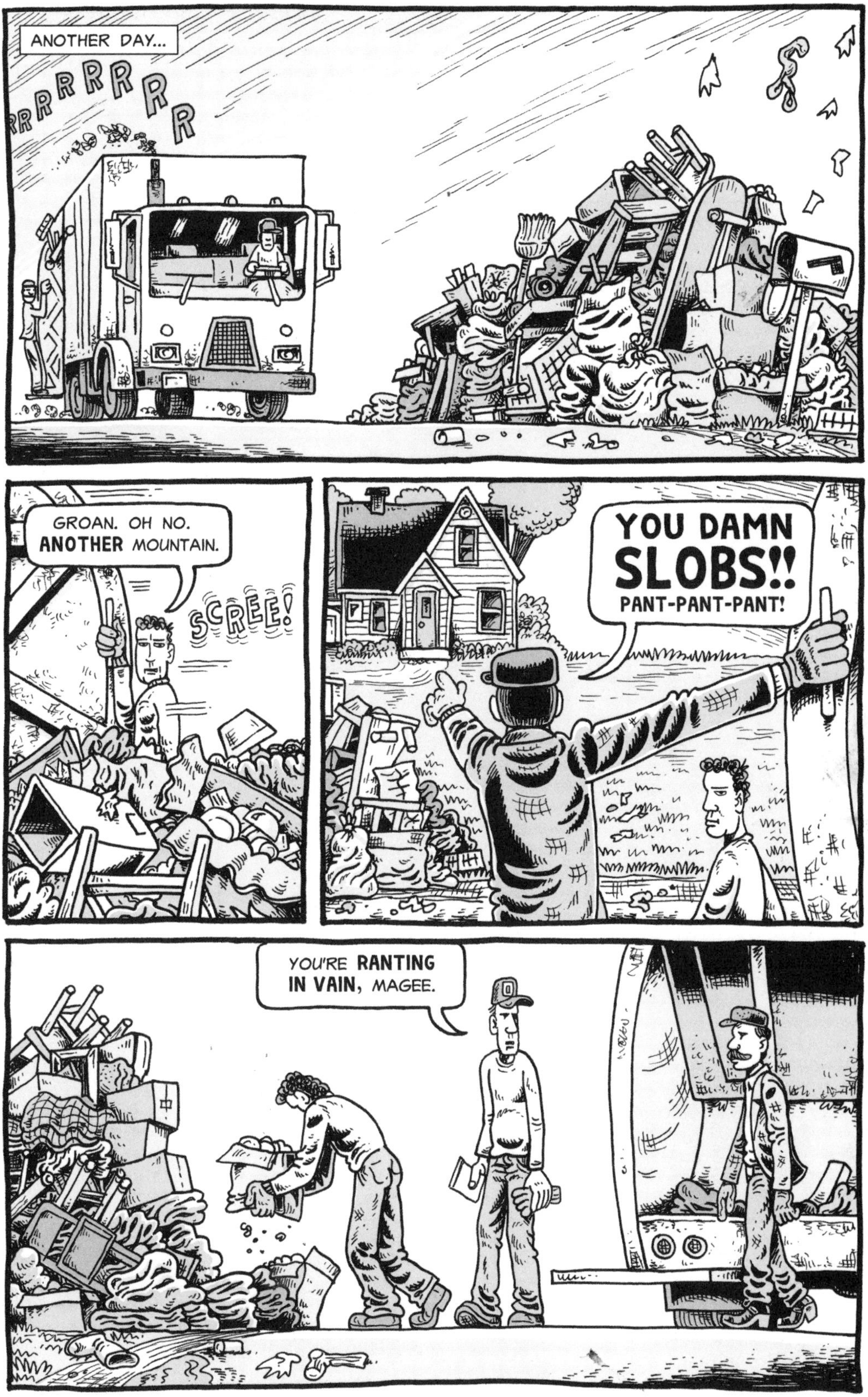
ANOTHER DAY...
RRRRRRRRR
GROAN. OH NO. ANOTHER MOUNTAIN.
SCREE!
YOU DAMN SLOBS!! PANT-PANT-PANT!
YOU'RE RANTING IN VAIN, MAGEE.

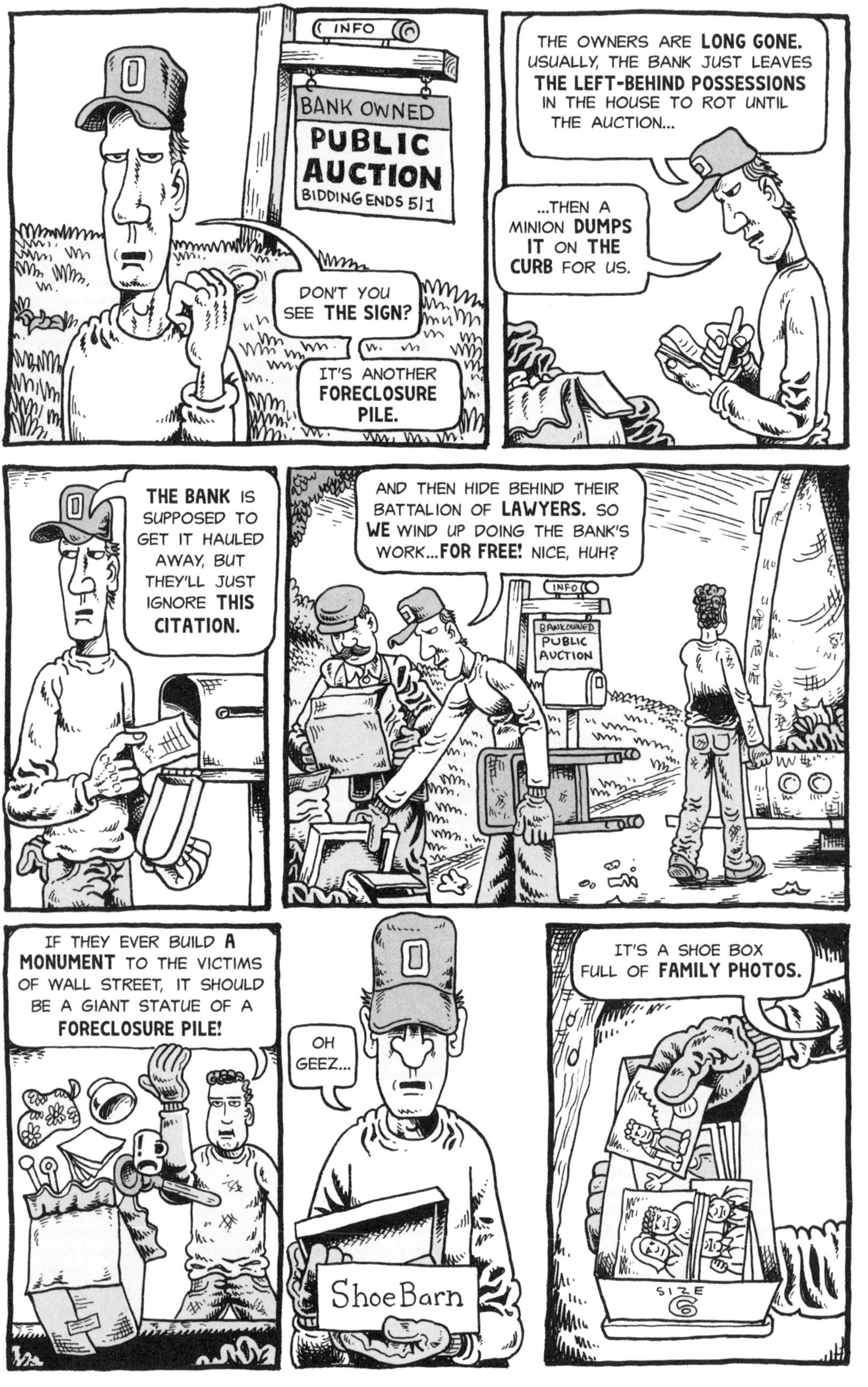
INFO
BANK OWNED
PUBLIC AUCTION
BIDDING ENDS 5/1
DON'T YOU SEE THE SIGN?
IT'S ANOTHER FORECLOSURE PILE.
THE OWNERS ARE LONG GONE. USUALLY, THE BANK JUST LEAVES THE LEFT-BEHIND POSSESSIONS IN THE HOUSE TO ROT UNTIL THE AUCTION...
...THEN A MINION DUMPS IT ON THE CURB FOR US.
THE BANK IS SUPPOSED TO GET IT HAULED AWAY, BUT THEY'LL JUST IGNORE THIS CITATION.
AND THEN HIDE BEHIND THEIR BATTALION OF LAWYERS. SO WE WIND UP DOING THE BANK'S WORK...FOR FREE! NICE, HUH?
INFO
BANK OWNED
PUBLIC AUCTION
IF THEY EVER BUILD A MONUMENT TO THE VICTIMS OF WALL STREET, IT SHOULD BE A GIANT STATUE OF A FORECLOSURE PILE!
OH GEEZ...
ShoeBarn
IT'S A SHOE BOX FULL OF FAMILY PHOTOS.
SIZE 6

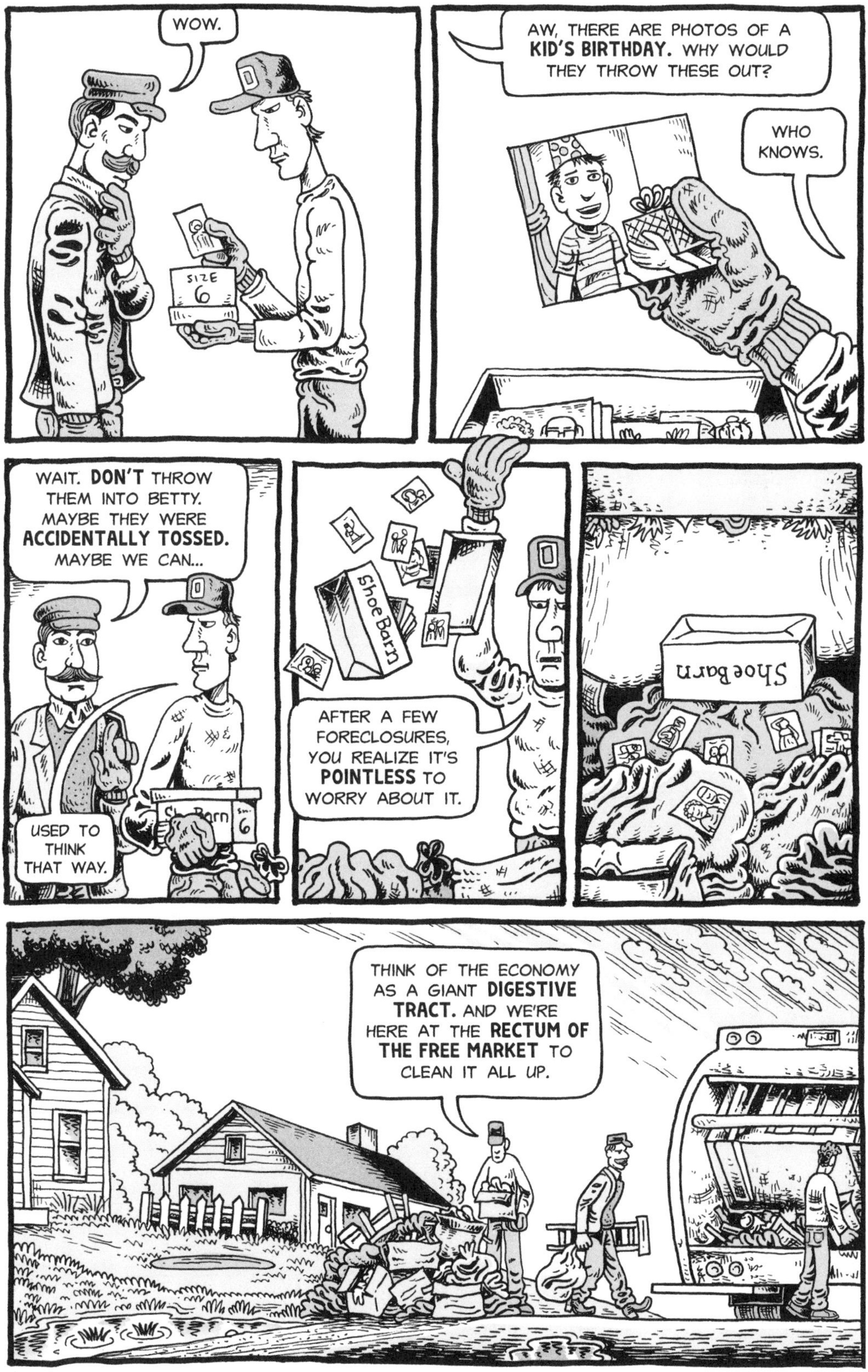
WOW.
SIZE 6
AW, THERE ARE PHOTOS OF A KID'S BIRTHDAY. WHY WOULD THEY THROW THESE OUT?
WHO KNOWS.
WAIT. DON'T THROW THEM INTO BETTY. MAYBE THEY WERE ACCIDENTALLY TOSSED. MAYBE WE CAN...
USED TO THINK THAT WAY.
Shoe Barn
AFTER A FEW FORECLOSURES, YOU REALIZE IT'S POINTLESS TO WORRY ABOUT IT.
Shoe Barn
THINK OF THE ECONOMY AS A GIANT DIGESTIVE TRACT. AND WE'RE HERE AT THE RECTUM OF THE FREE MARKET TO CLEAN IT ALL UP.

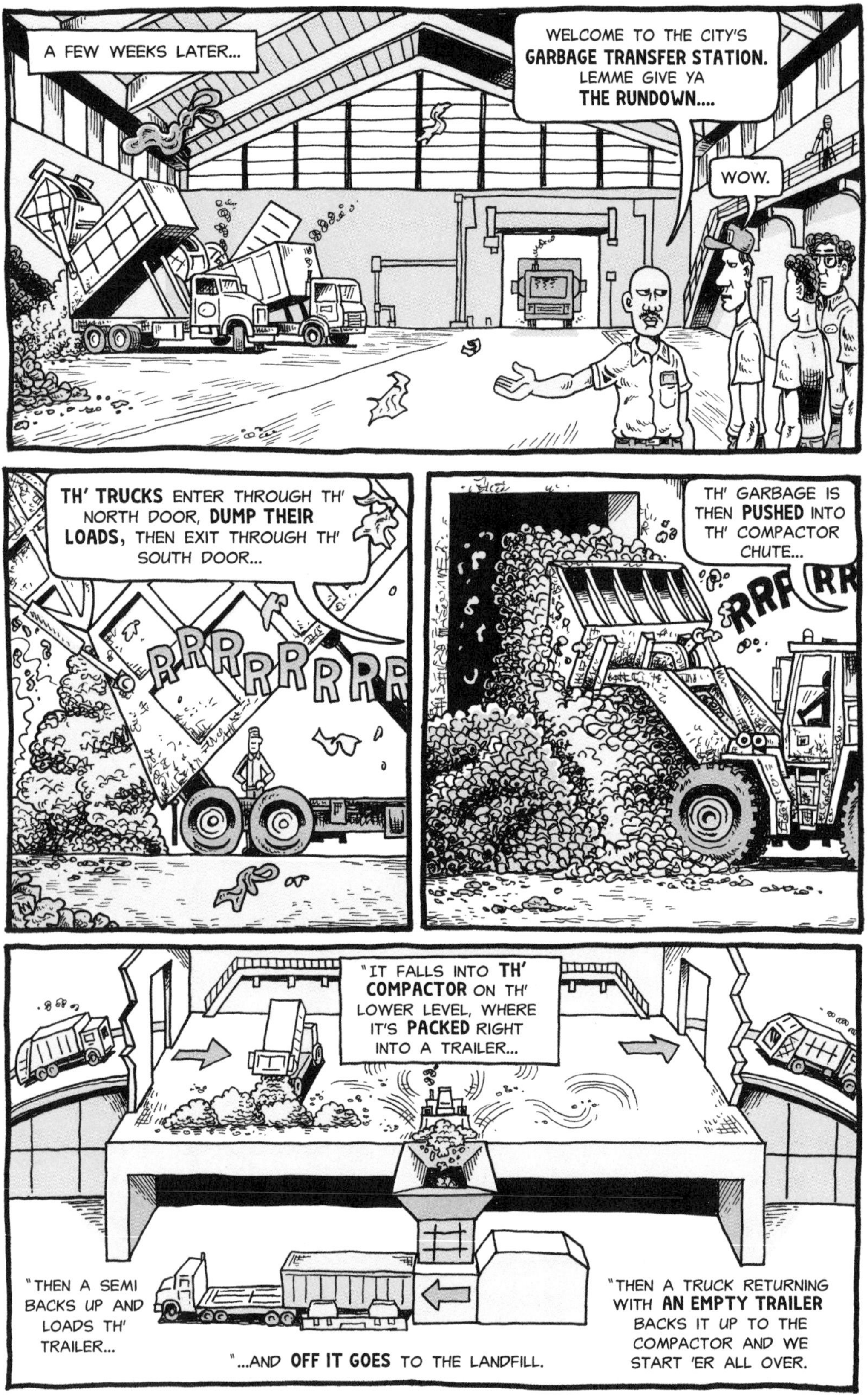
A FEW WEEKS LATER...
WELCOME TO THE CITY'S GARBAGE TRANSFER STATION. LEMME GIVE YA THE RUNDOWN....
WOW.
TH' TRUCKS ENTER THROUGH TH' NORTH DOOR, DUMP THEIR LOADS, THEN EXIT THROUGH TH' SOUTH DOOR...
RRRRRRRR
TH' GARBAGE IS THEN PUSHED INTO TH' COMPACTOR CHUTE...
RRR RR
"IT FALLS INTO TH' COMPACTOR ON TH' LOWER LEVEL, WHERE IT'S PACKED RIGHT INTO A TRAILER...
"THEN A SEMI BACKS UP AND LOADS TH' TRAILER...
"...AND OFF IT GOES TO THE LANDFILL.
"THEN A TRUCK RETURNING WITH AN EMPTY TRAILER BACKS IT UP TO THE COMPACTOR AND WE START 'ER ALL OVER.

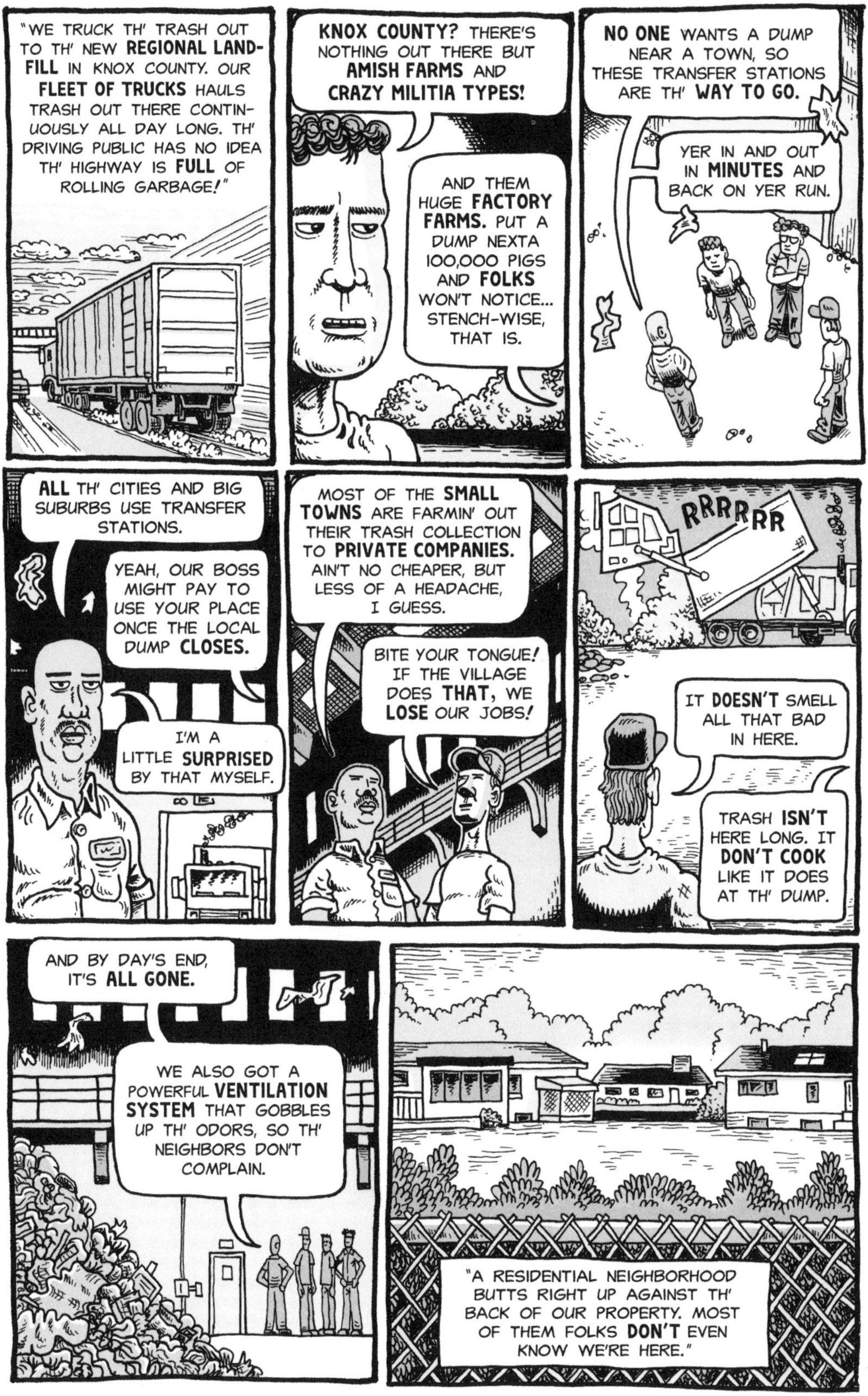
"WE TRUCK TH' TRASH OUT TO TH' NEW REGIONAL LANDFILL IN KNOX COUNTY. OUR FLEET OF TRUCKS HAULS TRASH OUT THERE CONTINUOUSLY ALL DAY LONG. TH' DRIVING PUBLIC HAS NO IDEA TH' HIGHWAY IS FULL OF ROLLING GARBAGE!"
KNOX COUNTY? THERE'S NOTHING OUT THERE BUT AMISH FARMS AND CRAZY MILITIA TYPES!
AND THEM HUGE FACTORY FARMS. PUT A DUMP NEXTA 100,000 PIGS AND FOLKS WON'T NOTICE... STENCH-WISE, THAT IS.
NO ONE WANTS A DUMP NEAR A TOWN, SO THESE TRANSFER STATIONS ARE TH' WAY TO GO.
YER IN AND OUT IN MINUTES AND BACK ON YER RUN.
ALL TH' CITIES AND BIG SUBURBS USE TRANSFER STATIONS.
YEAH, OUR BOSS MIGHT PAY TO USE YOUR PLACE ONCE THE LOCAL DUMP CLOSES.
I'M A LITTLE SURPRISED BY THAT MYSELF.
MOST OF THE SMALL TOWNS ARE FARMIN' OUT THEIR TRASH COLLECTION TO PRIVATE COMPANIES. AIN'T NO CHEAPER, BUT LESS OF A HEADACHE, I GUESS.
BITE YOUR TONGUE! IF THE VILLAGE DOES THAT, WE LOSE OUR JOBS!
RRRRRR
IT DOESN'T SMELL ALL THAT BAD IN HERE.
TRASH ISN'T HERE LONG. IT DON'T COOK LIKE IT DOES AT TH' DUMP.
AND BY DAY'S END, IT'S ALL GONE.
WE ALSO GOT A POWERFUL VENTILATION SYSTEM THAT GOBBLES UP TH' ODORS, SO TH' NEIGHBORS DON'T COMPLAIN.
"A RESIDENTIAL NEIGHBORHOOD BUTTS RIGHT UP AGAINST TH' BACK OF OUR PROPERTY. MOST OF THEM FOLKS DON'T EVEN KNOW WE'RE HERE."

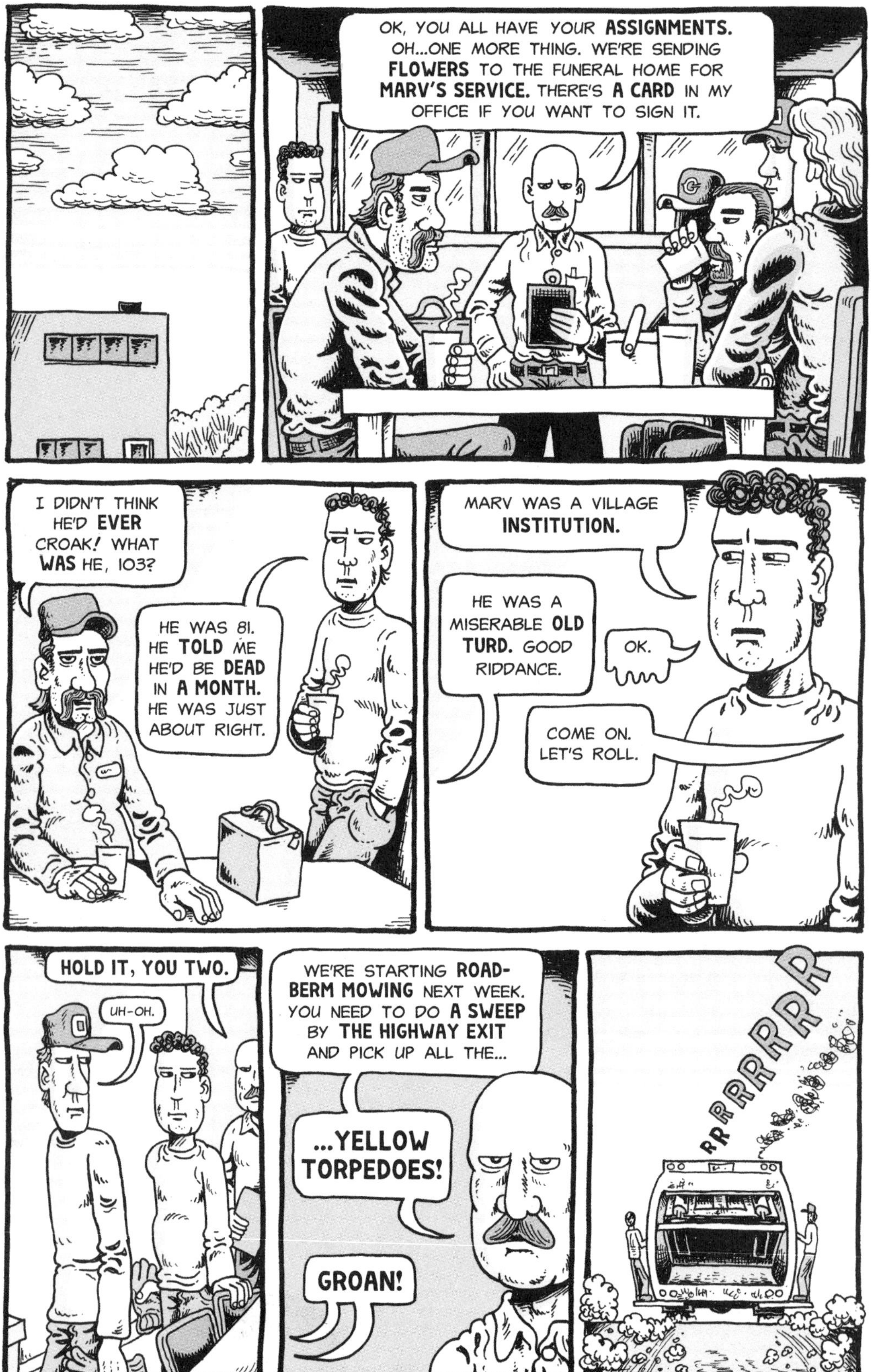
OK, YOU ALL HAVE YOUR ASSIGNMENTS. OH...ONE MORE THING. WE'RE SENDING FLOWERS TO THE FUNERAL HOME FOR MARV'S SERVICE. THERE'S A CARD IN MY OFFICE IF YOU WANT TO SIGN IT.
I DIDN'T THINK HE'D EVER CROAK! WHAT WAS HE, 103?
HE WAS 81. HE TOLD ME HE'D BE DEAD IN A MONTH. HE WAS JUST ABOUT RIGHT.
MARV WAS A VILLAGE INSTITUTION.
HE WAS A MISERABLE OLD TURD. GOOD RIDDANCE.
OK.
COME ON. LET'S ROLL.
HOLD IT, YOU TWO.
UH-OH.
WE'RE STARTING ROAD-BERM MOWING NEXT WEEK. YOU NEED TO DO A SWEEP BY THE HIGHWAY EXIT AND PICK UP ALL THE...
...YELLOW TORPEDOES!
GROAN!
RRRRRRRRR

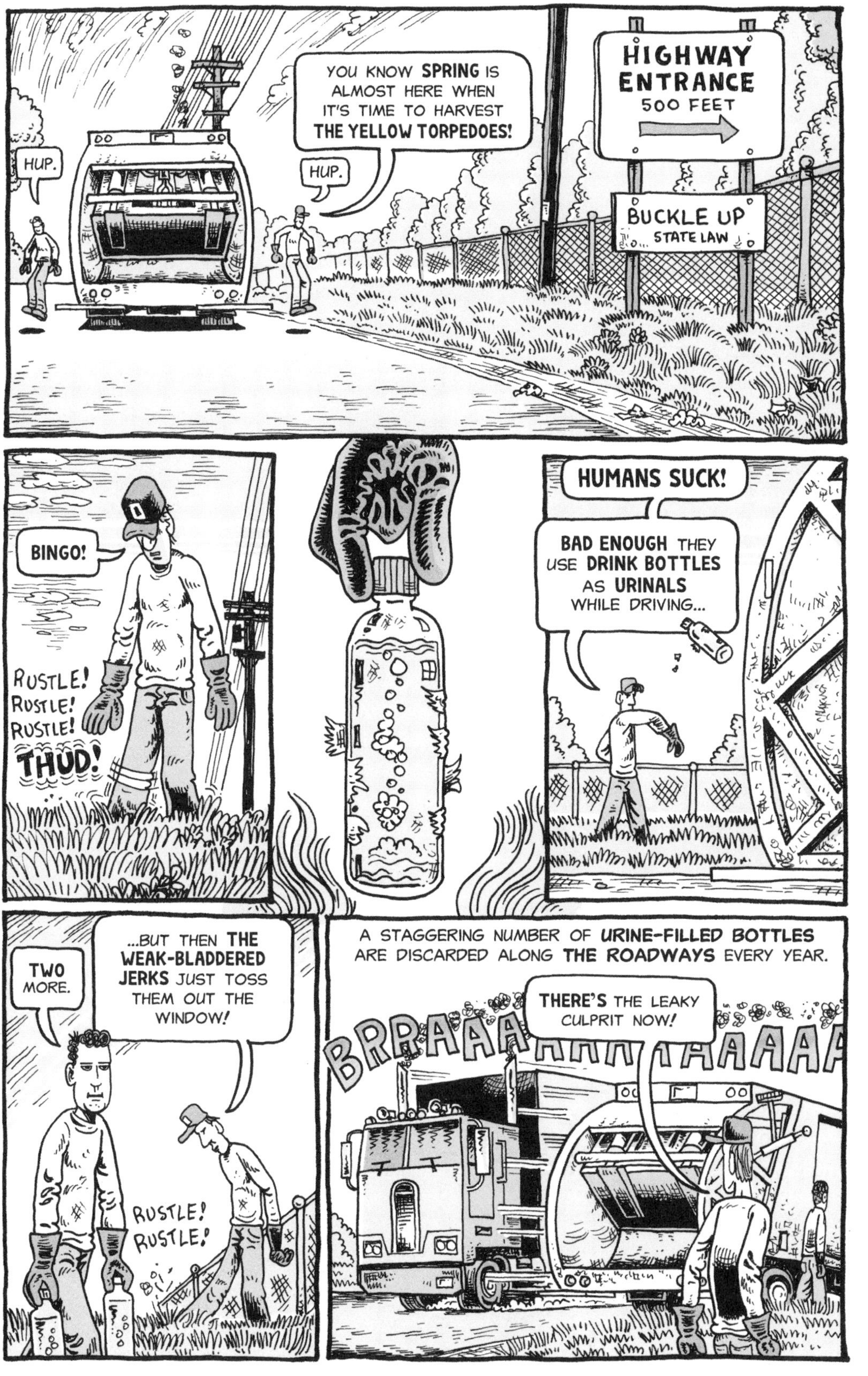
HUP.
YOU KNOW **SPRING** IS ALMOST HERE WHEN IT'S TIME TO HARVEST **THE YELLOW TORPEDOES!**
HUP.
HIGHWAY ENTRANCE 500 FEET
BUCKLE UP STATE LAW
BINGO!
RUSTLE! RUSTLE! RUSTLE!
THUD!
HUMANS SUCK!
BAD ENOUGH THEY USE **DRINK BOTTLES** AS **URINALS** WHILE DRIVING...
TWO MORE.
...BUT THEN **THE WEAK-BLADDERED JERKS** JUST TOSS THEM OUT THE WINDOW!
RUSTLE! RUSTLE!
A STAGGERING NUMBER OF **URINE-FILLED BOTTLES** ARE DISCARDED ALONG **THE ROADWAYS** EVERY YEAR.
BRRAAAAAAAAAAAAA
THERE'S THE LEAKY CULPRIT NOW!

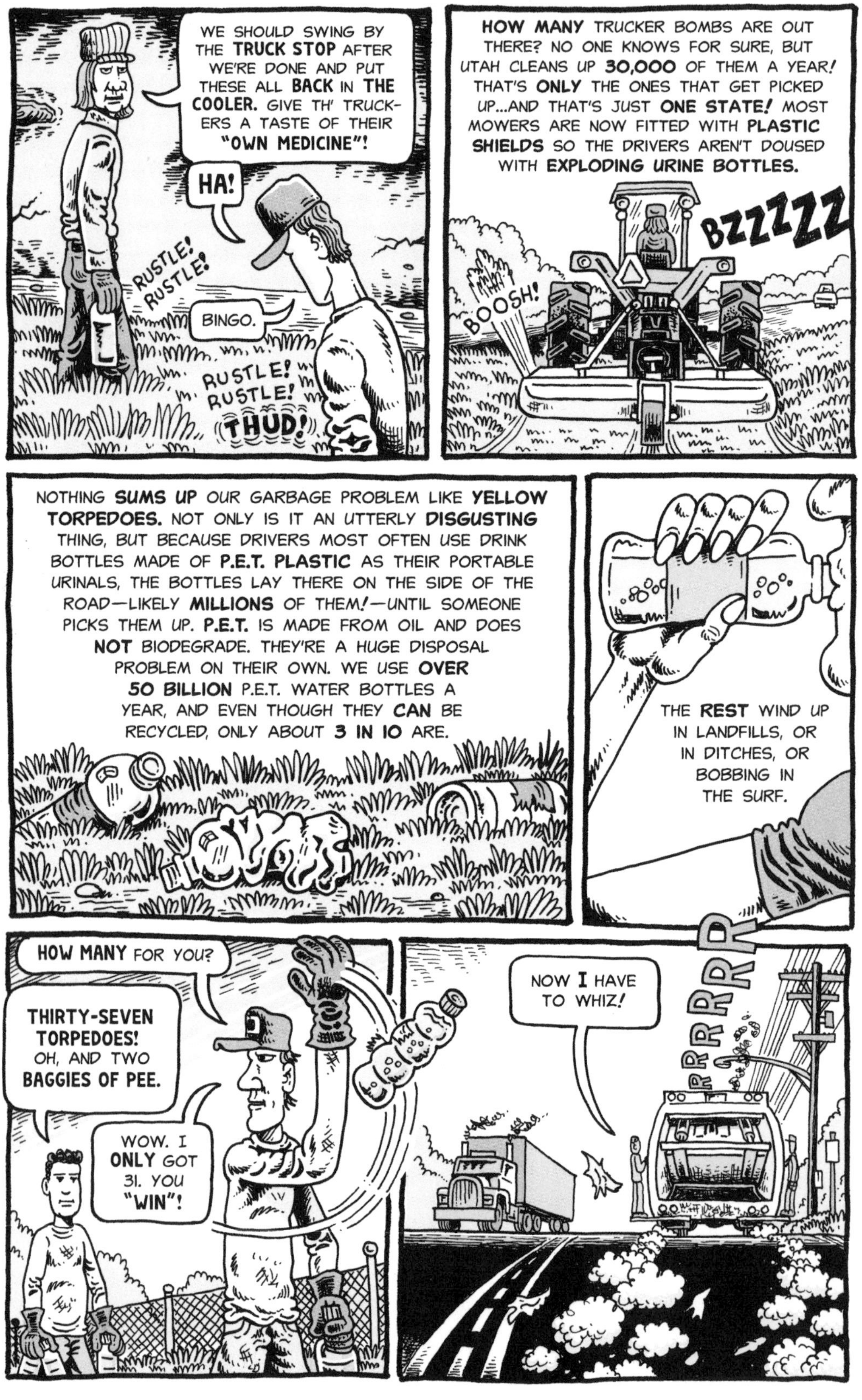
WE SHOULD SWING BY THE TRUCK STOP AFTER WE'RE DONE AND PUT THESE ALL BACK IN THE COOLER. GIVE TH' TRUCKERS A TASTE OF THEIR "OWN MEDICINE"!
HA!
RUSTLE! RUSTLE!
BINGO.
RUSTLE! RUSTLE!
THUD!
HOW MANY TRUCKER BOMBS ARE OUT THERE? NO ONE KNOWS FOR SURE, BUT UTAH CLEANS UP 30,000 OF THEM A YEAR! THAT'S ONLY THE ONES THAT GET PICKED UP...AND THAT'S JUST ONE STATE! MOST MOWERS ARE NOW FITTED WITH PLASTIC SHIELDS SO THE DRIVERS AREN'T DOUSED WITH EXPLODING URINE BOTTLES.
BZZZZZ
BOOSH!
NOTHING SUMS UP OUR GARBAGE PROBLEM LIKE YELLOW TORPEDOES. NOT ONLY IS IT AN UTTERLY DISGUSTING THING, BUT BECAUSE DRIVERS MOST OFTEN USE DRINK BOTTLES MADE OF P.E.T. PLASTIC AS THEIR PORTABLE URINALS, THE BOTTLES LAY THERE ON THE SIDE OF THE ROAD—LIKELY MILLIONS OF THEM!—UNTIL SOMEONE PICKS THEM UP. P.E.T. IS MADE FROM OIL AND DOES NOT BIODEGRADE. THEY'RE A HUGE DISPOSAL PROBLEM ON THEIR OWN. WE USE OVER 50 BILLION P.E.T. WATER BOTTLES A YEAR, AND EVEN THOUGH THEY CAN BE RECYCLED, ONLY ABOUT 3 IN 10 ARE.
THE REST WIND UP IN LANDFILLS, OR IN DITCHES, OR BOBBING IN THE SURF.
HOW MANY FOR YOU?
THIRTY-SEVEN TORPEDOES! OH, AND TWO BAGGIES OF PEE.
WOW. I ONLY GOT 31. YOU "WIN"!
NOW I HAVE TO WHIZ!
RRRRRR

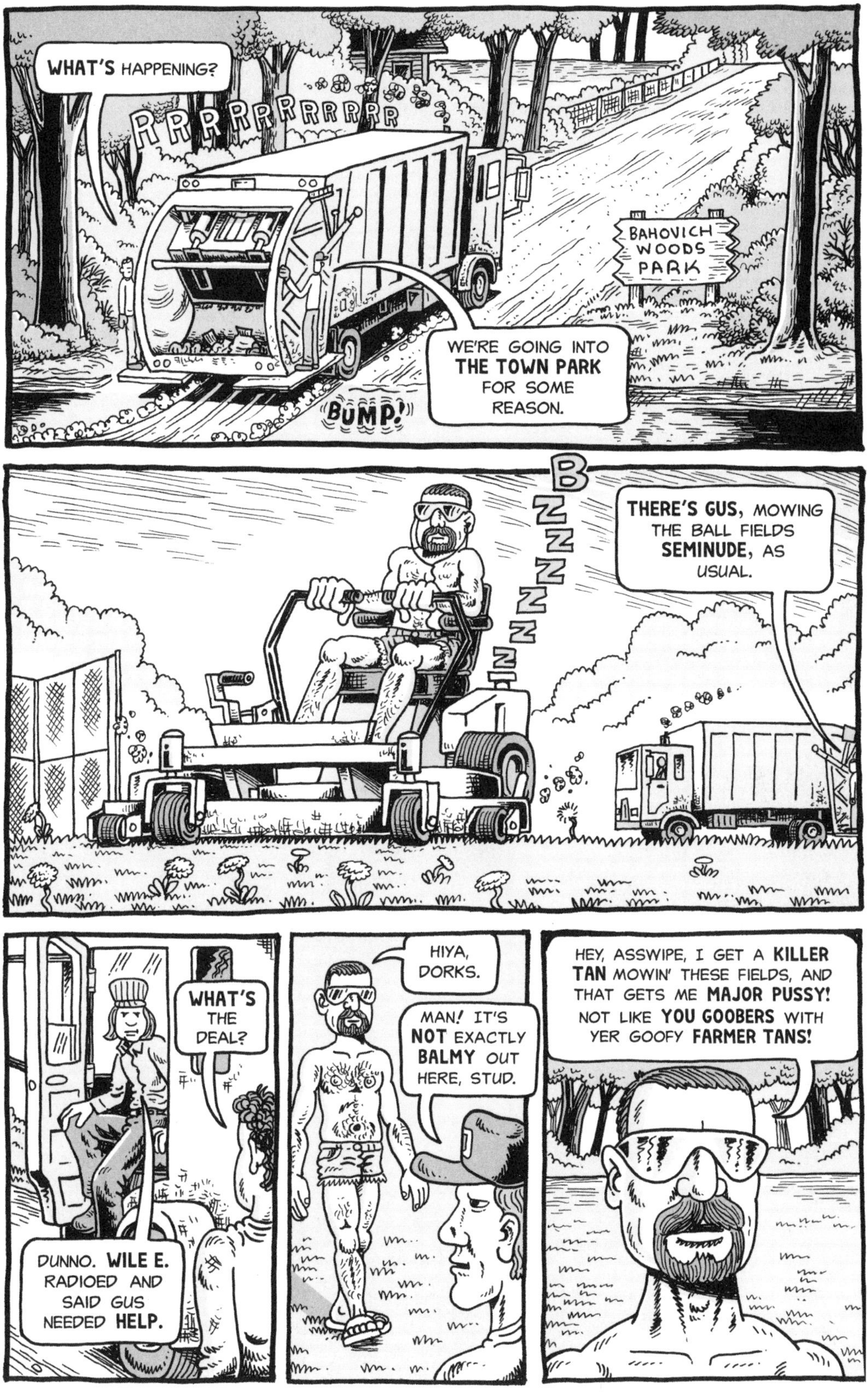

WHAT'S HAPPENING?
RRRRRRRRRRRRR
BAHOVICH WOODS PARK
WE'RE GOING INTO THE TOWN PARK FOR SOME REASON.
BUMP!
BZZZZZZ
THERE'S GUS, MOWING THE BALL FIELDS SEMINUDE, AS USUAL.
WHAT'S THE DEAL?
DUNNO. WILE E. RADIOED AND SAID GUS NEEDED HELP.
HIYA, DORKS.
MAN! IT'S NOT EXACTLY BALMY OUT HERE, STUD.
HEY, ASSWIPE, I GET A KILLER TAN MOWIN' THESE FIELDS, AND THAT GETS ME MAJOR PUSSY! NOT LIKE YOU GOOBERS WITH YER GOOFY FARMER TANS!

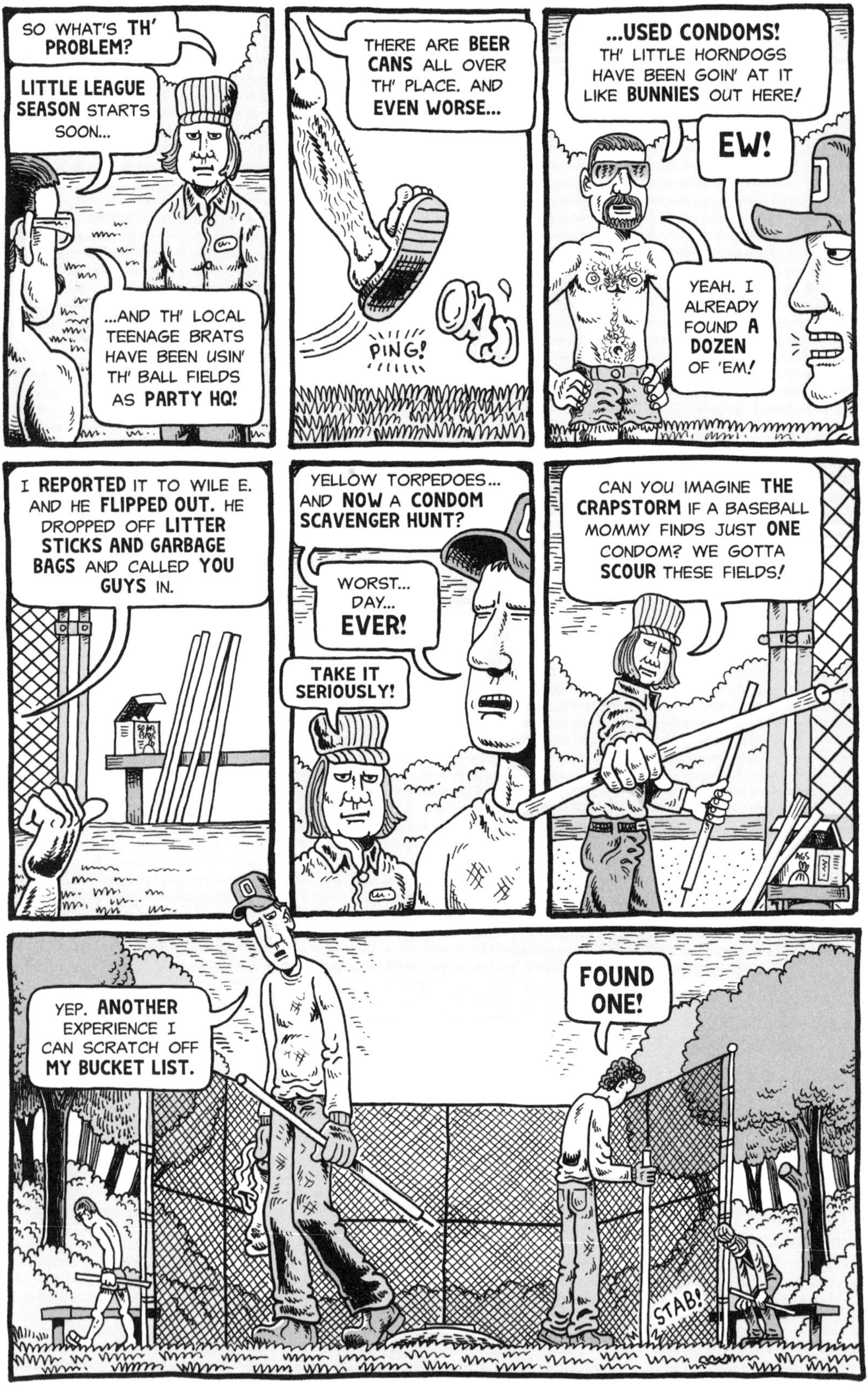
SO WHAT'S TH' PROBLEM?
LITTLE LEAGUE SEASON STARTS SOON...
...AND TH' LOCAL TEENAGE BRATS HAVE BEEN USIN' TH' BALL FIELDS AS PARTY HQ!
THERE ARE BEER CANS ALL OVER TH' PLACE. AND EVEN WORSE...
PING!
...USED CONDOMS! TH' LITTLE HORNDOGS HAVE BEEN GOIN' AT IT LIKE BUNNIES OUT HERE!
EW!
YEAH. I ALREADY FOUND A DOZEN OF 'EM!
I REPORTED IT TO WILE E. AND HE FLIPPED OUT. HE DROPPED OFF LITTER STICKS AND GARBAGE BAGS AND CALLED YOU GUYS IN.
YELLOW TORPEDOES... AND NOW A CONDOM SCAVENGER HUNT?
WORST... DAY... EVER!
TAKE IT SERIOUSLY!
CAN YOU IMAGINE THE CRAPSTORM IF A BASEBALL MOMMY FINDS JUST ONE CONDOM? WE GOTTA SCOUR THESE FIELDS!
YEP. ANOTHER EXPERIENCE I CAN SCRATCH OFF MY BUCKET LIST.
FOUND ONE!
STAB!

WELL, **THAT** WAS AN HOUR OF MY LIFE I'LL ALWAYS TREASURE.

LATER.

RRRRRR

HUH. WHAT'S **DIRK** LOOKING AT HERE?

WHAZZUP?

TWENTY LANDSCAPE **BOULDERS** GOT PINCHED.

YOU MEAN TH' ONES YOU AND I **JUST** PUT IN LAST WEEK?

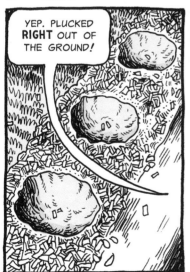

YEP. PLUCKED **RIGHT** OUT OF THE GROUND!

THAT **JERK** IS MAKING WORK FOR **US** NOW.

YEAH. IT'S **TIME.**

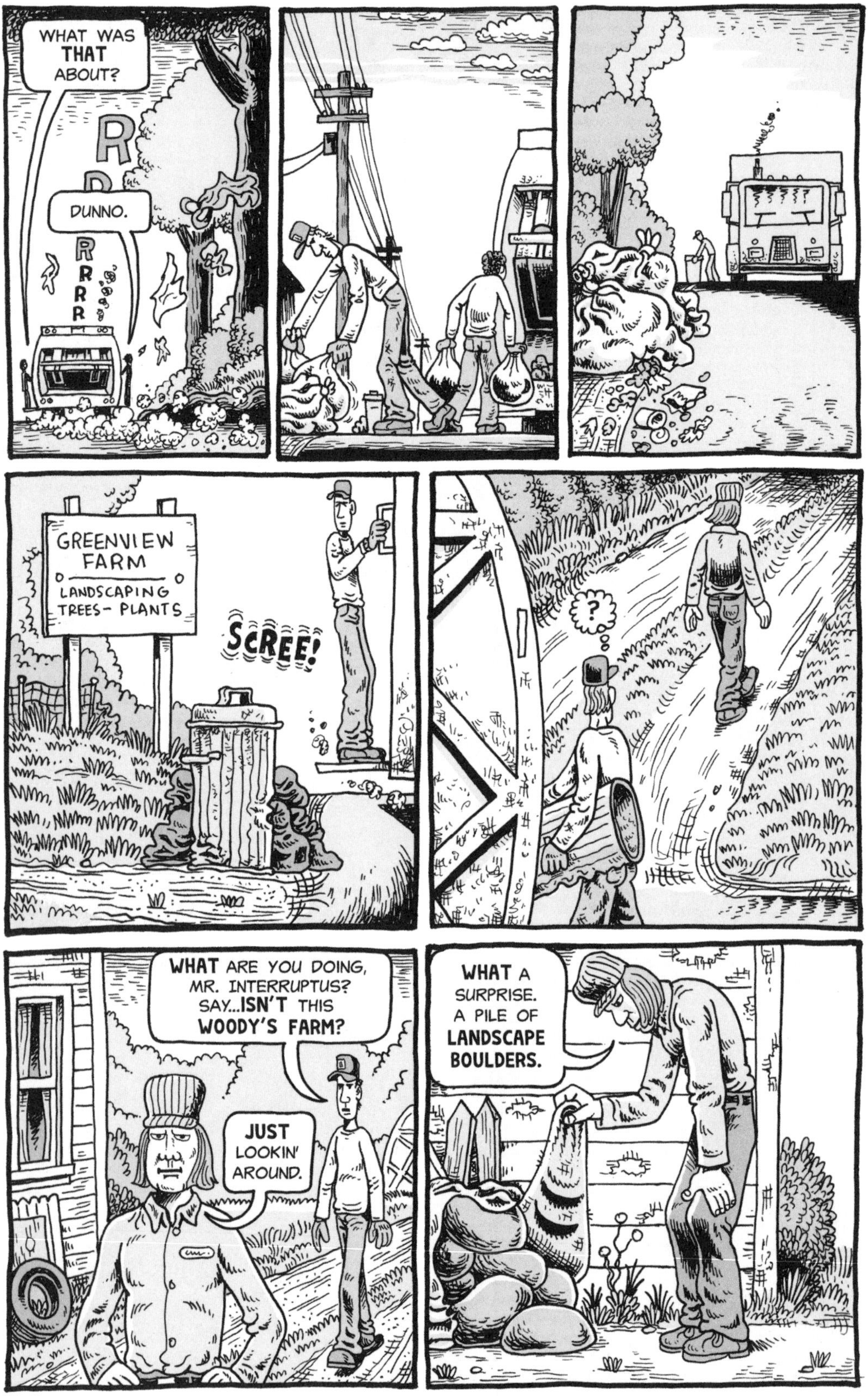
WHAT WAS **THAT** ABOUT?
R
R
DUNNO.
R R R R
GREENVIEW FARM
LANDSCAPING TREES- PLANTS
SCREE!
?
WHAT ARE YOU DOING, MR. INTERRUPTUS? SAY...**ISN'T** THIS **WOODY'S FARM?**
JUST LOOKIN' AROUND.
WHAT A SURPRISE. A PILE OF **LANDSCAPE BOULDERS.**

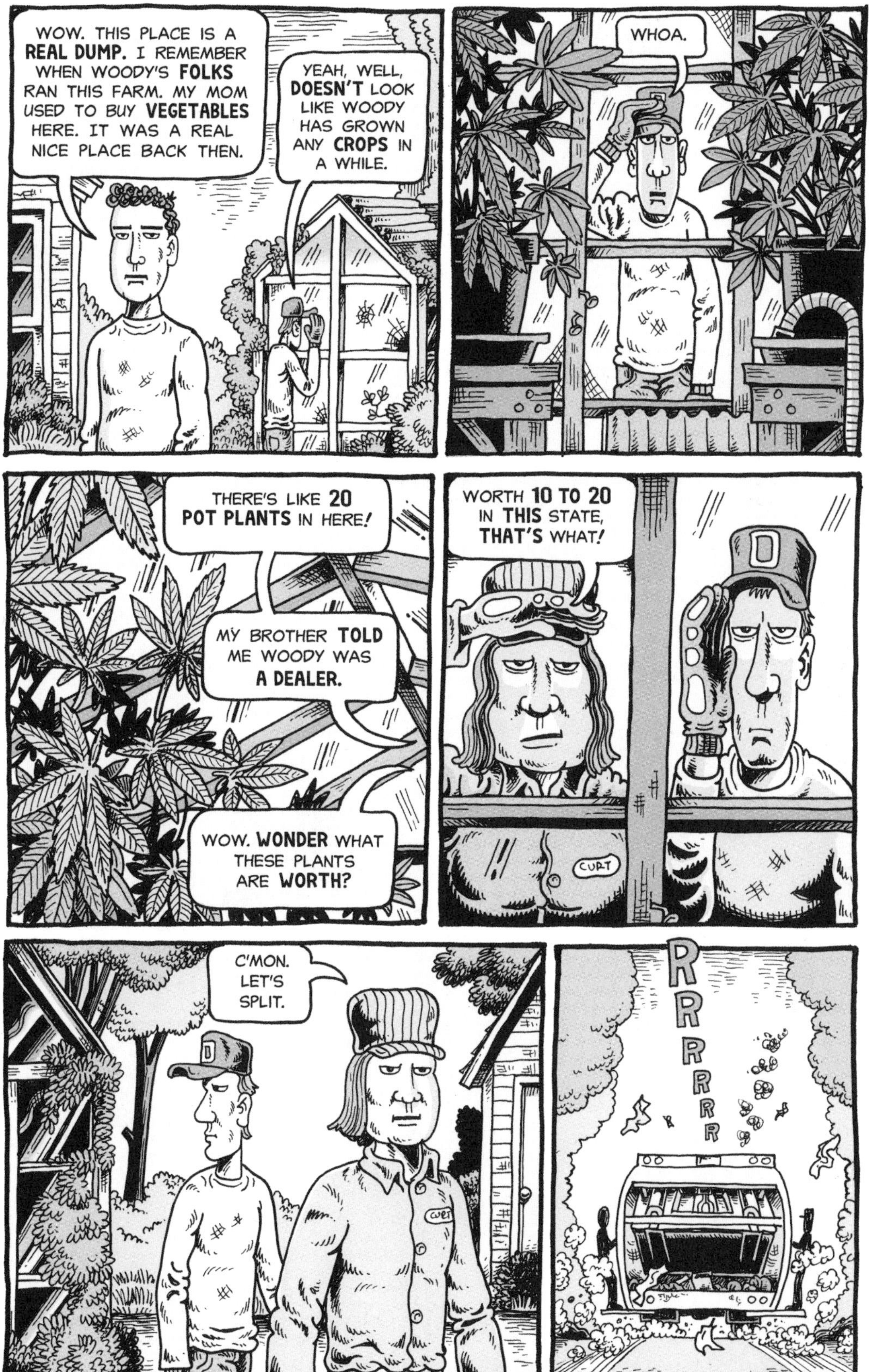

WOW. THIS PLACE IS A REAL DUMP. I REMEMBER WHEN WOODY'S FOLKS RAN THIS FARM. MY MOM USED TO BUY VEGETABLES HERE. IT WAS A REAL NICE PLACE BACK THEN.
YEAH, WELL, DOESN'T LOOK LIKE WOODY HAS GROWN ANY CROPS IN A WHILE.
WHOA.
THERE'S LIKE 20 POT PLANTS IN HERE!
MY BROTHER TOLD ME WOODY WAS A DEALER.
WOW. WONDER WHAT THESE PLANTS ARE WORTH?
WORTH 10 TO 20 IN THIS STATE, THAT'S WHAT!
CURT
C'MON. LET'S SPLIT.
RRRRRR

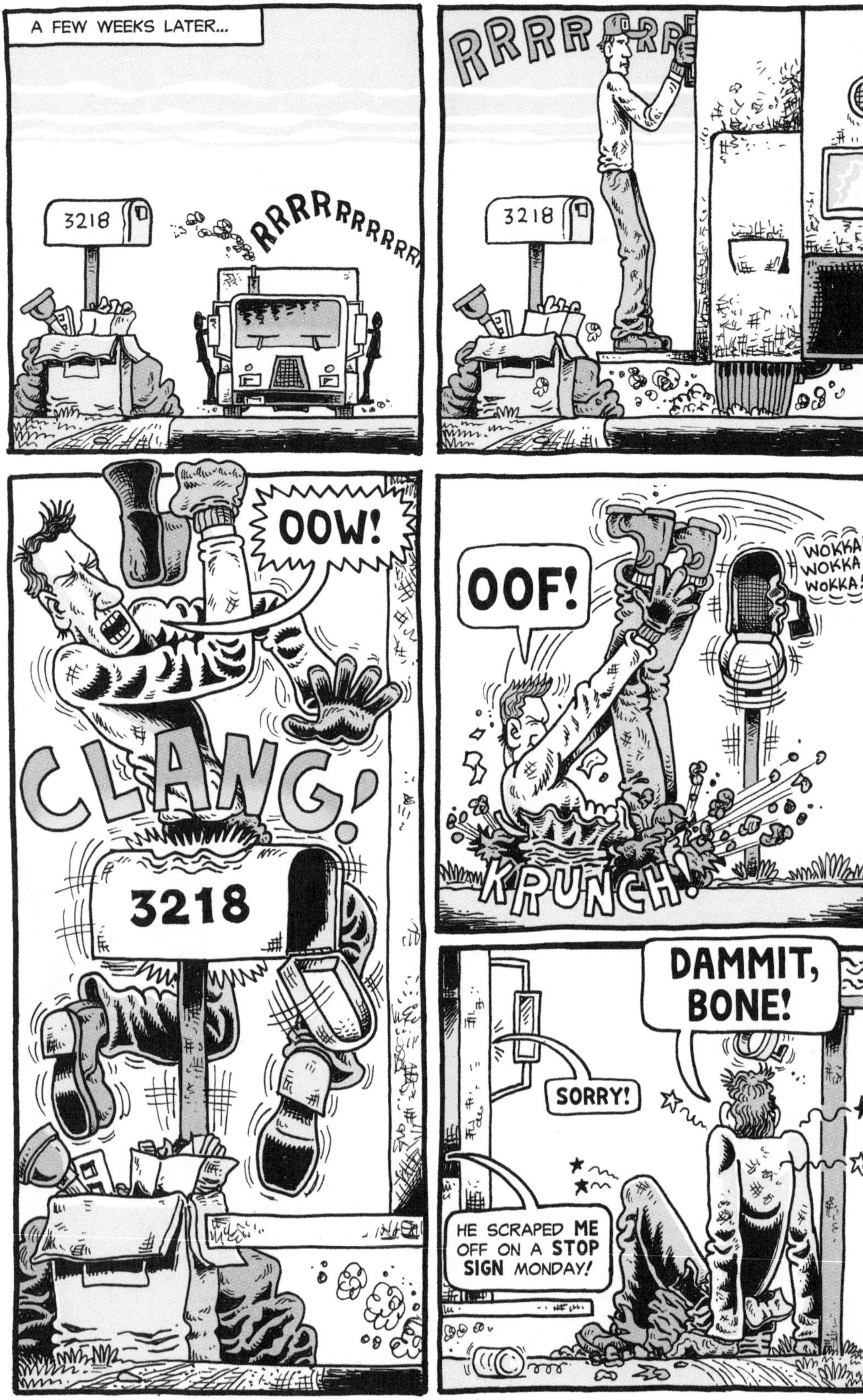
A FEW WEEKS LATER...
3218
RRRRRRRRRRR
RRRRRRRRR
3218
OOW!
CLANG!
3218
OOF!
WOKKA! WOKKA! WOKKA!
KRUNCH!
DAMMIT, BONE!
SORRY!
HE SCRAPED ME OFF ON A STOP SIGN MONDAY!

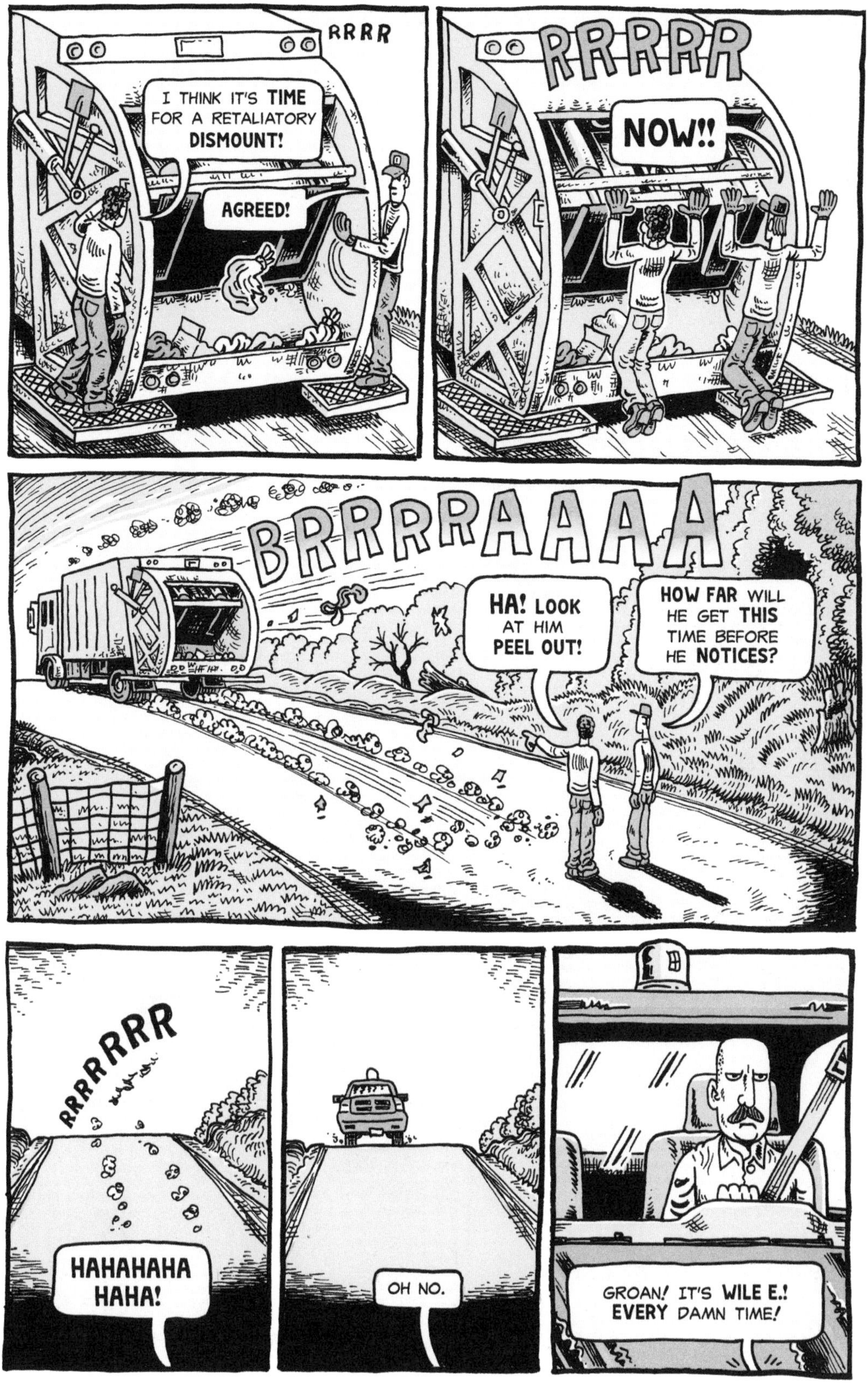
RRRR
I THINK IT'S TIME FOR A RETALIATORY DISMOUNT!
AGREED!
RRRRR
NOW!!
BRRRRAAAA
HA! LOOK AT HIM PEEL OUT!
HOW FAR WILL HE GET THIS TIME BEFORE HE NOTICES?
RRRRRRR
HAHAHAHA HAHA!
OH NO.
GROAN! IT'S WILE E.! EVERY DAMN TIME!

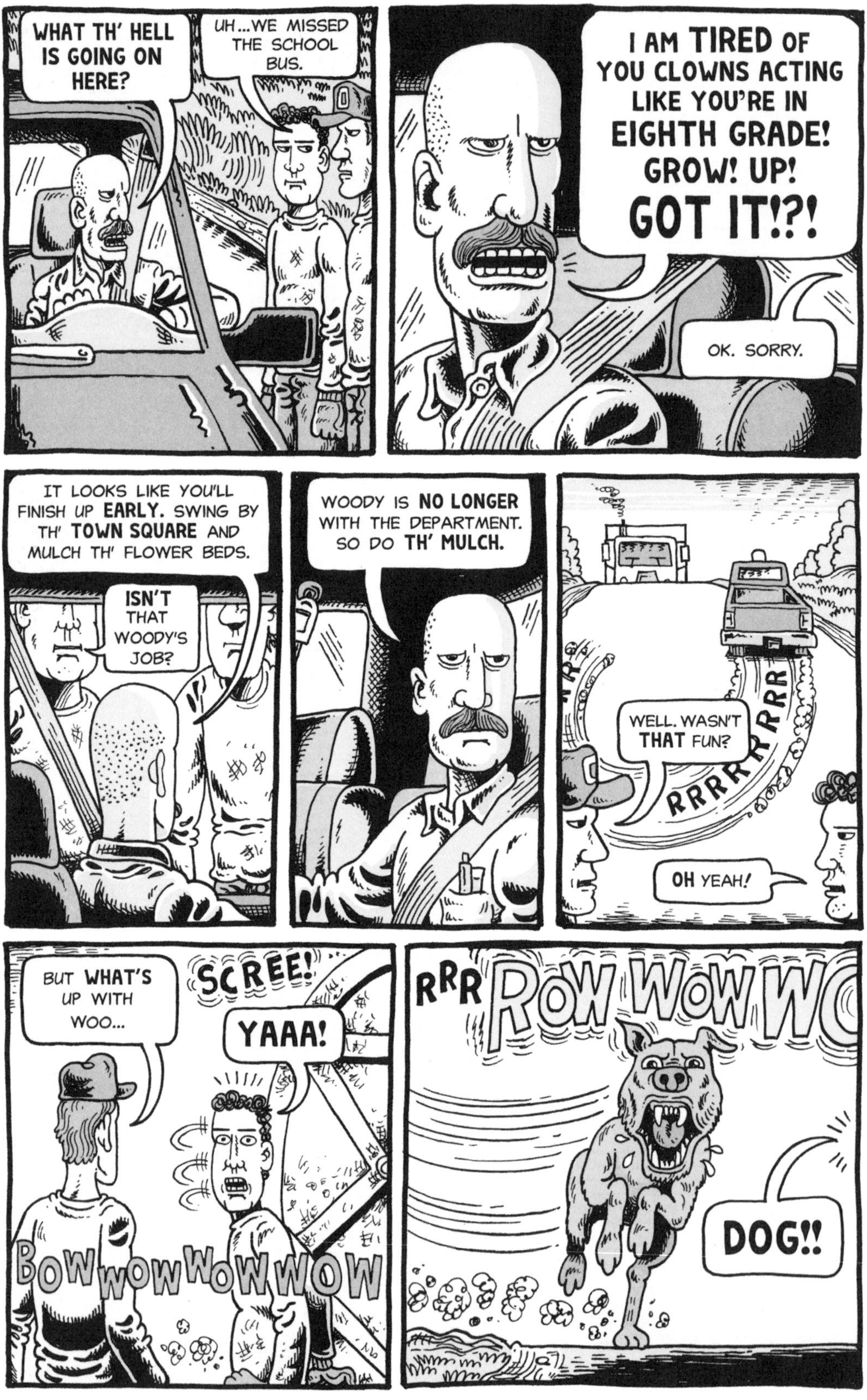
WHAT TH' HELL IS GOING ON HERE?
UH...WE MISSED THE SCHOOL BUS.
I AM TIRED OF YOU CLOWNS ACTING LIKE YOU'RE IN EIGHTH GRADE! GROW! UP! GOT IT!?!
OK. SORRY.
IT LOOKS LIKE YOU'LL FINISH UP EARLY. SWING BY TH' TOWN SQUARE AND MULCH TH' FLOWER BEDS.
ISN'T THAT WOODY'S JOB?
WOODY IS NO LONGER WITH THE DEPARTMENT. SO DO TH' MULCH.
RRRRRRRRRR
WELL. WASN'T THAT FUN?
OH YEAH!
BUT WHAT'S UP WITH WOO...
SCREE!
YAAA!
BOWWOWWOWWOW
RRR ROW WOW WO
DOG!!

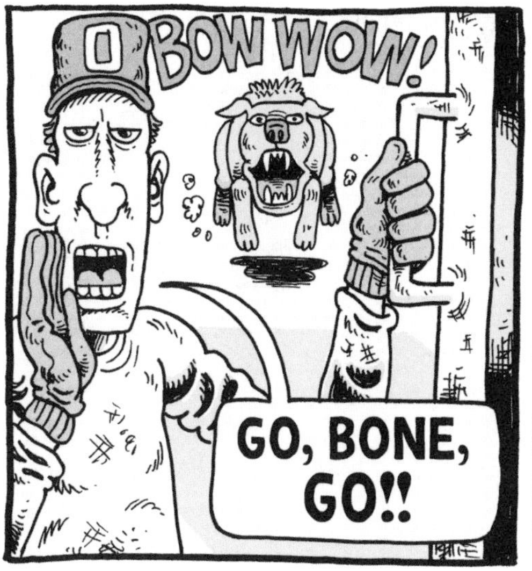
BOW WOW!
GO, BONE, GO!!
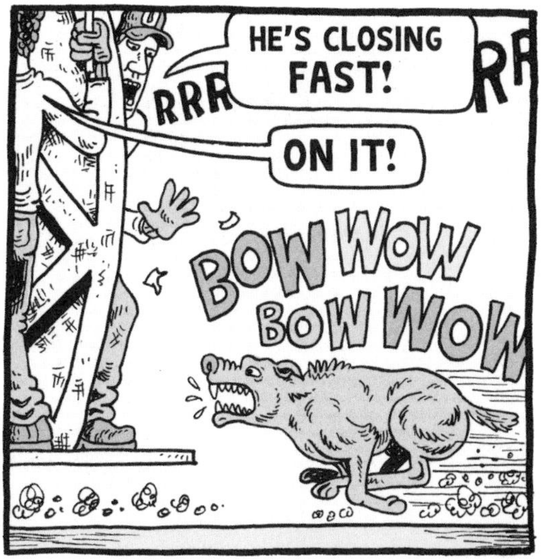
HE'S CLOSING FAST!
RRR
RR
ON IT!
BOW WOW
BOW WOW
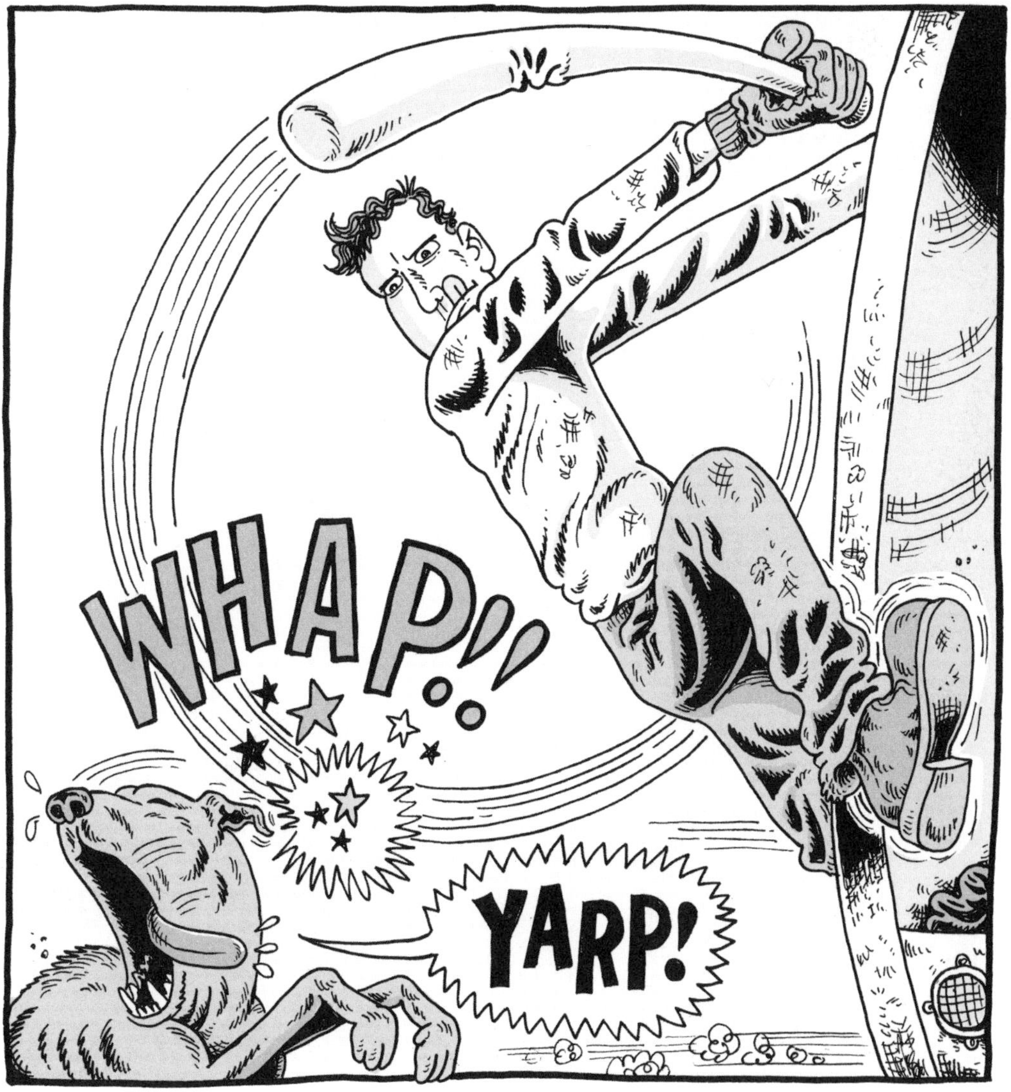
WHAP!!
YARP!

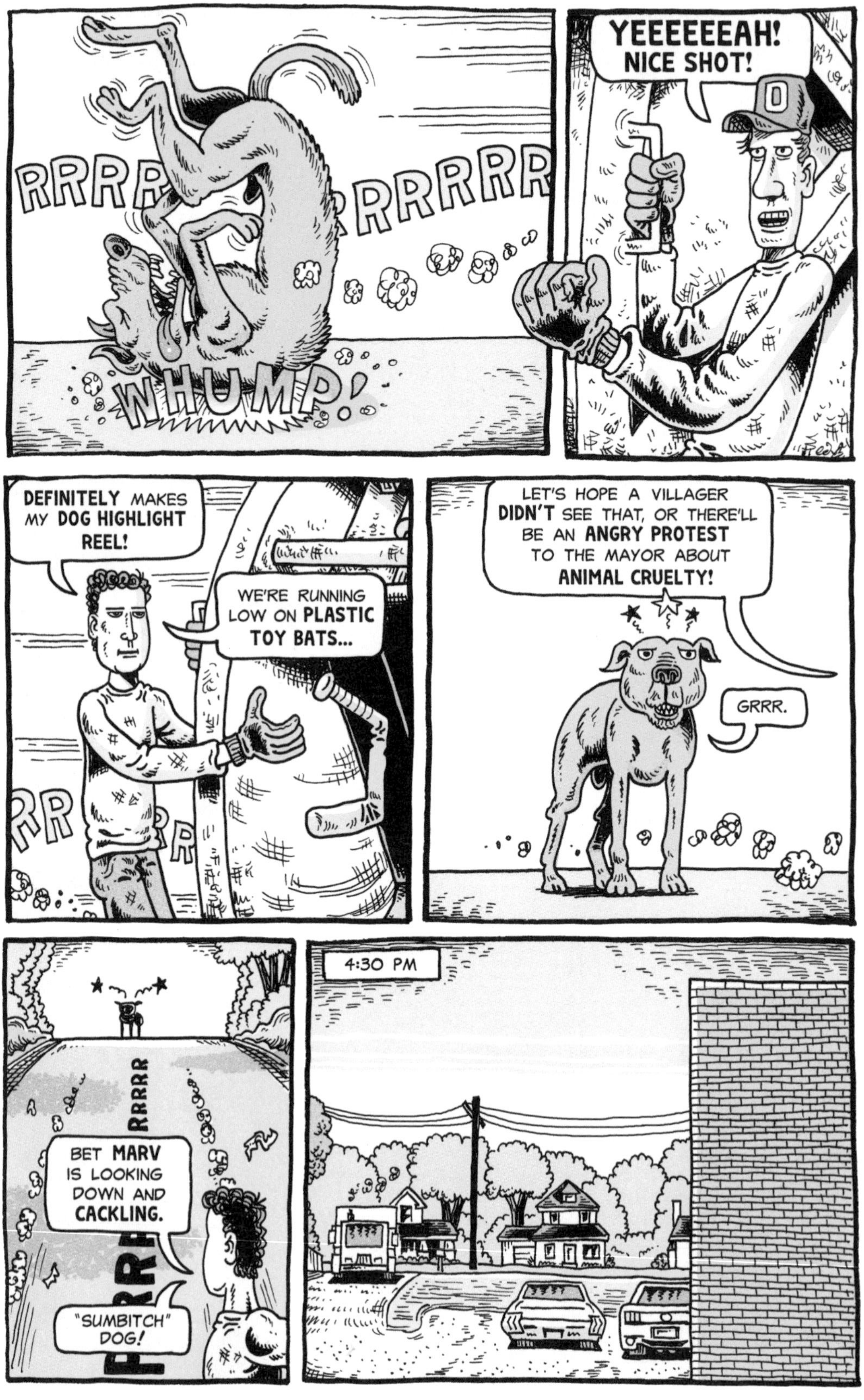
RRRRRRRRRRR
WHUMP!
YEEEEEEAH! NICE SHOT!
DEFINITELY MAKES MY DOG HIGHLIGHT REEL!
WE'RE RUNNING LOW ON PLASTIC TOY BATS...
RR
RR
LET'S HOPE A VILLAGER DIDN'T SEE THAT, OR THERE'LL BE AN ANGRY PROTEST TO THE MAYOR ABOUT ANIMAL CRUELTY!
GRRR.
RRRRR
BET MARV IS LOOKING DOWN AND CACKLING.
"SUMBITCH" DOG!
4:30 PM

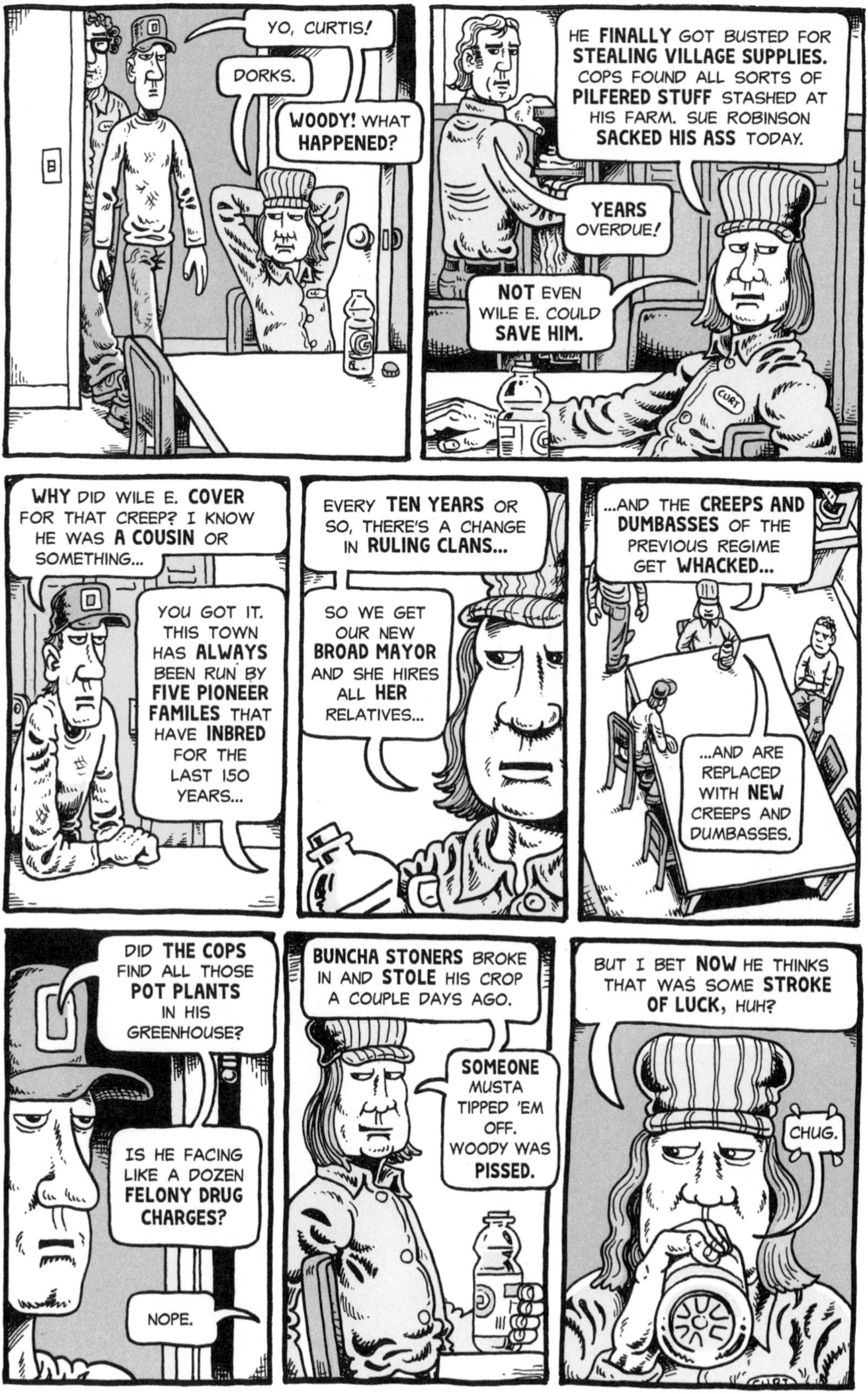
YO, CURTIS!
DORKS.
WOODY! WHAT HAPPENED?
HE FINALLY GOT BUSTED FOR STEALING VILLAGE SUPPLIES. COPS FOUND ALL SORTS OF PILFERED STUFF STASHED AT HIS FARM. SUE ROBINSON SACKED HIS ASS TODAY.
YEARS OVERDUE!
NOT EVEN WILE E. COULD SAVE HIM.
CURT
WHY DID WILE E. COVER FOR THAT CREEP? I KNOW HE WAS A COUSIN OR SOMETHING...
YOU GOT IT. THIS TOWN HAS ALWAYS BEEN RUN BY FIVE PIONEER FAMILES THAT HAVE INBRED FOR THE LAST 150 YEARS...
EVERY TEN YEARS OR SO, THERE'S A CHANGE IN RULING CLANS...
SO WE GET OUR NEW BROAD MAYOR AND SHE HIRES ALL HER RELATIVES...
...AND THE CREEPS AND DUMBASSES OF THE PREVIOUS REGIME GET WHACKED...
...AND ARE REPLACED WITH NEW CREEPS AND DUMBASSES.
DID THE COPS FIND ALL THOSE POT PLANTS IN HIS GREENHOUSE?
IS HE FACING LIKE A DOZEN FELONY DRUG CHARGES?
NOPE.
BUNCHA STONERS BROKE IN AND STOLE HIS CROP A COUPLE DAYS AGO.
SOMEONE MUSTA TIPPED 'EM OFF. WOODY WAS PISSED.
BUT I BET NOW HE THINKS THAT WAS SOME STROKE OF LUCK, HUH?
CHUG.

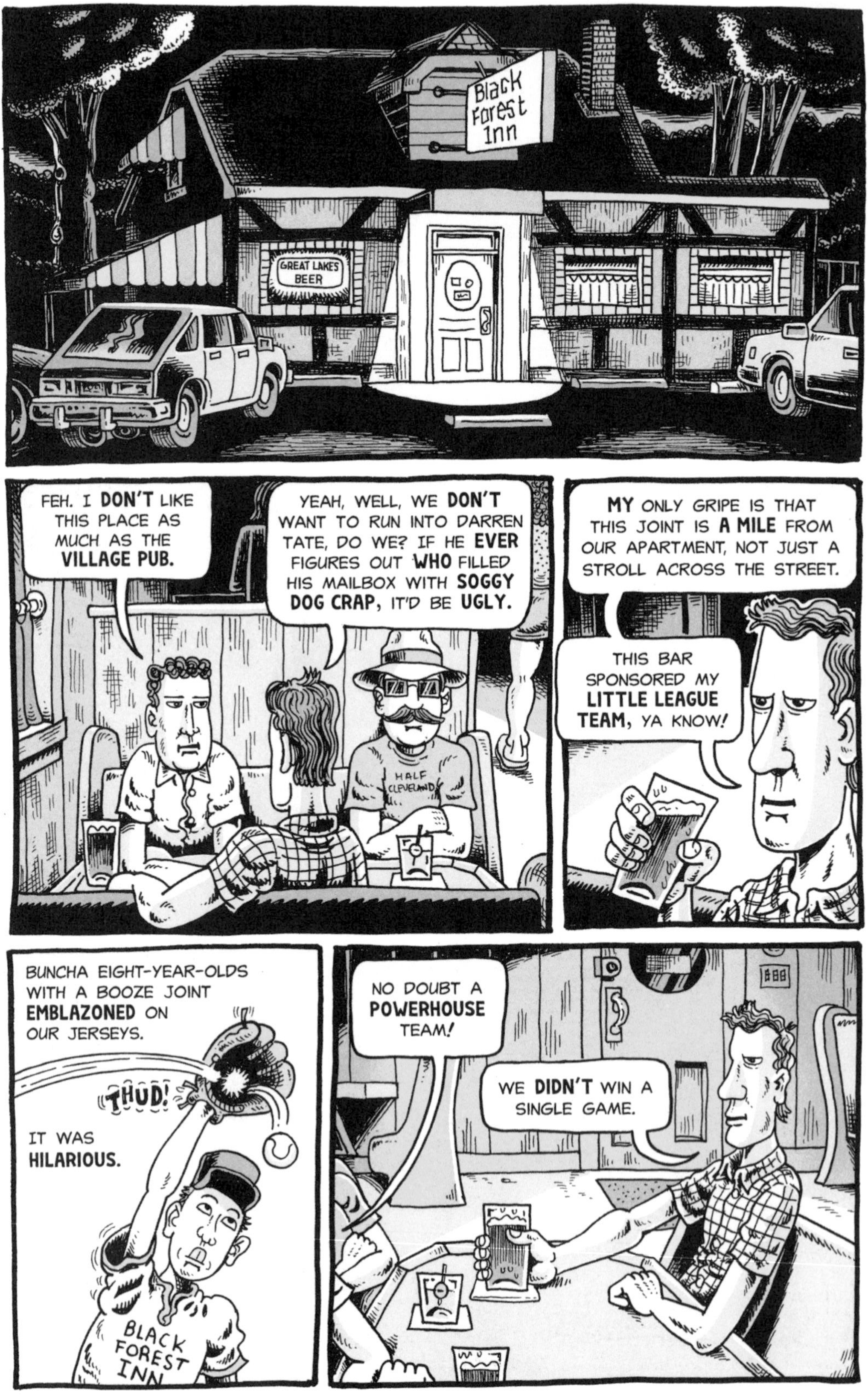
Black Forest Inn
GREAT LAKES BEER
FEH. I DON'T LIKE THIS PLACE AS MUCH AS THE VILLAGE PUB.
YEAH, WELL, WE DON'T WANT TO RUN INTO DARREN TATE, DO WE? IF HE EVER FIGURES OUT WHO FILLED HIS MAILBOX WITH SOGGY DOG CRAP, IT'D BE UGLY.
HALF CLEVELAND
MY ONLY GRIPE IS THAT THIS JOINT IS A MILE FROM OUR APARTMENT, NOT JUST A STROLL ACROSS THE STREET.
THIS BAR SPONSORED MY LITTLE LEAGUE TEAM, YA KNOW!
BUNCHA EIGHT-YEAR-OLDS WITH A BOOZE JOINT EMBLAZONED ON OUR JERSEYS.
THUD!
IT WAS HILARIOUS.
BLACK FOREST INN
NO DOUBT A POWERHOUSE TEAM!
WE DIDN'T WIN A SINGLE GAME.

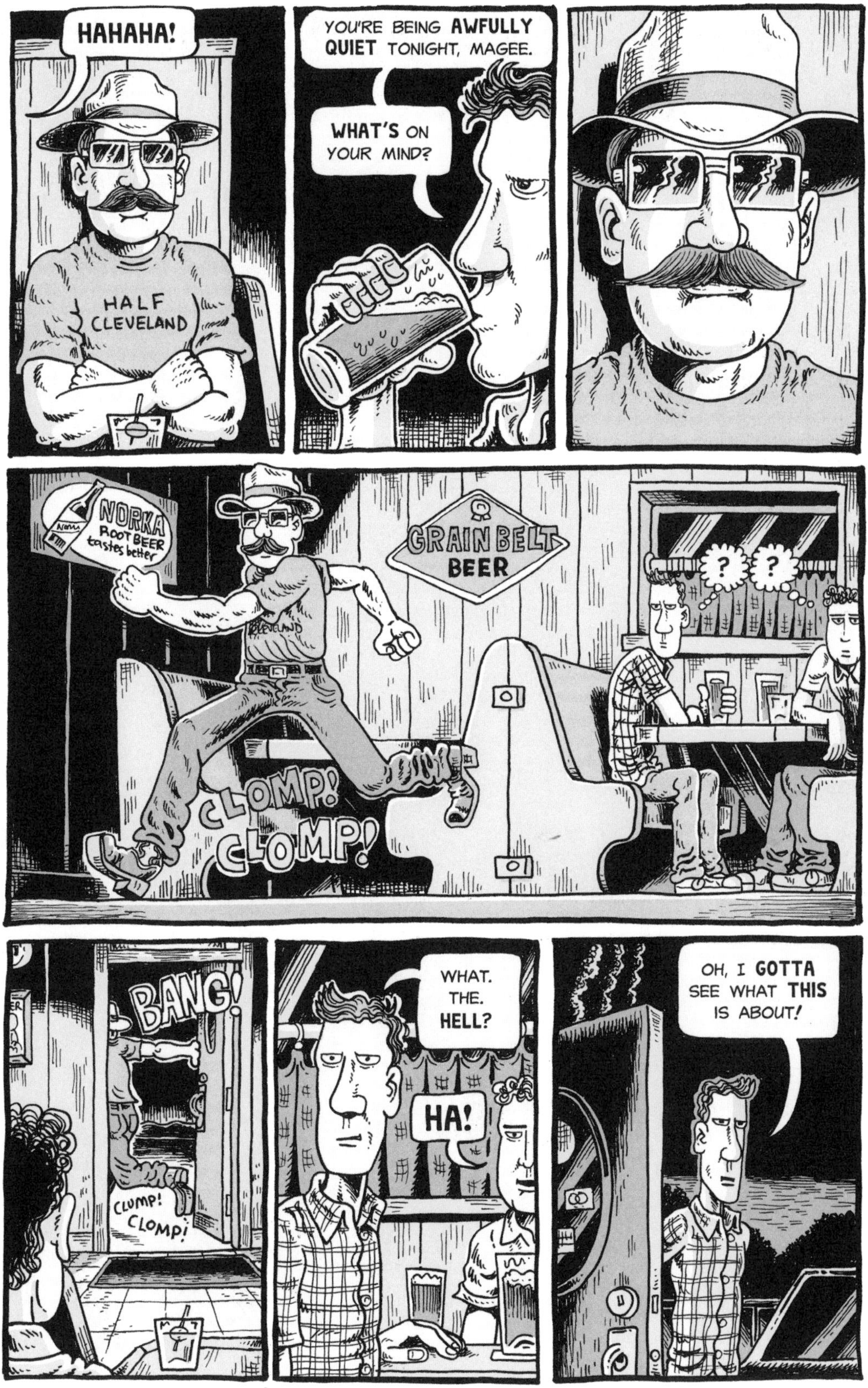
HAHAHA!
HALF CLEVELAND
YOU'RE BEING AWFULLY QUIET TONIGHT, MAGEE.
WHAT'S ON YOUR MIND?
NORKA ROOT BEER tastes better
GRAIN BELT BEER
CLEVELAND
?
?
CLOMP!
CLOMP!
BANG!
CLUMP!
CLOMP!
WHAT. THE. HELL?
HA!
OH, I GOTTA SEE WHAT THIS IS ABOUT!

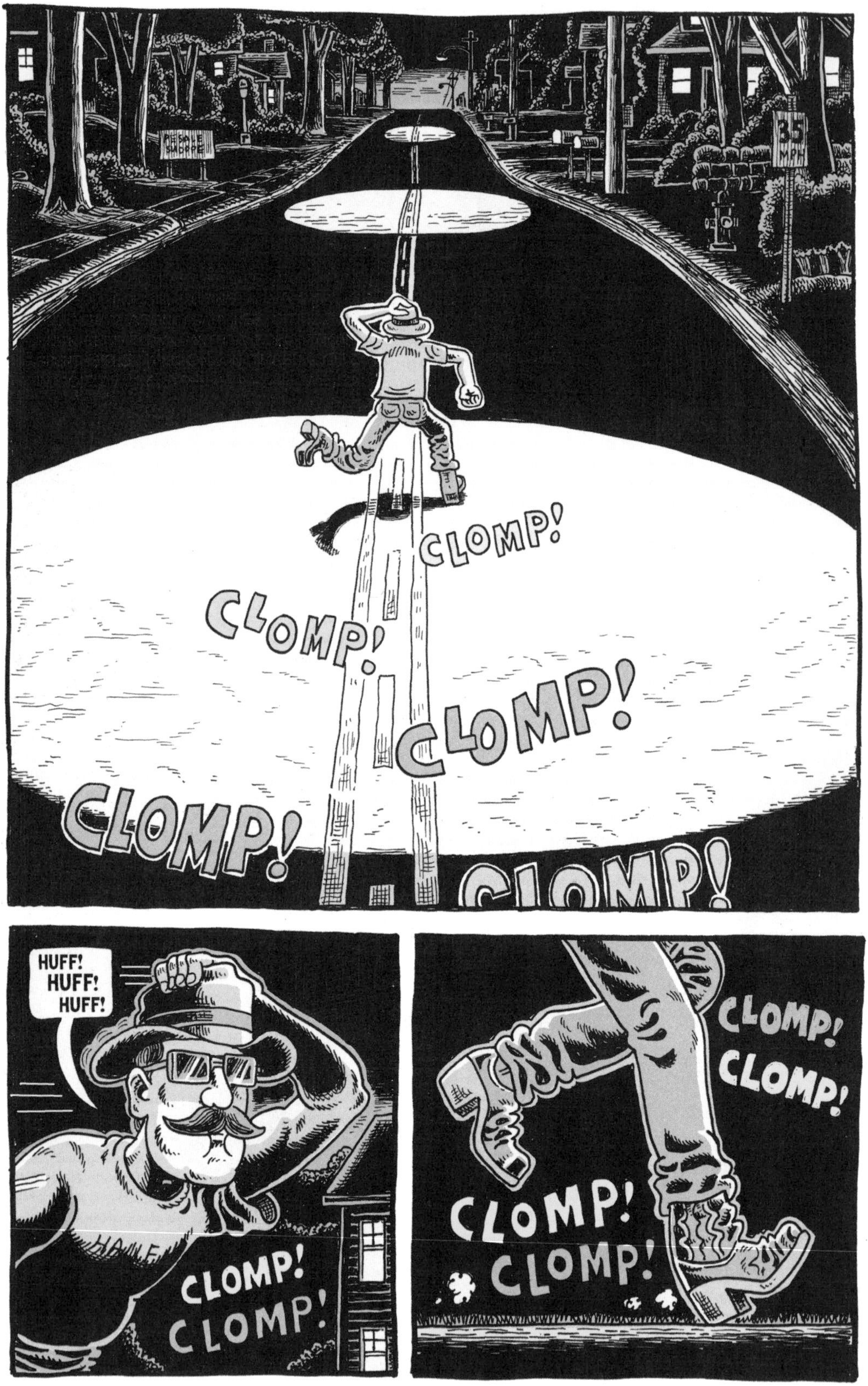
CLOMP!
CLOMP!
CLOMP!
CLOMP!
CLOMP!
HUFF! HUFF! HUFF!
CLOMP! CLOMP!
CLOMP! CLOMP!
CLOMP! CLOMP!

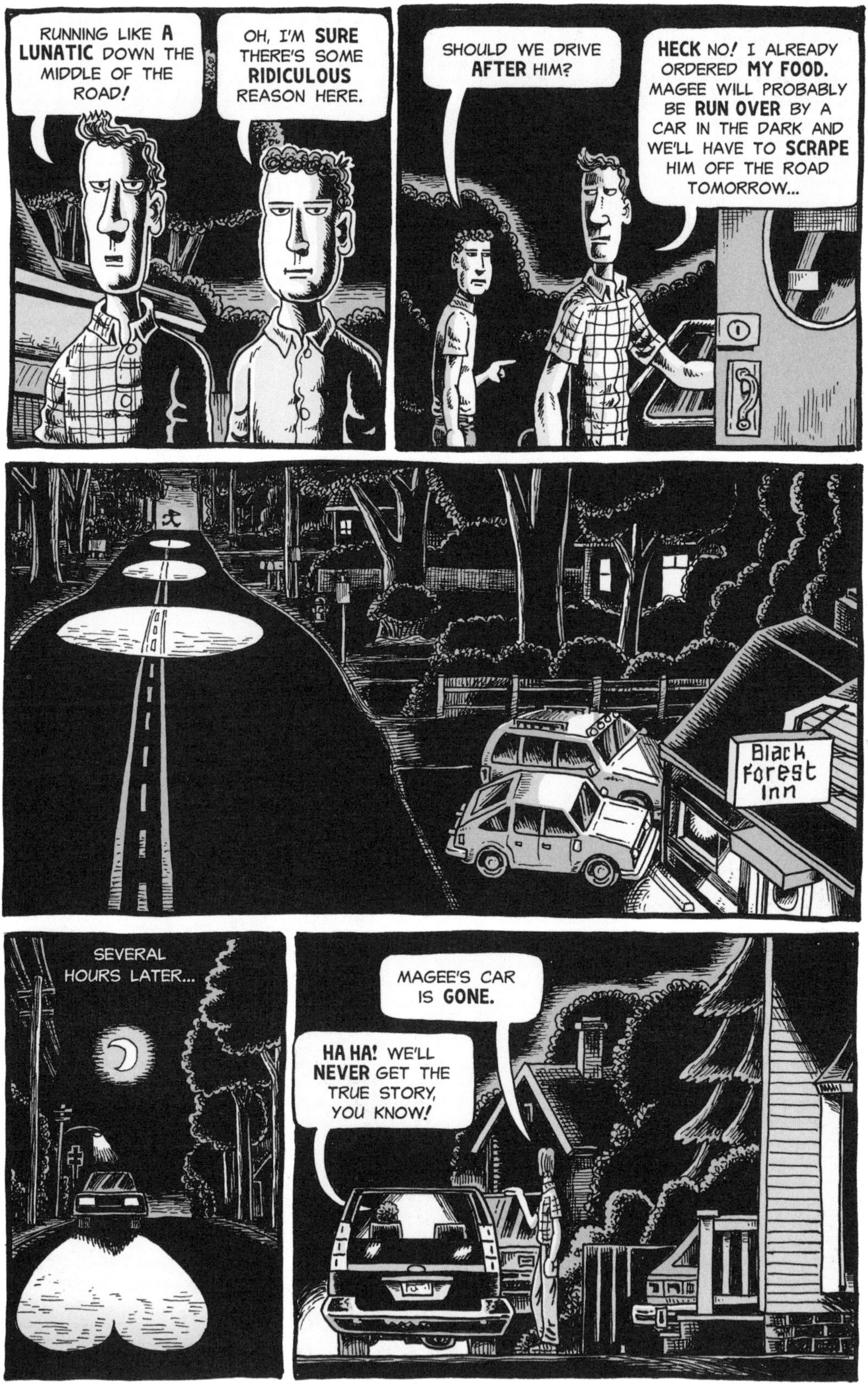
RUNNING LIKE A LUNATIC DOWN THE MIDDLE OF THE ROAD!
OH, I'M SURE THERE'S SOME RIDICULOUS REASON HERE.
SHOULD WE DRIVE AFTER HIM?
HECK NO! I ALREADY ORDERED MY FOOD. MAGEE WILL PROBABLY BE RUN OVER BY A CAR IN THE DARK AND WE'LL HAVE TO SCRAPE HIM OFF THE ROAD TOMORROW...
Black Forest Inn
SEVERAL HOURS LATER...
MAGEE'S CAR IS GONE.
HA HA! WE'LL NEVER GET THE TRUE STORY, YOU KNOW!

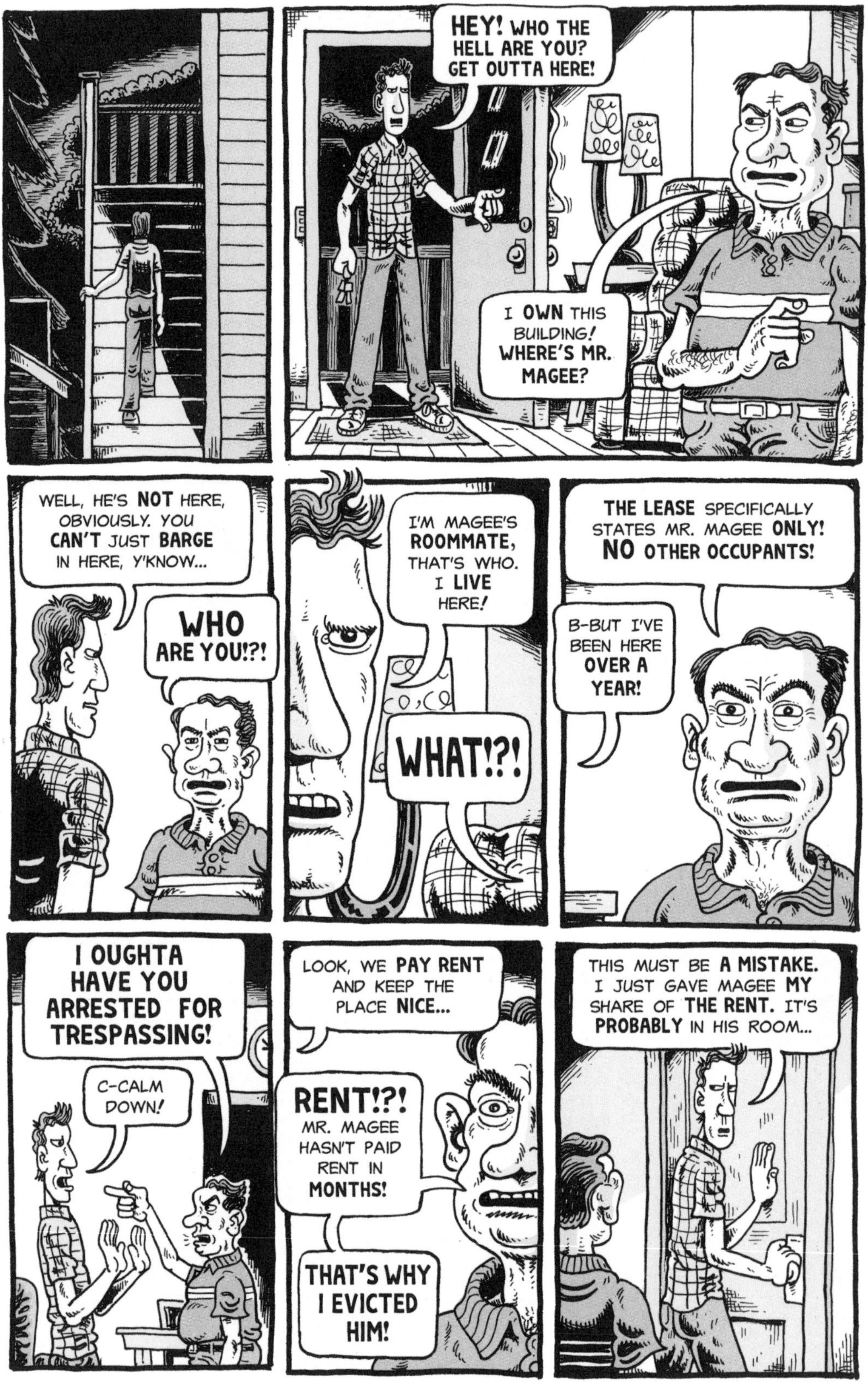
HEY! WHO THE HELL ARE YOU? GET OUTTA HERE!
I OWN THIS BUILDING! WHERE'S MR. MAGEE?
WELL, HE'S NOT HERE, OBVIOUSLY. YOU CAN'T JUST BARGE IN HERE, Y'KNOW...
WHO ARE YOU!?!
I'M MAGEE'S ROOMMATE, THAT'S WHO. I LIVE HERE!
WHAT!?!
THE LEASE SPECIFICALLY STATES MR. MAGEE ONLY! NO OTHER OCCUPANTS!
B-BUT I'VE BEEN HERE OVER A YEAR!
I OUGHTA HAVE YOU ARRESTED FOR TRESPASSING!
C-CALM DOWN!
LOOK, WE PAY RENT AND KEEP THE PLACE NICE...
RENT!?! MR. MAGEE HASN'T PAID RENT IN MONTHS!
THAT'S WHY I EVICTED HIM!
THIS MUST BE A MISTAKE. I JUST GAVE MAGEE MY SHARE OF THE RENT. IT'S PROBABLY IN HIS ROOM...

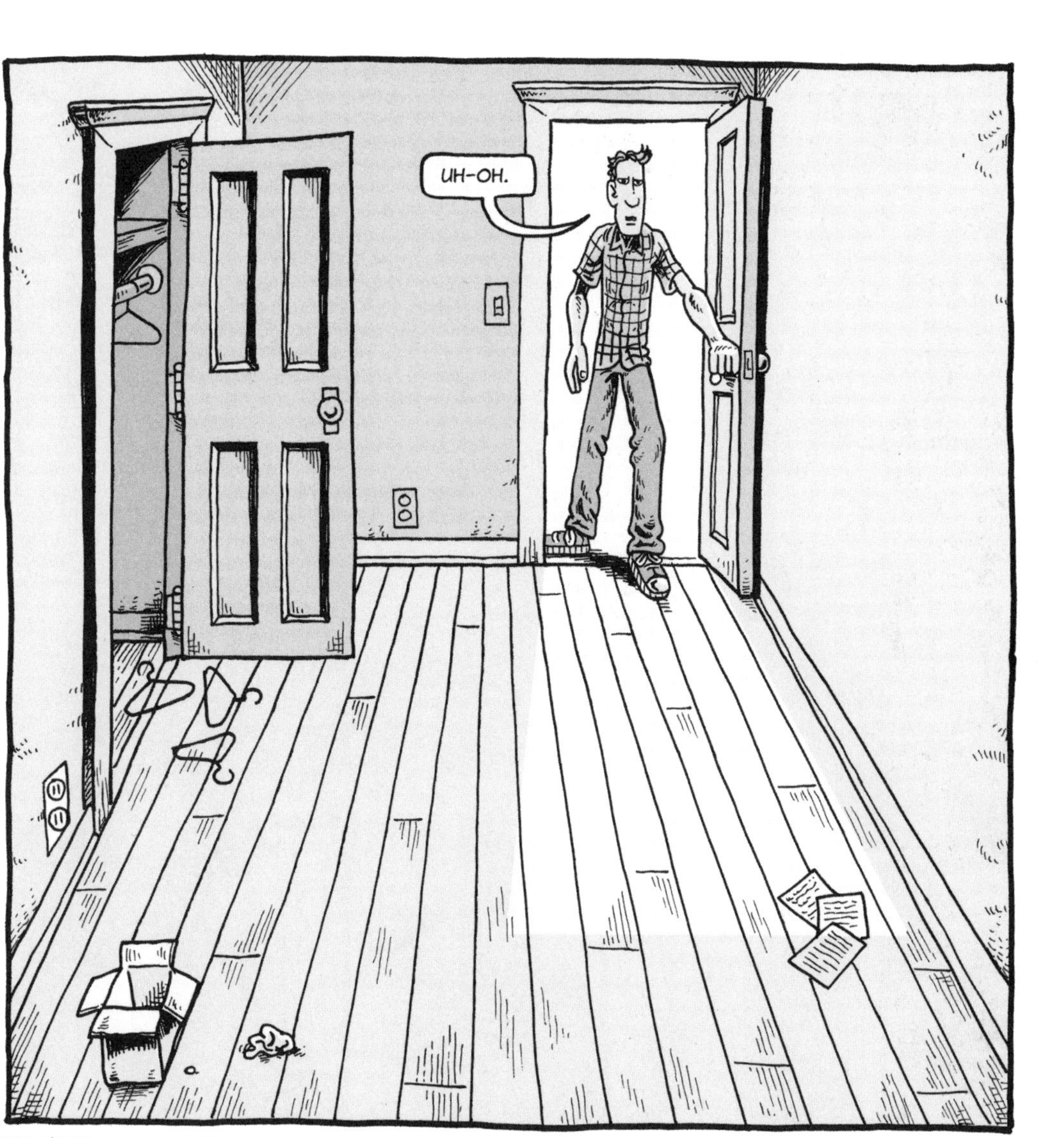
UH-OH.

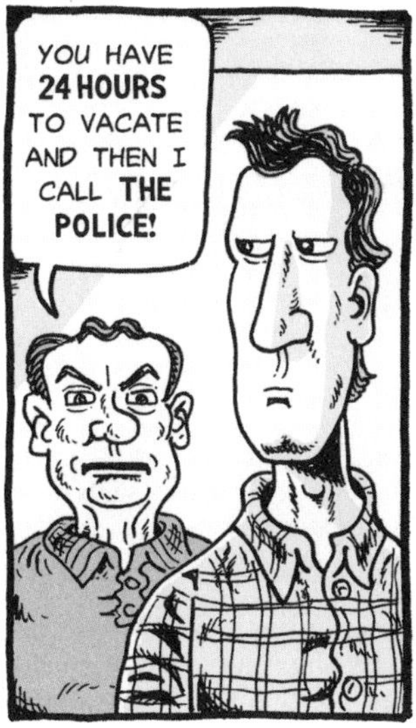
YOU HAVE **24 HOURS** TO VACATE AND THEN I CALL **THE POLICE!**

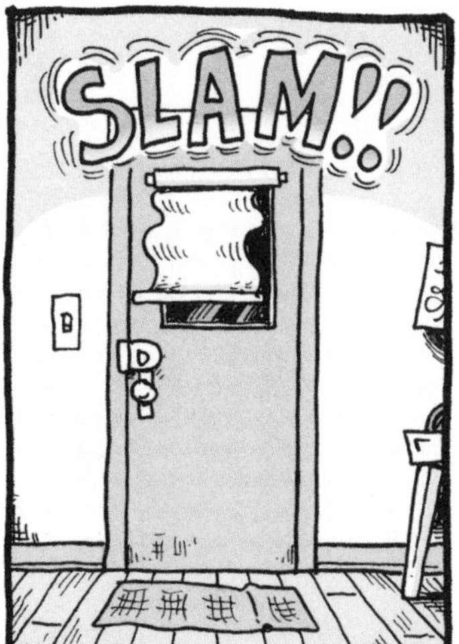
SLAM!!

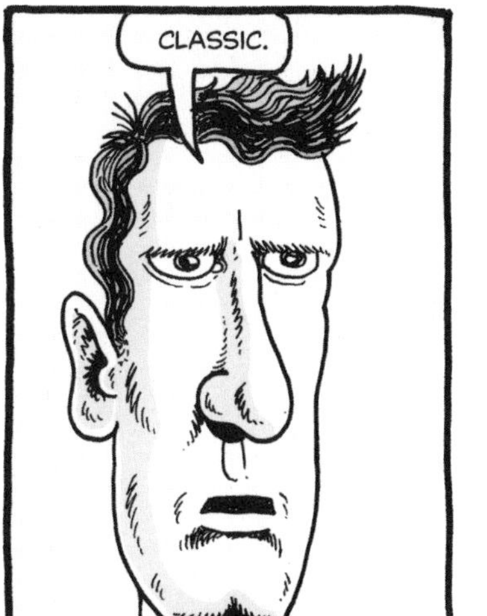
CLASSIC.

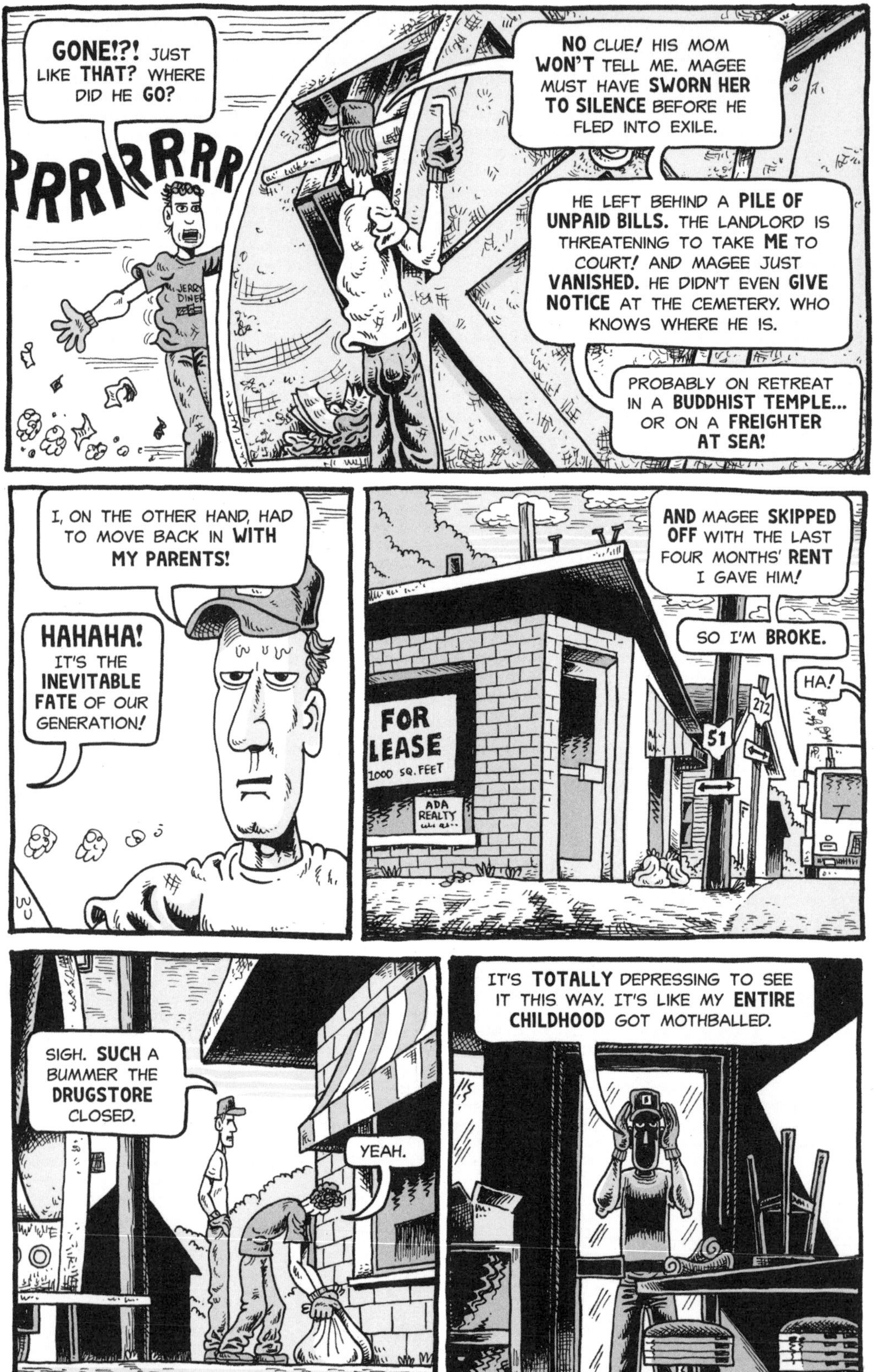
GONE!?! JUST LIKE THAT? WHERE DID HE GO?
RRRRRRR
NO CLUE! HIS MOM WON'T TELL ME. MAGEE MUST HAVE SWORN HER TO SILENCE BEFORE HE FLED INTO EXILE.
HE LEFT BEHIND A PILE OF UNPAID BILLS. THE LANDLORD IS THREATENING TO TAKE ME TO COURT! AND MAGEE JUST VANISHED. HE DIDN'T EVEN GIVE NOTICE AT THE CEMETERY. WHO KNOWS WHERE HE IS.
PROBABLY ON RETREAT IN A BUDDHIST TEMPLE... OR ON A FREIGHTER AT SEA!
JERRY DINER
I, ON THE OTHER HAND, HAD TO MOVE BACK IN WITH MY PARENTS!
HAHAHA! IT'S THE INEVITABLE FATE OF OUR GENERATION!
AND MAGEE SKIPPED OFF WITH THE LAST FOUR MONTHS' RENT I GAVE HIM!
SO I'M BROKE.
HA!
FOR LEASE
1000 SQ. FEET
ADA REALTY
51
272
SIGH. SUCH A BUMMER THE DRUGSTORE CLOSED.
YEAH.
IT'S TOTALLY DEPRESSING TO SEE IT THIS WAY. IT'S LIKE MY ENTIRE CHILDHOOD GOT MOTHBALLED.

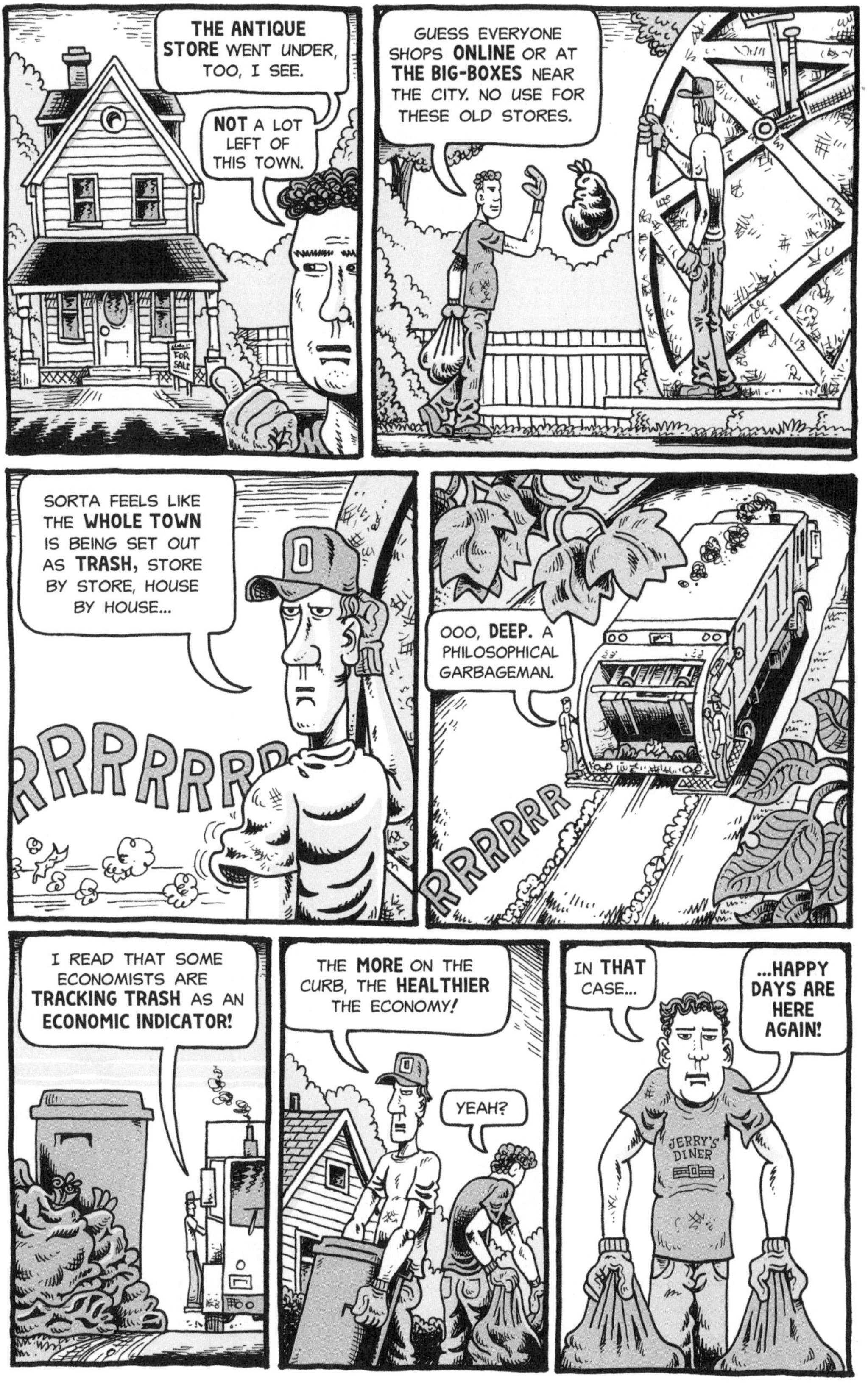
THE ANTIQUE STORE WENT UNDER, TOO, I SEE.
NOT A LOT LEFT OF THIS TOWN.
FOR SALE
GUESS EVERYONE SHOPS ONLINE OR AT THE BIG-BOXES NEAR THE CITY. NO USE FOR THESE OLD STORES.
SORTA FEELS LIKE THE WHOLE TOWN IS BEING SET OUT AS TRASH, STORE BY STORE, HOUSE BY HOUSE...
RRRRRRR
OOO, DEEP. A PHILOSOPHICAL GARBAGEMAN.
RRRRRR
I READ THAT SOME ECONOMISTS ARE TRACKING TRASH AS AN ECONOMIC INDICATOR!
THE MORE ON THE CURB, THE HEALTHIER THE ECONOMY!
YEAH?
IN THAT CASE...
...HAPPY DAYS ARE HERE AGAIN!
JERRY'S DINER

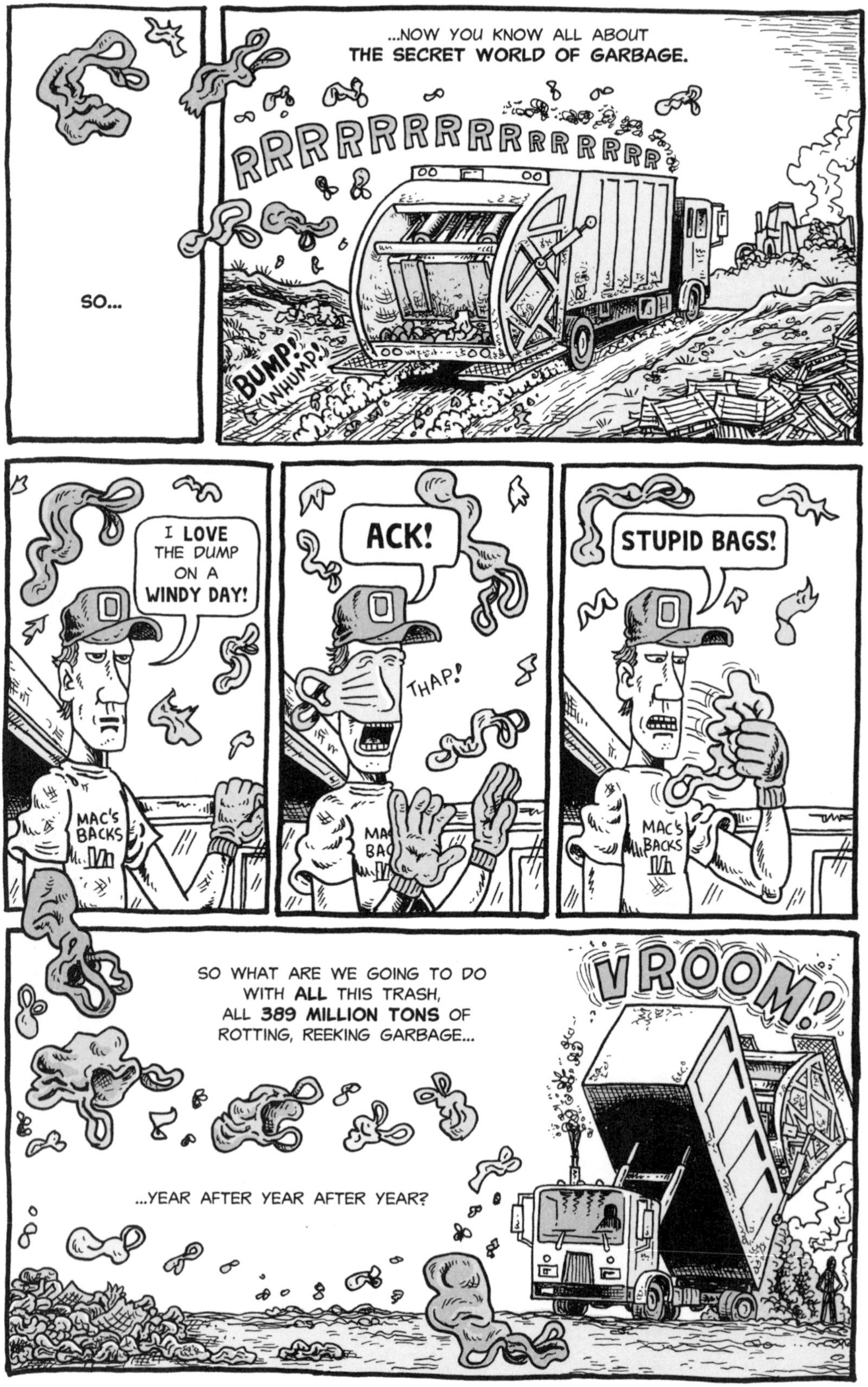
SO...
...NOW YOU KNOW ALL ABOUT THE SECRET WORLD OF GARBAGE.
RRRRRRRRRRRRRRR
BUMP! WHUMP!
I LOVE THE DUMP ON A WINDY DAY!
MAC'S BACKS
ACK!
THAP!
STUPID BAGS!
MAC'S BACKS
SO WHAT ARE WE GOING TO DO WITH ALL THIS TRASH, ALL 389 MILLION TONS OF ROTTING, REEKING GARBAGE...
VROOM!
...YEAR AFTER YEAR AFTER YEAR?

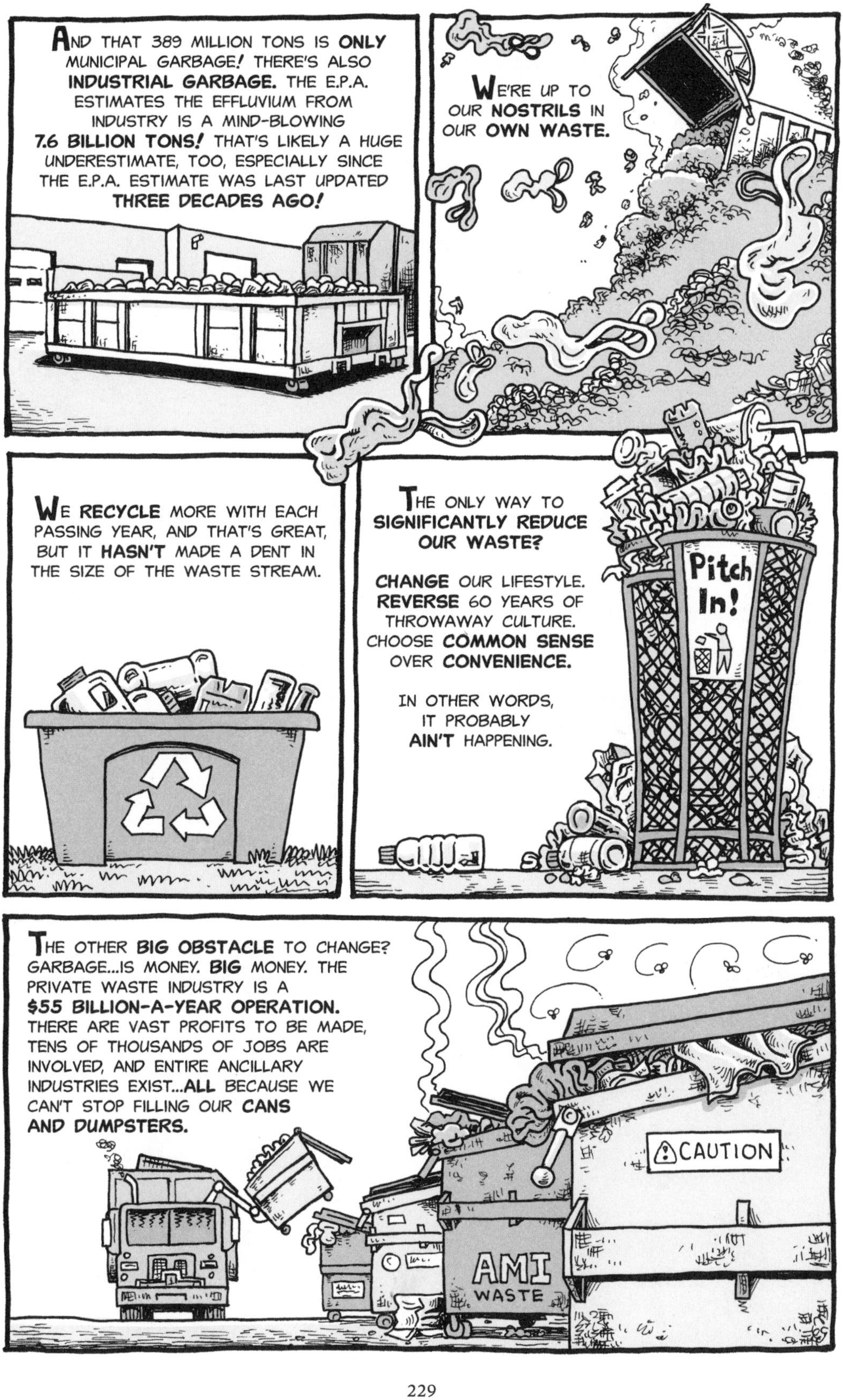
AND THAT 389 MILLION TONS IS ONLY MUNICIPAL GARBAGE! THERE'S ALSO INDUSTRIAL GARBAGE. THE E.P.A. ESTIMATES THE EFFLUVIUM FROM INDUSTRY IS A MIND-BLOWING 7.6 BILLION TONS! THAT'S LIKELY A HUGE UNDERESTIMATE, TOO, ESPECIALLY SINCE THE E.P.A. ESTIMATE WAS LAST UPDATED THREE DECADES AGO!
WE'RE UP TO OUR NOSTRILS IN OUR OWN WASTE.
WE RECYCLE MORE WITH EACH PASSING YEAR, AND THAT'S GREAT, BUT IT HASN'T MADE A DENT IN THE SIZE OF THE WASTE STREAM.
THE ONLY WAY TO SIGNIFICANTLY REDUCE OUR WASTE?
CHANGE OUR LIFESTYLE. REVERSE 60 YEARS OF THROWAWAY CULTURE. CHOOSE COMMON SENSE OVER CONVENIENCE.
IN OTHER WORDS, IT PROBABLY AIN'T HAPPENING.
Pitch In!
THE OTHER BIG OBSTACLE TO CHANGE? GARBAGE...IS MONEY. BIG MONEY. THE PRIVATE WASTE INDUSTRY IS A $55 BILLION-A-YEAR OPERATION. THERE ARE VAST PROFITS TO BE MADE, TENS OF THOUSANDS OF JOBS ARE INVOLVED, AND ENTIRE ANCILLARY INDUSTRIES EXIST...ALL BECAUSE WE CAN'T STOP FILLING OUR CANS AND DUMPSTERS.
CAUTION
AMI WASTE

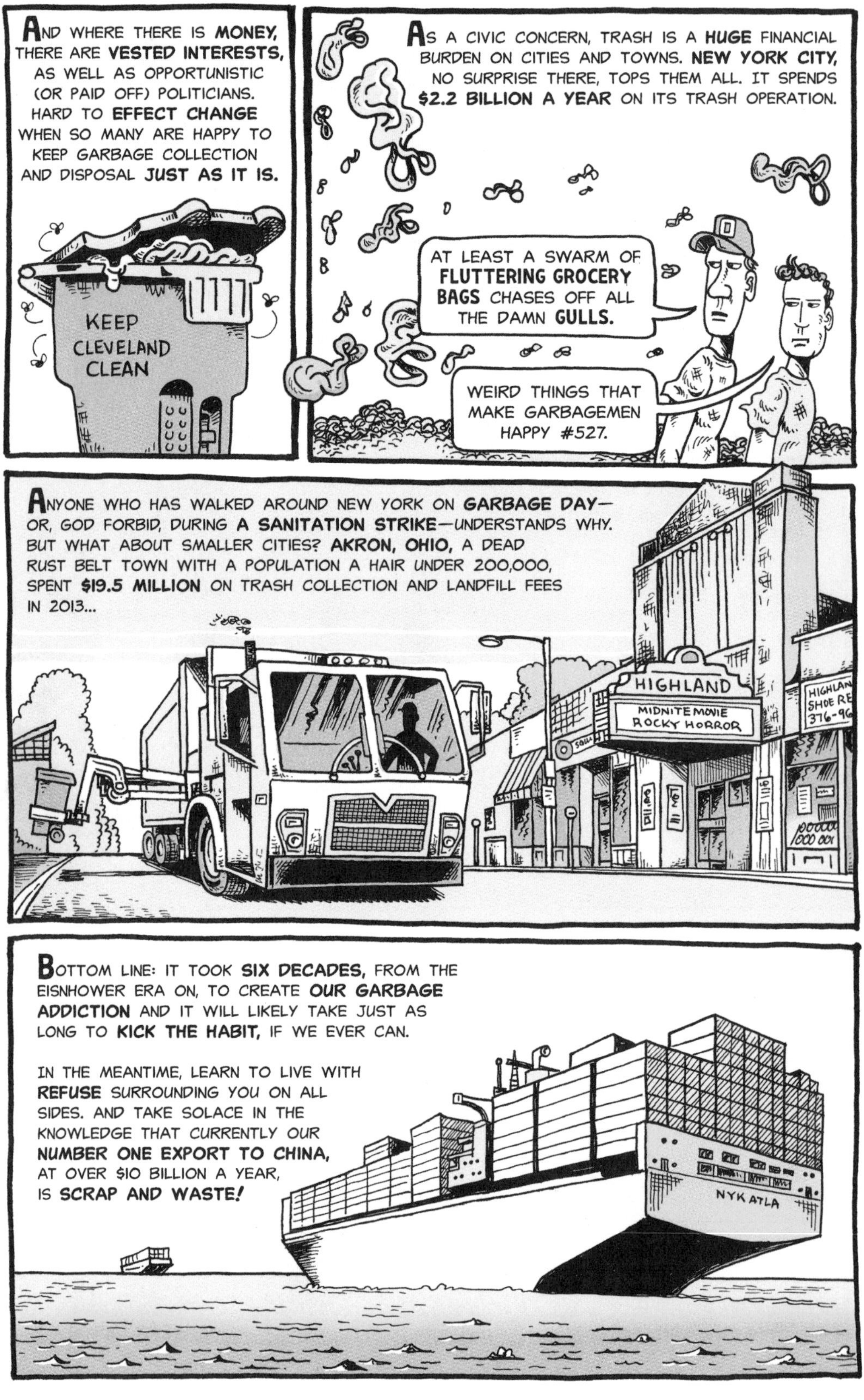
AND WHERE THERE IS MONEY, THERE ARE VESTED INTERESTS, AS WELL AS OPPORTUNISTIC (OR PAID OFF) POLITICIANS. HARD TO EFFECT CHANGE WHEN SO MANY ARE HAPPY TO KEEP GARBAGE COLLECTION AND DISPOSAL JUST AS IT IS.
KEEP CLEVELAND CLEAN
AS A CIVIC CONCERN, TRASH IS A HUGE FINANCIAL BURDEN ON CITIES AND TOWNS. NEW YORK CITY, NO SURPRISE THERE, TOPS THEM ALL. IT SPENDS $2.2 BILLION A YEAR ON ITS TRASH OPERATION.
AT LEAST A SWARM OF FLUTTERING GROCERY BAGS CHASES OFF ALL THE DAMN GULLS.
WEIRD THINGS THAT MAKE GARBAGEMEN HAPPY #527.
ANYONE WHO HAS WALKED AROUND NEW YORK ON GARBAGE DAY—OR, GOD FORBID, DURING A SANITATION STRIKE—UNDERSTANDS WHY. BUT WHAT ABOUT SMALLER CITIES? AKRON, OHIO, A DEAD RUST BELT TOWN WITH A POPULATION A HAIR UNDER 200,000, SPENT $19.5 MILLION ON TRASH COLLECTION AND LANDFILL FEES IN 2013...
HIGHLAND
MIDNITE MOVIE
ROCKY HORROR
BOTTOM LINE: IT TOOK SIX DECADES, FROM THE EISNHOWER ERA ON, TO CREATE OUR GARBAGE ADDICTION AND IT WILL LIKELY TAKE JUST AS LONG TO KICK THE HABIT, IF WE EVER CAN.
IN THE MEANTIME, LEARN TO LIVE WITH REFUSE SURROUNDING YOU ON ALL SIDES. AND TAKE SOLACE IN THE KNOWLEDGE THAT CURRENTLY OUR NUMBER ONE EXPORT TO CHINA, AT OVER $10 BILLION A YEAR, IS SCRAP AND WASTE!
NYK ATLA

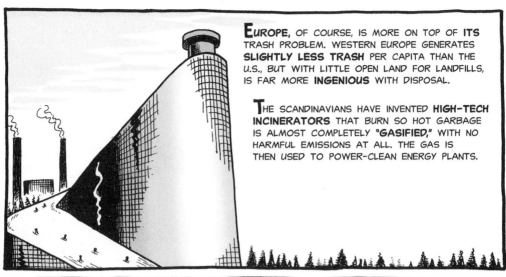

EUROPE, OF COURSE, IS MORE ON TOP OF **ITS** TRASH PROBLEM. WESTERN EUROPE GENERATES **SLIGHTLY LESS TRASH** PER CAPITA THAN THE U.S., BUT WITH LITTLE OPEN LAND FOR LANDFILLS, IS FAR MORE **INGENIOUS** WITH DISPOSAL.

THE SCANDINAVIANS HAVE INVENTED **HIGH-TECH INCINERATORS** THAT BURN SO HOT GARBAGE IS ALMOST COMPLETELY **"GASIFIED,"** WITH NO HARMFUL EMISSIONS AT ALL. THE GAS IS THEN USED TO POWER-CLEAN ENERGY PLANTS.

WE DON'T HAVE TO BE **CLEVER** IN THE U.S. BECAUSE WE HAVE **VAST TRACTS** OF BARREN LAND. WE **BURY** OUR GARBAGE BECAUSE WE **CAN.** IT'S NO COINCIDENCE THAT **ALASKA** IS NUMBER ONE IN LANDFILLS.

SOME **GARBOLOGISTS** (YES, THERE ARE SUCH PEOPLE) THINK WE'D BE BETTER OFF DUMPING **ALL** THE NATION'S TRASH IN **A FEW**—OR MAYBE JUST **ONE!**— **VAST LANDFILL** SOMEWHERE OUT IN THE WASTELAND.

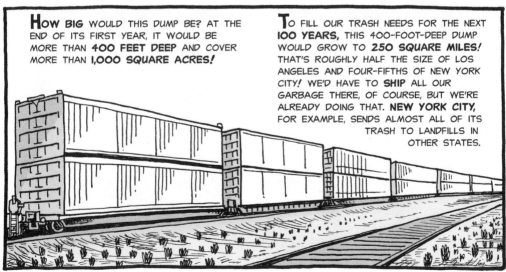

HOW **BIG** WOULD THIS DUMP BE? AT THE END OF ITS FIRST YEAR, IT WOULD BE MORE THAN **400 FEET DEEP** AND COVER MORE THAN **1,000 SQUARE ACRES!**

TO FILL OUR TRASH NEEDS FOR THE NEXT **100 YEARS,** THIS 400-FOOT-DEEP DUMP WOULD GROW TO **250 SQUARE MILES!** THAT'S ROUGHLY HALF THE SIZE OF LOS ANGELES AND FOUR-FIFTHS OF NEW YORK CITY! WE'D HAVE TO **SHIP** ALL OUR GARBAGE THERE, OF COURSE, BUT WE'RE ALREADY DOING THAT. **NEW YORK CITY,** FOR EXAMPLE, SENDS ALMOST ALL OF ITS TRASH TO LANDFILLS IN OTHER STATES.

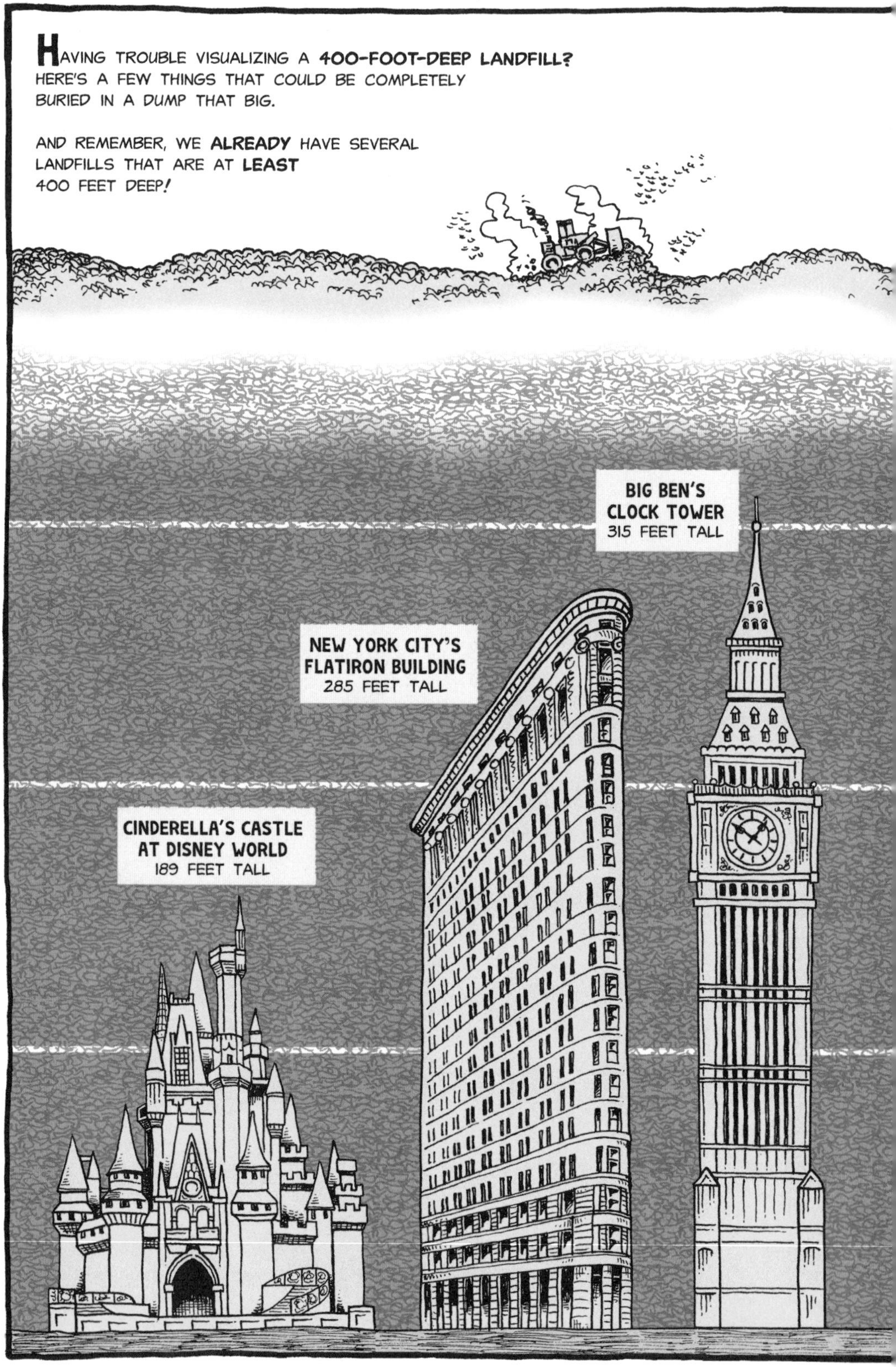
HAVING TROUBLE VISUALIZING A **400-FOOT-DEEP LANDFILL?** HERE'S A FEW THINGS THAT COULD BE COMPLETELY BURIED IN A DUMP THAT BIG.
AND REMEMBER, WE **ALREADY** HAVE SEVERAL LANDFILLS THAT ARE AT **LEAST** 400 FEET DEEP!
BIG BEN'S CLOCK TOWER 315 FEET TALL
NEW YORK CITY'S FLATIRON BUILDING 285 FEET TALL
CINDERELLA'S CASTLE AT DISNEY WORLD 189 FEET TALL

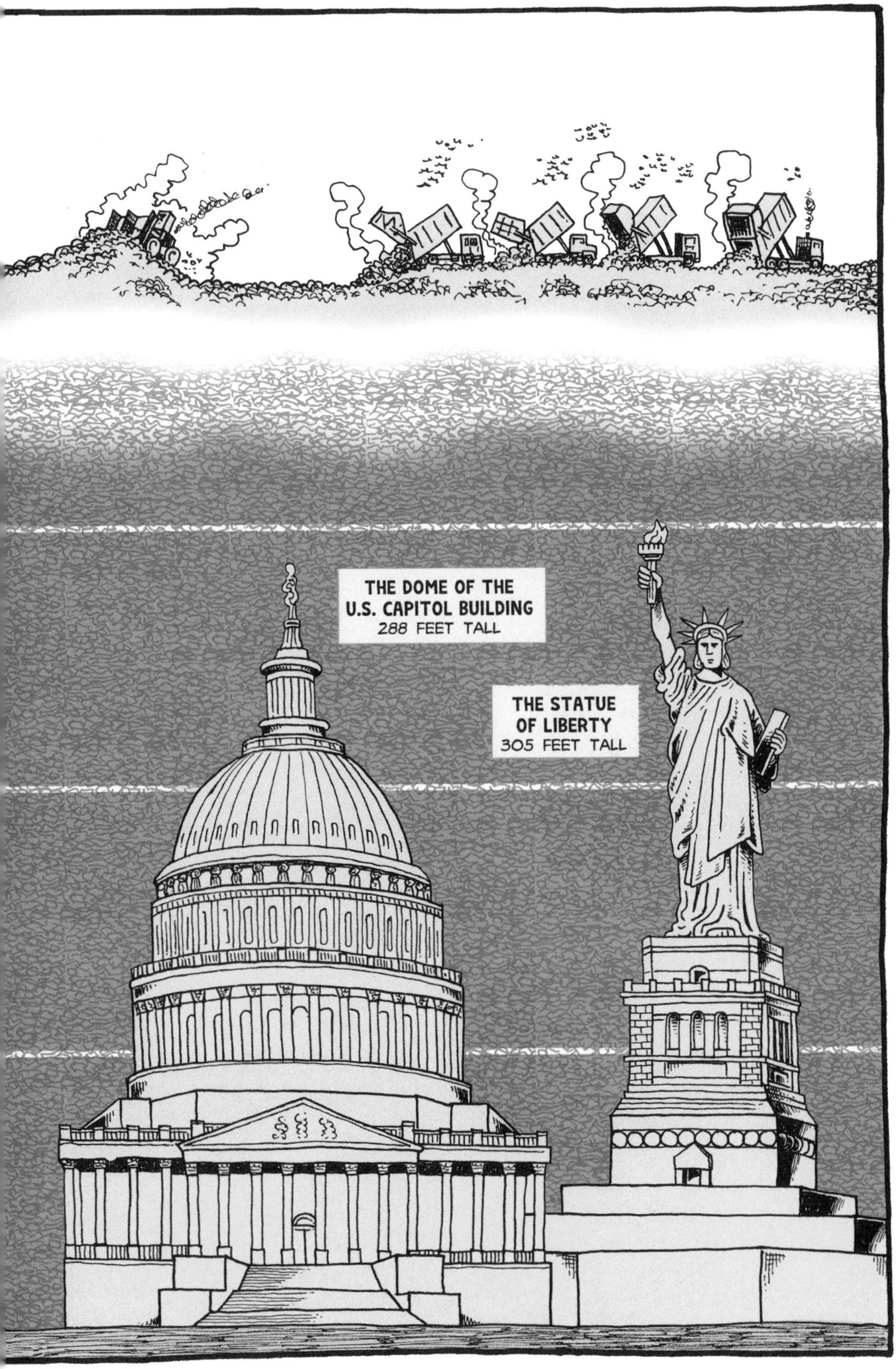
THE DOME OF THE
U.S. CAPITOL BUILDING
288 FEET TALL
THE STATUE
OF LIBERTY
305 FEET TALL

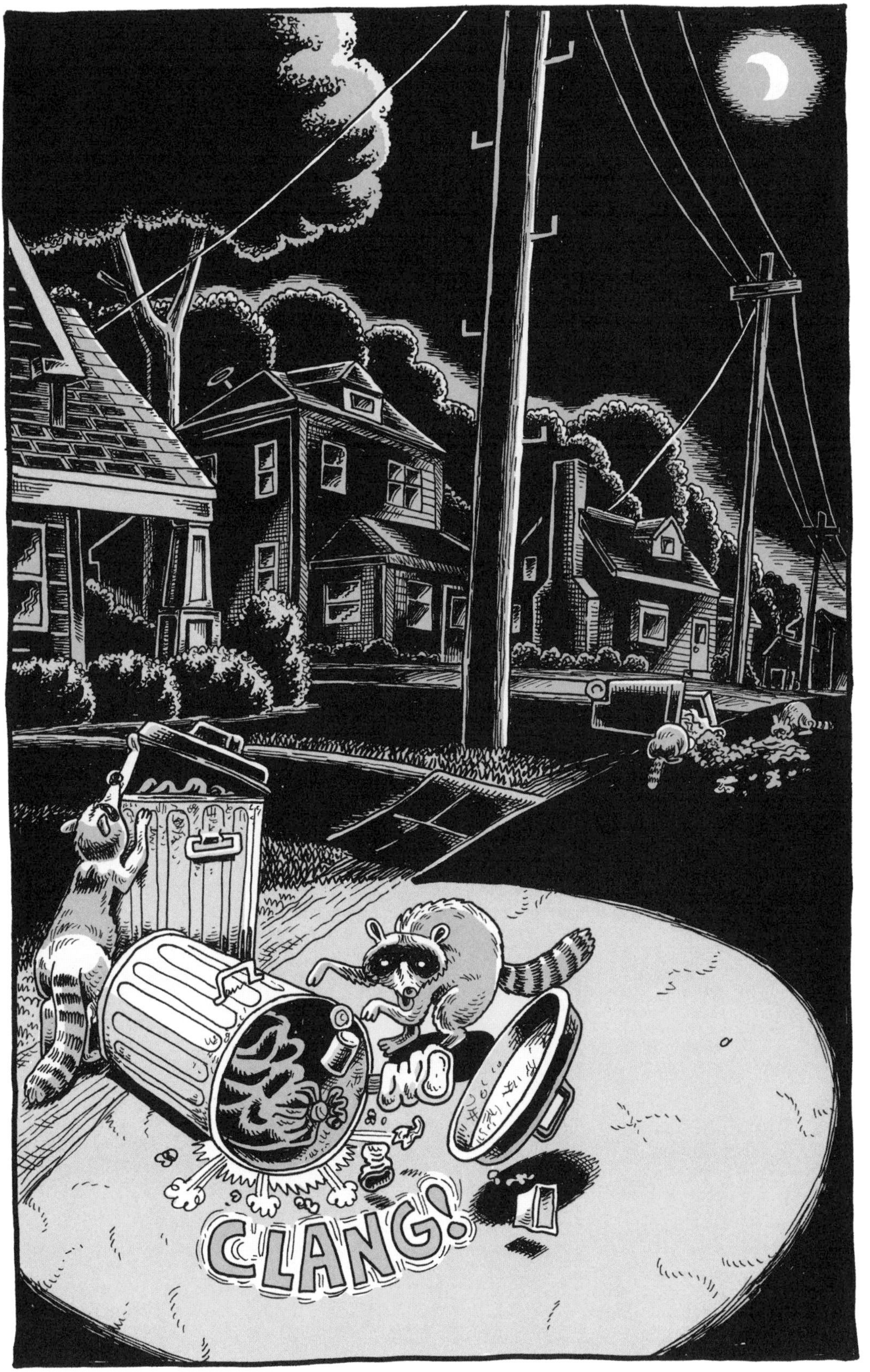
CLANG!

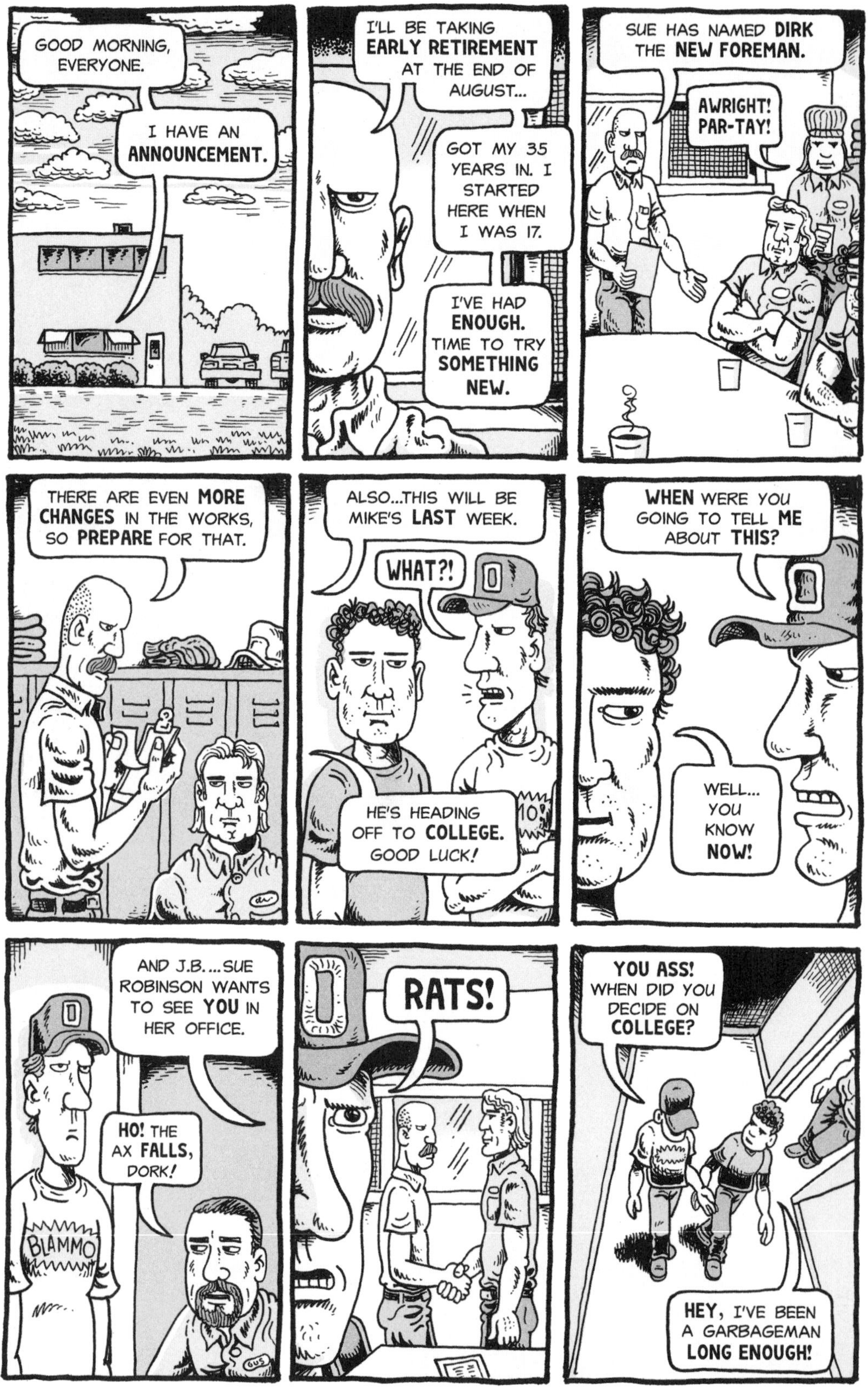
GOOD MORNING, EVERYONE.
I HAVE AN ANNOUNCEMENT.
I'LL BE TAKING EARLY RETIREMENT AT THE END OF AUGUST...
GOT MY 35 YEARS IN. I STARTED HERE WHEN I WAS 17.
I'VE HAD ENOUGH. TIME TO TRY SOMETHING NEW.
SUE HAS NAMED DIRK THE NEW FOREMAN.
AWRIGHT! PAR-TAY!
THERE ARE EVEN MORE CHANGES IN THE WORKS, SO PREPARE FOR THAT.
ALSO...THIS WILL BE MIKE'S LAST WEEK.
WHAT?!
HE'S HEADING OFF TO COLLEGE. GOOD LUCK!
WHEN WERE YOU GOING TO TELL ME ABOUT THIS?
WELL... YOU KNOW NOW!
AND J.B. ...SUE ROBINSON WANTS TO SEE YOU IN HER OFFICE.
HO! THE AX FALLS, DORK!
BLAMMO
GUS
RATS!
YOU ASS! WHEN DID YOU DECIDE ON COLLEGE?
HEY, I'VE BEEN A GARBAGEMAN LONG ENOUGH!

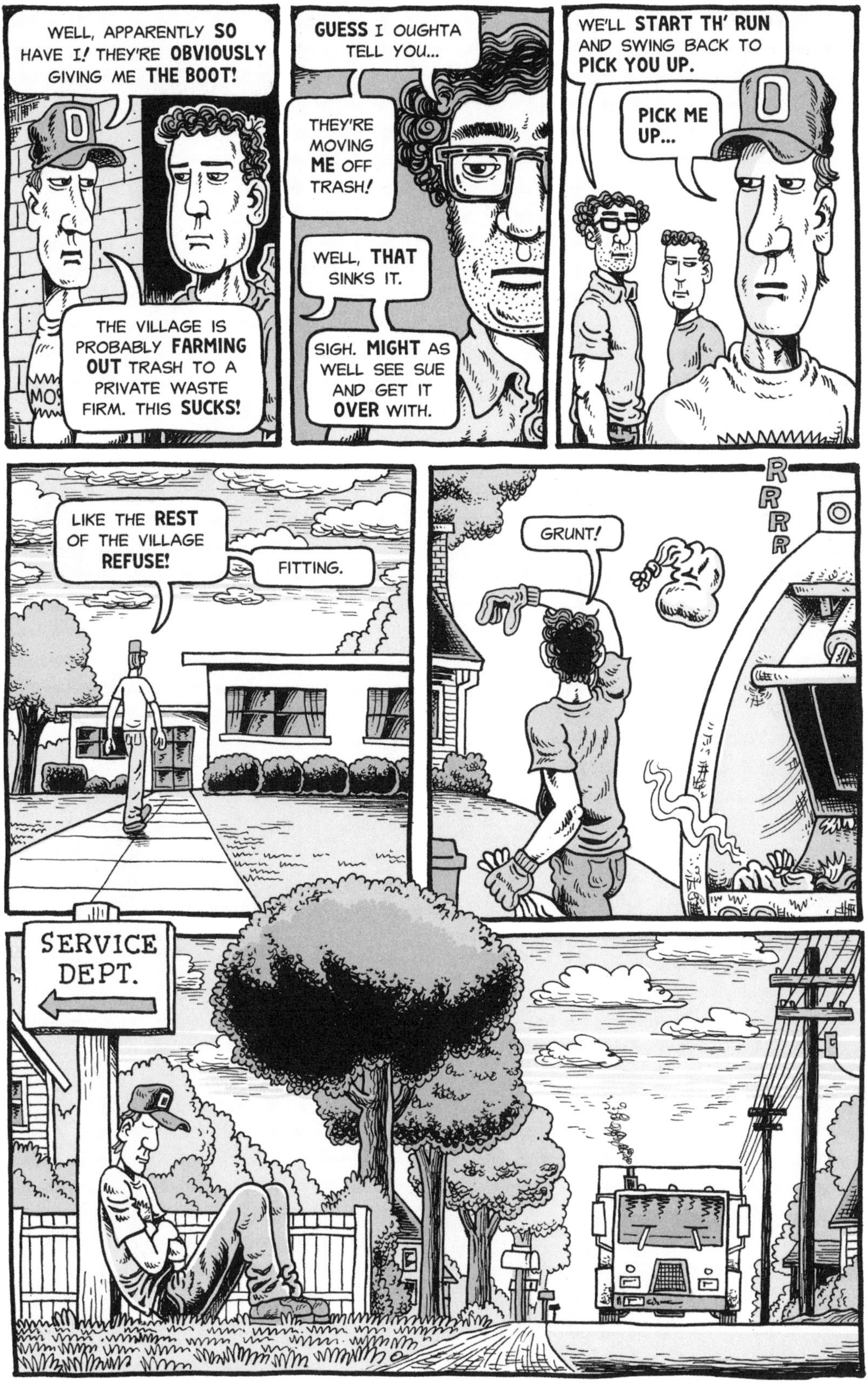

WELL, APPARENTLY SO HAVE I! THEY'RE OBVIOUSLY GIVING ME THE BOOT!
THE VILLAGE IS PROBABLY FARMING OUT TRASH TO A PRIVATE WASTE FIRM. THIS SUCKS!
GUESS I OUGHTA TELL YOU...
THEY'RE MOVING ME OFF TRASH!
WELL, THAT SINKS IT.
SIGH. MIGHT AS WELL SEE SUE AND GET IT OVER WITH.
WE'LL START TH' RUN AND SWING BACK TO PICK YOU UP.
PICK ME UP...
LIKE THE REST OF THE VILLAGE REFUSE!
FITTING.
GRUNT!
RRRR
SERVICE DEPT.

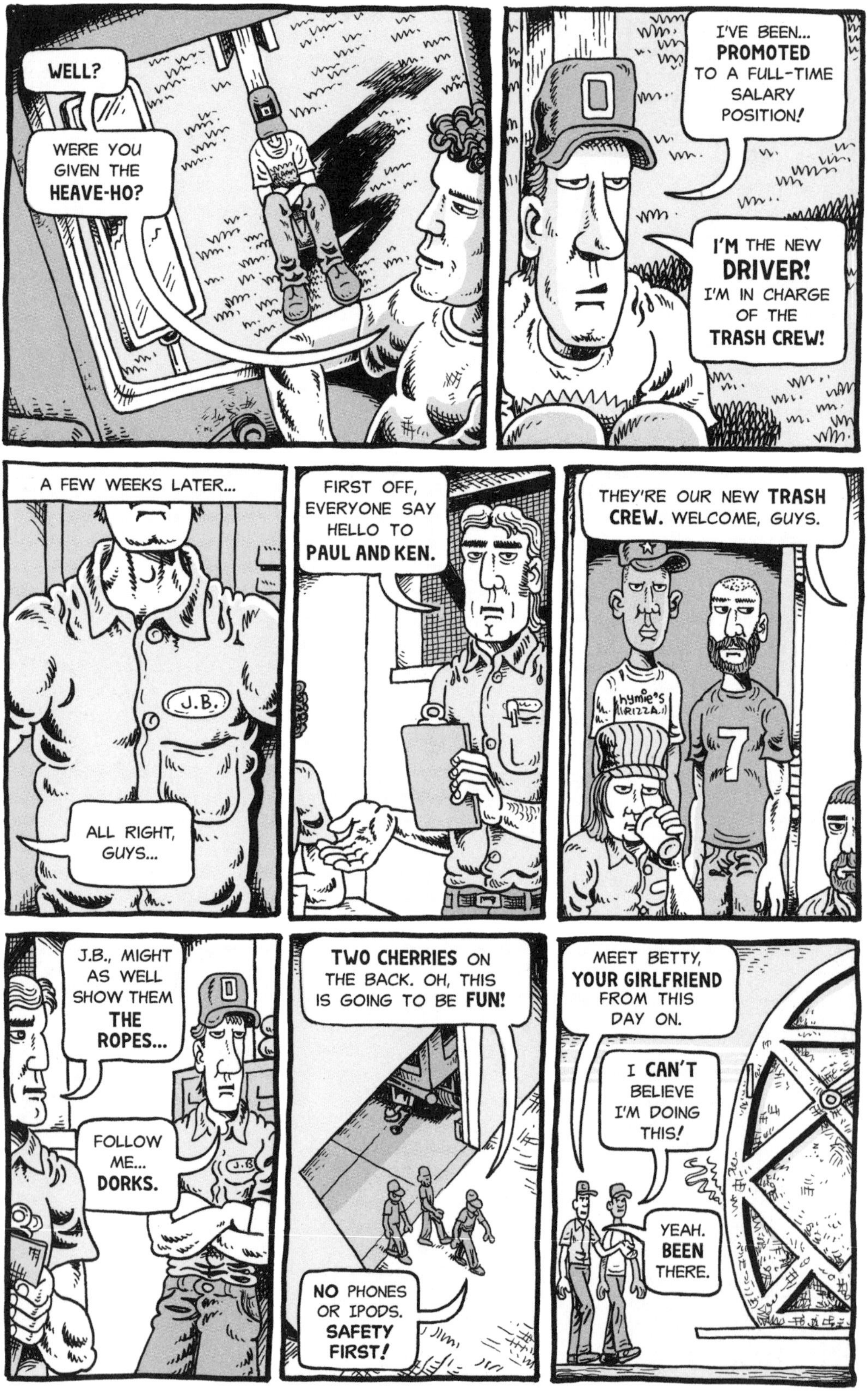
WELL?
WERE YOU GIVEN THE HEAVE-HO?
I'VE BEEN... PROMOTED TO A FULL-TIME SALARY POSITION!
I'M THE NEW DRIVER! I'M IN CHARGE OF THE TRASH CREW!
A FEW WEEKS LATER...
J.B.
ALL RIGHT, GUYS...
FIRST OFF, EVERYONE SAY HELLO TO PAUL AND KEN.
THEY'RE OUR NEW TRASH CREW. WELCOME, GUYS.
hymie's PIZZA
7
J.B., MIGHT AS WELL SHOW THEM THE ROPES...
FOLLOW ME... DORKS.
J.B.
TWO CHERRIES ON THE BACK. OH, THIS IS GOING TO BE FUN!
NO PHONES OR IPODS. SAFETY FIRST!
MEET BETTY, YOUR GIRLFRIEND FROM THIS DAY ON.
I CAN'T BELIEVE I'M DOING THIS!
YEAH. BEEN THERE.

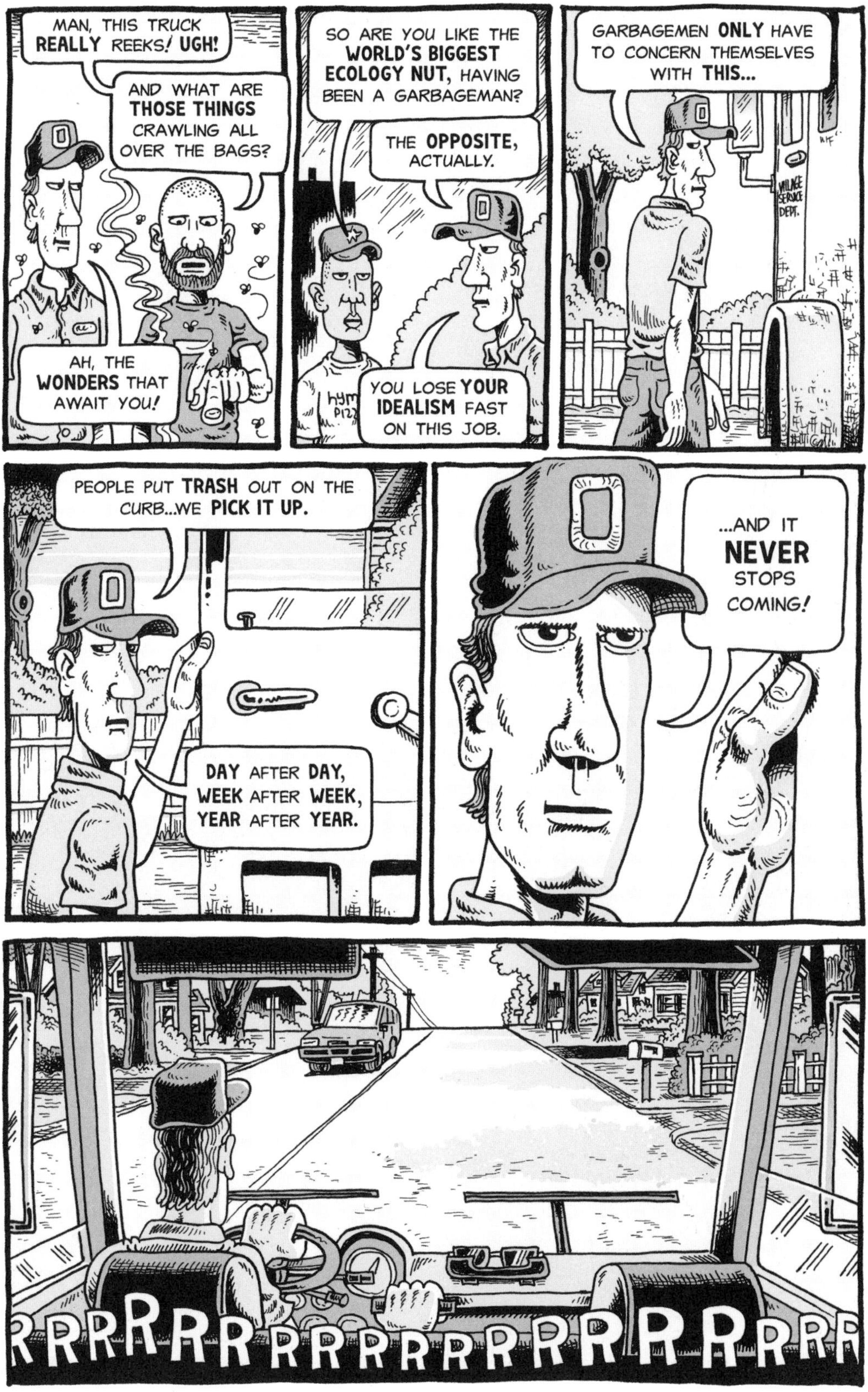

MAN, THIS TRUCK **REALLY** REEKS! **UGH!**
AND WHAT ARE **THOSE THINGS** CRAWLING ALL OVER THE BAGS?
AH, THE **WONDERS** THAT AWAIT YOU!
SO ARE YOU LIKE THE **WORLD'S BIGGEST ECOLOGY NUT,** HAVING BEEN A GARBAGEMAN?
THE **OPPOSITE,** ACTUALLY.
YOU LOSE **YOUR IDEALISM** FAST ON THIS JOB.
GARBAGEMEN **ONLY** HAVE TO CONCERN THEMSELVES WITH **THIS...**
VILLAGE SERVICE DEPT.
PEOPLE PUT **TRASH** OUT ON THE CURB...WE **PICK IT UP.**
DAY AFTER **DAY,** **WEEK** AFTER **WEEK,** **YEAR** AFTER **YEAR.**
...AND IT **NEVER** STOPS COMING!
RRRRRRRRRRRRRRRRRR

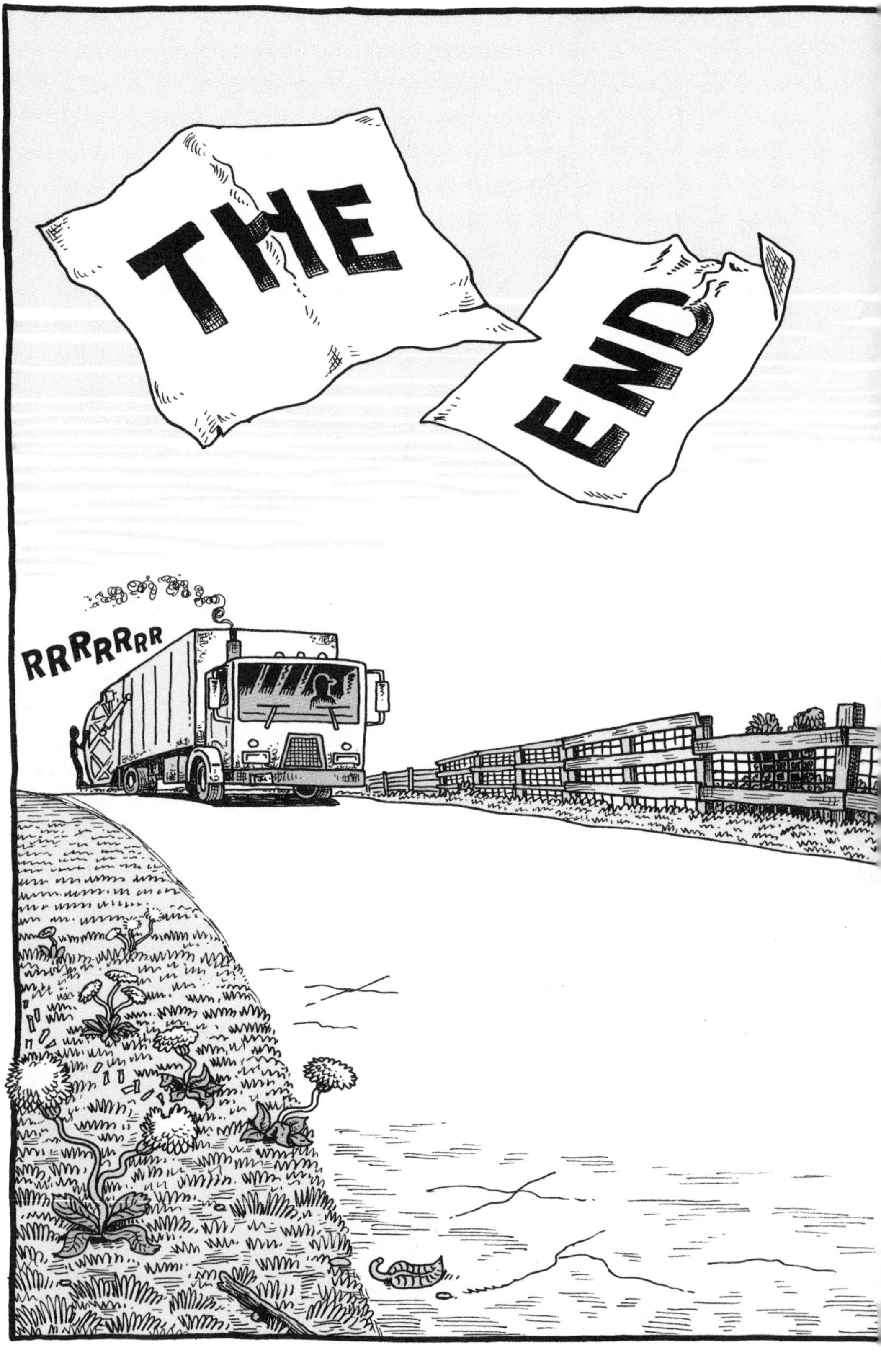
THE
END
RRRRRRR

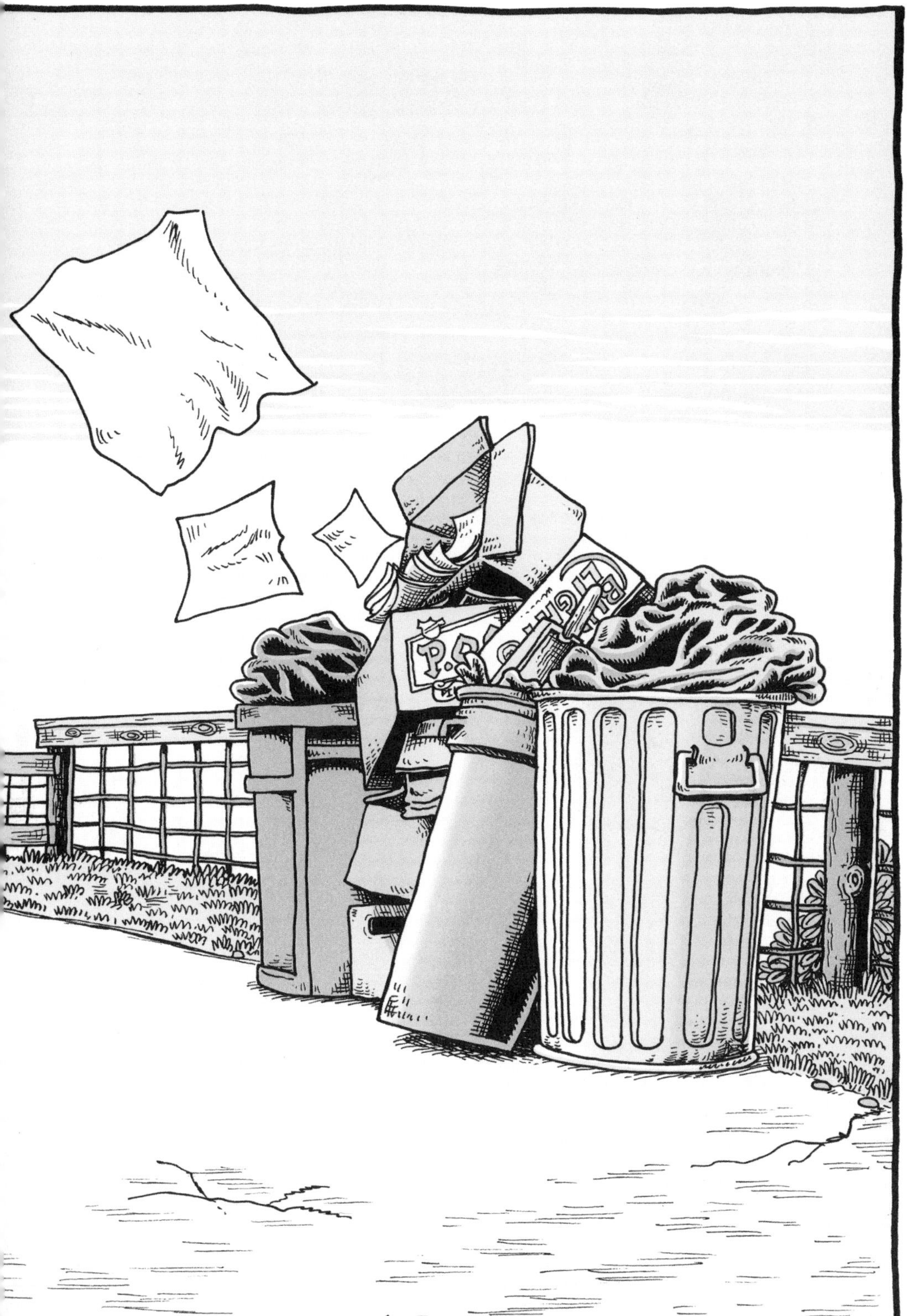

NOTES

The main sources for this book are derived from the following:

Personal Experience
As I wrote in the preface, I was a garbageman in 1979 and 1980. That's a long time ago, but surprisingly little has changed in garbage collection since then. The main difference is recycling. We recycled very little back then. In 2014, we recycle almost 30 percent of our trash.

Though fictional, all the episodes in this book are inspired by personal experience.

For you trivia buffs, I started on the truck roughly six months after the events in *My Friend Dahmer*. Jeff butchered the body of his first victim, a teenage hitchhiker he had murdered the previous June. He bagged up the remains and set them out for the trash collectors! It was just a few months before I first climbed on the back of the truck. Creepy, huh? Welcome to my world.

I've supplemented my experience with recent fact-finding visits to landfills, both closed and active; to garbage transfer stations; and to recycling facilities.

I also relied on the following reports on municipal garbage.

The Columbia University Biannual Report on Municipal Waste, 2014
This report, conducted every other year since 2002 by the Earth Engineering Center at Columbia University in New York City, is the go-to source for an accurate view of our municipal waste. Until 2010, this report was done in conjunction with *BioCycle* magazine and was titled *The State of Garbage in America*. There was no 2012 report, and Columbia now creates the report on its own, with the more academically cumbersome title *Generation and Disposition of Municipal Solid Waste in the United States—A National Survey*. Columbia gets its info directly from the states, unlike the E.P.A. in its biannual survey. Columbia's report was my main source for the numbers in this book. Garbologists and the private-waste industry also cite Columbia's findings. Even the E.P.A. uses some of the Columbia numbers in its report.

The E.P.A. Report on Municipal Solid Waste
For its report on municipal solid waste in the United States, *Advancing Sustainable Material Management: Facts and Figures 2013*, the E.P.A. doesn't use hard data like Columbia, but rather a byzantine formula of projections

and estimates. Its totals are ridiculously low. For example, its 2013 munici-pal waste total is a jaw-dropping 135 million tons *less* than that of Columbia University's. Why? Probably a mix of the usual bureaucratic dogma and a healthy dose of politics. For example, the E.P.A. last conducted a survey of the nation's landfills in 1995! The agency doesn't even know how many munici-pal landfills are out there in the United States! Thirty years without a detailed look at our massive dumps? Incredible. Obviously, there are political forces that don't want this information documented or publicized.

Still, because the E.P.A. has used the same formulas since 1960, its reports are useful in showing trends, even if its totals are laughably inaccurate. The 2013 report was the most recent one available when this book was written.

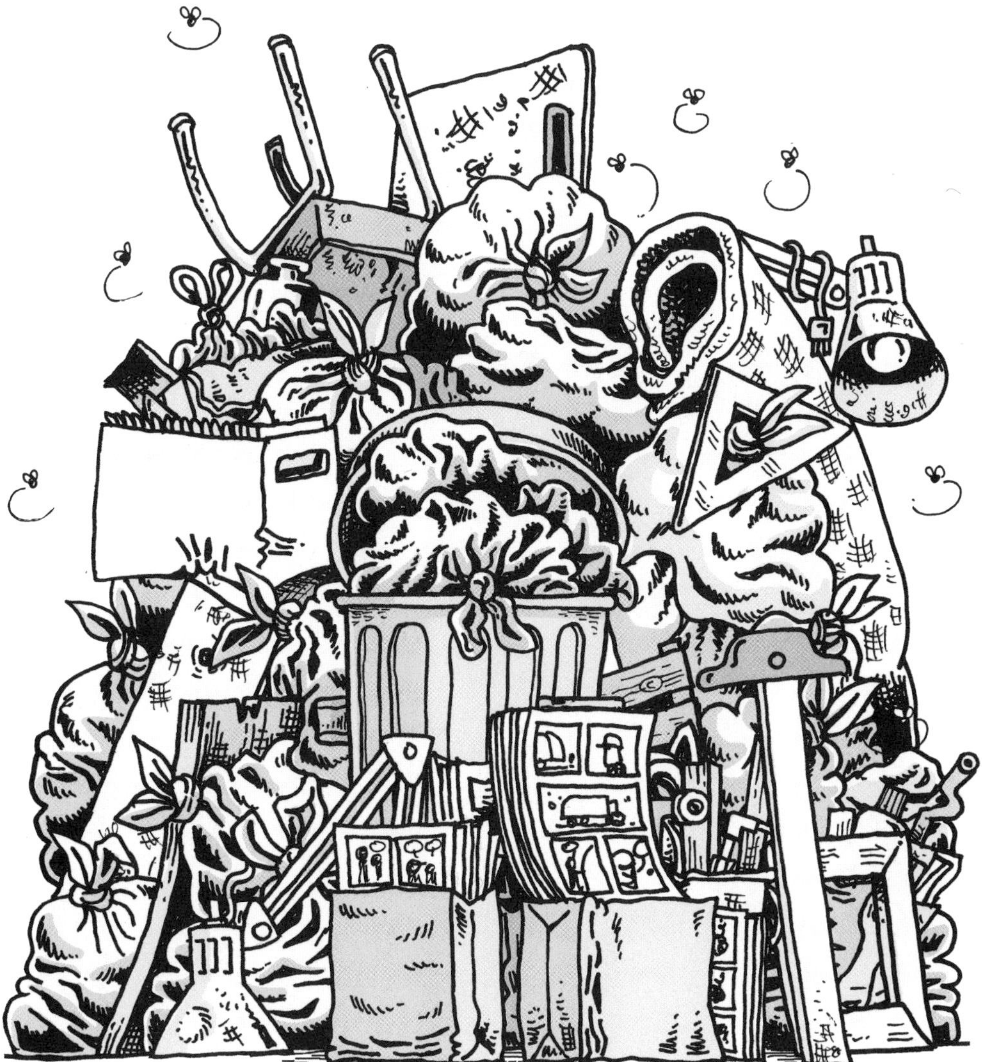

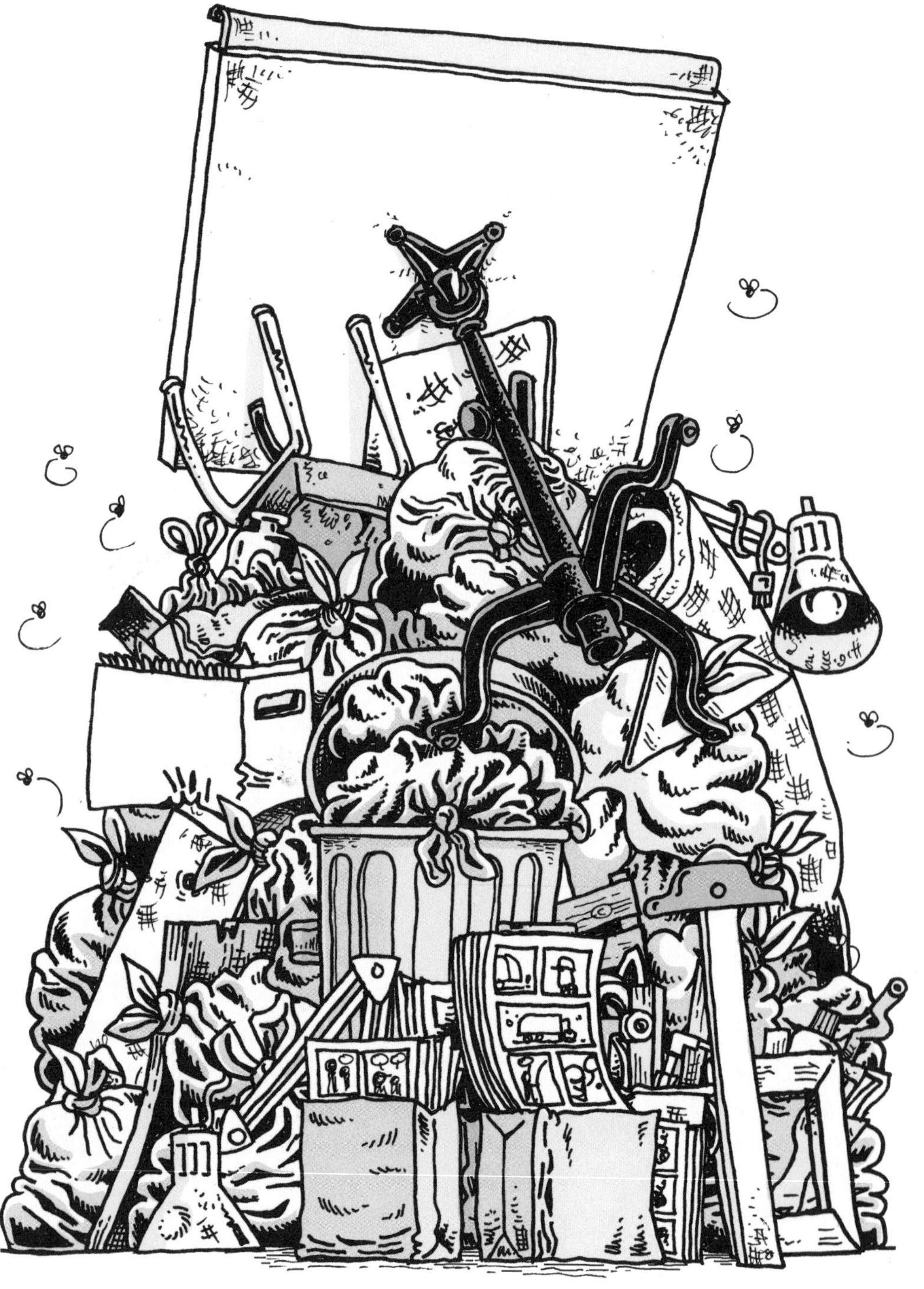